KEY ENGINEERING MATERIALS

Volume II: Interdisciplinary Concepts and Research

KEY ENGINEERING MATERIALS

Volume II: Interdisciplinary Concepts and Research

Edited by

François Kajzar, PhD, Eli M. Pearce, PhD,
Nikolai A. Turovskij, PhD, and Omari Vasilii Mukbaniani, DSc

A. K. Haghi, PhD, and Gennady E. Zaikov, DSc
Reviewers and Advisory Board Members

Apple Academic Press

TORONTO NEW JERSEY

Apple Academic Press Inc.
3333 Mistwell Crescent
Oakville, ON L6L 0A2
Canada

Apple Academic Press Inc.
9 Spinnaker Way
Waretown, NJ 08758
USA

Exclusive worldwide distribution by CRC Press, a member of Taylor & Francis Group

Printed in the United States of America on acid-free paper

International Standard Book Number-13: 978-1-926895-74-1 (Hardcover)

Library of Congress Control Number: 2013956108

Library and Archives Canada Cataloguing in Publication

Key engineering materials.

Includes bibliographical references and index.
Contents: v. 1. Current state of the art on novel materials/edited by Devrim Balköse, PhD, Daniel Horak, PhD, and Ladislav Šoltés, DSc; A. K. Haghi, PhD, and G. E. Zaikov, DSc, reviewers and advisory board members -- v. 2. Interdisciplinary concepts and research/edited by François Kajzar, PhD, Eli M. Pearce, PhD, Nikolai A. Turovskij, PhD, and Omari Vasilii Mukbaniani, DSc; A. K. Haghi, PhD, and Gennady E. Zaikov, DSc, reviewers and advisory board members.
ISBN ISBN 978-1-926895-74-1 (v. 2: bound)

1. Materials. I. Balköse, Devrim, editor of compilation II. Horák, D. (Daniel), editor of compilation III. Šoltés, Ladislav, editor of compilation IV. Title: Current state of the art on novel materials. V. Title: Interdisciplinary concepts and research.

TA403.K49 2014 620.1'1 C2013-908007-4

Apple Academic Press also publishes its books in a variety of electronic formats. Some content that appears in print may not be available in electronic format. For information about Apple Academic Press products, visit our website at **www.appleacademicpress.com** and the CRC Press website at **www.crcpress.com**

ABOUT THE EDITORS

François Kajzar, PhD

François Kajzar, PhD, is currently Associate Research Director at the University of Angers in France. He has taught and lectured at Jagellonian University, Kraków, Poland; the Academy of Mining and Metallurgy, Kraków; and the National Research Council-Institute of Structure of Matter (ISM-CNR), Bolognia, Italy, among other places. He was the Research Director and Senior Scientist at the Atomic Energy and Alternative Energies Commission, France. He has won numerous awards for his work and has written over 450 papers, books and book chapters, and conference presentations. He is also the editor of several journals and is on the editorial review boards of several others. His specialization fields include nonlinear optics, materials research, biomaterials, and biophotonics.

Eli M. Pearce, PhD

Dr. Eli M. Pearce was President of the American Chemical Society. He served as Dean of the Faculty of Science and Art at the Polytechnic Institute of New York University, as well as a professor of chemistry and chemical engineering. He was Director of the Polymer Research Institute, also in Brooklyn. At present, he consults for the Polymer Research Institute. A prolific author and researcher, he edited the *Journal of Polymer Science* (Chemistry Edition) for 25 years and was an active member of many professional organizations.

Nikolai A. Turovskij, PhD

Nikolai A. Turovskij, PhD, is currently Associate Professor, Physical Chemistry Department, Donetsk National University, Donetsk, Ukraine. He is the author of more than 200 scientific works and six author's certificates on invention. He is a supervisor of four candidates' theses, and the head of two scientific projects of the International Scientific Fund and two projects of the International Soros Science Education Program. Dr. Turovskij has worked in the L. M. Litvinenko Institute of Physical Organic and Coal Chemistry National Academy of Sciences of Ukraine in job titles of junior research fellow and senior scientific employee. His research interests include kinetics, structural chemistry, and molecular modeling of the supramolecular reactions of organic peroxides.

Omari Vasilii Mukbaniani, DSc

Omari Vasilii Mukbaniani, DSc, is Professor and Director of the Macromolecular Chemistry Department of the I. Javakhishvili Tbilisi State University, Tbilisi, Georgia. He is also the Director of the Institute of Macromolecular Chemistry of Academy of

Sciences of the Czech Republic. For several years he was a member of advisory board of the *Journal Proceedings of Iv. Javakhishvili Tbilisi State University* (Chemical Series), contributing editor of the journal *Polymer News* and the *Polymers Research Journal*. His research interests include polymer chemistry, polymeric materials, and chemistry of organosilicon compounds. He is an author more than 360 publication, 8 books, 3 monographs, and 10 inventions.

REVIEWERS AND ADVISORY BOARD MEMBERS

A. K. Haghi, PhD

A. K. Haghi, PhD, holds a BSc in urban and environmental engineering from University of North Carolina (USA); a MSc in mechanical engineering from North Carolina A&T State University (USA); a DEA in applied mechanics, acoustics and materials from Université de Technologie de Compiègne (France); and a PhD in engineering sciences from Université de Franche-Comté (France). He is the author and editor of 65 books as well as 1000 published papers in various journals and conference proceedings. Dr. Haghi has received several grants, consulted for a number of major corporations, and is a frequent speaker to national and international audiences. Since 1983, he served as a professor at several universities. He is currently Editor-in-Chief of the International *Journal of Chemoinformatics and Chemical Engineering* and *Polymers Research Journal* and on the editorial boards of many international journals. He is a member of the Canadian Research and Development Center of Sciences and Cultures (CRDCSC), Montreal, Quebec, Canada.

Gennady E. Zaikov, DSc

Gennady E. Zaikov, DSc, is Head of the Polymer Division at the N. M. Emanuel Institute of Biochemical Physics, Russian Academy of Sciences, Moscow, Russia, and Professor at Moscow State Academy of Fine Chemical Technology, Russia, as well as Professor at Kazan National Research Technological University, Kazan, Russia. He is also a prolific author, researcher, and lecturer. He has received several awards for his work, including the Russian Federation Scholarship for Outstanding Scientists. He has been a member of many professional organizations and on the editorial boards of many international science journals.

CONTENTS

LIST OF CONTRIBUTORS

O. R. Abdurakhmonov
The Institute of General and Inorganic Chemistry, National Academy of Sciences, Uzbekistan.

Y. A. Antonov
N. M. Emanuel Institute of Biochemical Physics, Russian Academy of Sciences, Kosigin str. 4. Moscow-119334, Russia.

S. N. Bondarenko
Volzhsky Polytechnical Institute, branch of Federal State Budgetary Educational Institution of Higher Professional Education, Volgograd State Technical University, Engels str. 42a, Volzhsky-404121, Volgograd Region, Russia.
E-mail: d.provotorova@gmail.com; www.volpi.ru

A. V. Bychkova
Federal State Budgetary Institution of Science Emanuel Institute of Biochemical Physics of Russian Academy of Sciences, Kosygina str. 4, Moscow-119334, Russia.
E-mail: annb0005@yandex.ru

A. A. Dalinkevich
Central Scientific Research Institute of Special Mashine-Building, Zavodskaya str. 1, Khotkovo-141371, Moscow region, Russia.
E-mail: dalinckevich@yandex.ru

K. S. Dibirova
Dagestan State Pedagogical University, Yaragskii str. 57, Makhachkala-367003, Russian Federation.

S. N. Gaidamaka
Moscow State University, Chemistry Faculty, Department of Chemical Enzymology, Leninskye gory MSU 1/11, Moscow-119991, Russia, Phone: +7(495) 939-5083, Fax: +7(495) 939-5417.
E-mail: vp_murygina@mail.ru, vpm@enzyme.chem.msu.ru

K. Z. Gumargalieva
N. N. Semenov Institute of Chemical Physics, RAS, Kosygin str. 4, Moscow-119991, Russia.
E-mail: guklara@yandex.ru

A. K. Haghi
University of Guilan, Rasht, Iran.

N. V. Ilyashenko
Tver State University, Zheliabova str. 33, Tver–170100, Russia.

A. I. Ivanova
Tver State University, Zheliabova str. 33, Tver–170100, Russia.

V. F. Kablov
Volzhsky Polytechnical Institute, Branch of Federal State Budgetary Educational Institution of Higher Professional Education, Volgograd State Technical University, Engels str. 42a, Volzhsky-404121, Volgograd Region, Russia.
E-mail: d.provotorova@gmail.com; www.volpi.ru

N. A. Keibal
Volzhsky Polytechnical Institute, Branch of Federal State Budgetary Educational Institution of Higher Professional Education, Volgograd State Technical University, Engels str. 42a, Volzhsky-404121, Volgograd Region, Russia.
E-mail: d.provotorova@gmail.com; www.volpi.ru

G. V. Kozlov
N. M. Emanuel Institute of Biochemical Physics of Russian Academy of Sciences, Kosygin str. 4, Moscow-119334, Russian Federation.

S. M. Lomakin
N. M. Emanuel Institute of Biochemical Physics, Russian Academy of Sciences, Kosygin str. 4, Moscow-119334, Russia.
E-mail: Chembio@sky.chph.ras.ru

G. M. Magomedov
Dagestan State Pedagogical University, Makhachkala-367003, Yaragskii str. 57, Russian Federation.

S. S. Marakhovsky
«Armoproject» Company» LLC, dominion 27 Malakhitovaya str., Moscow-129128,
E-mail: cmc@aproject.ton.ru

T. I. Medintseva
Semenov Institute of Chemical Physics of Russian Academy of Sciences, Kosygin str. 4, Moscow-119991.

Y. P. Miroshnikov
Lomonosov Moscow State University for Fine Chemical Technology.
E-mail: ypm@mail.ru

P. Moldenaers
K. U. of Leuven, Department Chemical Engineering, Willem de Croylaan 46, Leuven- B-3001, Belgium.

V. P. Murygina
Moscow State University, Chemistry Faculty, Department of Chemical Enzymology, Leninskye gory MSU, 1/11, Moscow-119991, Russia, Phone: +7(495) 939-5083, Fax: +7(495) 939-5417.
E-mail: vp_murygina@mail.ru, vpm@enzyme.chem.msu.ru

B. E. Murzabekov
JSC “Sea oil company”, “KazMunayTeniz”, Kazakhstan.

Yu. G. Oleneva
Tver State University, Zheliabova str., 33, Tver-170100, Russia.

Elena L. Pekhtasheva
G.V. Plekhanov Russian Economic University, 36 Stremyannyi way, Moscow-117997 Russia.
E-mail: pekhtashevael@mail.ru

M. A. Poldushov
Lomonosov Moscow State University for Fine Chemical Technology
E-mail: poldushov@mail.ru

D. A. Provotorova
Volzhsky Polytechnical Institute, branch of Federal State Budgetary Educational Institution of Higher Professional Education, Volgograd State Technical University, Engels str. 42a, Volzhsky-404121, Volgograd Region, Russia.
E-mail: d.provotorova@gmail.com; www.volpi.ru

E. V. Prut
Semenov Institute of Chemical Physics of Russian Academy of Sciences, Kosygin str. 4, Moscow-119991.

M. A. Rosenfeld
Federal state budgetary institution of science Emanuel Institute of Biochemical Physics of Russian Academy of Sciences, Kosygina str. 4, Moscow-119334, Russia.

P. A. Sakharov
N. M. Emanuel Institute of Biochemical Physics, Russian Academy of Sciences, Kosygin str. 4, Moscow-119334, Russia.
E-mail: Chembio@sky.chph.ras.ru

Z. S. Salimov
The Institute of General and Inorganic Chemistry, National Academy of Sciences, Uzbekistan.

Sh. M. Saydakhmedov
The Institute of General and Inorganic Chemistry, National Academy of Sciences, Uzbekistan.

F. B. Shevlyakov
Ufa State Petroleum Technological University, Russia.

A. V. Soukhanov
«Armoproject» Company» LLC, dominion 27 Malakhitovaya str., Moscow-129128,
E-mail: cmc@aproject.ton.ru

S. Ya. Trofimov
Moscow State University, Chemistry Faculty, Department of Chemical Enzymology,Leninskye gory MSU, 1/11, Moscow-19991, Russia, Phone: +7(495) 939-5083, Fax: +7(495) 939-5417.
E-mail: vp_murygina@mail.ru, vpm@enzyme.chem.msu.ru

T. G. Umergalin
Ufa State Petroleum Technological University, Russia.

G. E. Zaikov
N. M. Emanuel Institute of Biochemical Physics, Russian Academy of Sciences, Kosygin str. 4, Moscow-119334, Russia.
E-mail: Chembio@sky.chph.ras.ru

V. P. Zakharov
Bashkir State University, Russia.
E-mail: zaharovvp@mail.ru

LIST OF ABBREVIATIONS

AFM	Atomic force microscopy
ATP	Adenosine triphosphate
BGAF	Benzoguanamine formaldehyde resins
BSA	Bovine serum albumin
CNR	Chlorinated natural rubber
CP	Critical point
DC	Dupli-color
DEX	Dextran
DLS	Dynamic light scattering
DM	Dry matter
DSS	Dextran sulfate sodium
DTGS	Deuterated triglycine sulfate
EDX	Energy-dispersive X-ray
ELPP	Elastomeric polypropylene
EPDM	Ethylene propylene diene monomer
EPR	Ethylene propylene rubber
ESEM	Environment scanning electron microscope
ESR	Electron spin resonance
FG	Fibrinogen
FID	Flame ionization detector
FMR	Ferromagnetic resonance
FPLC	Fast protein liquid chromatography
FTIR	Fourier transform infrared
GAG	Glycosaminoglycan
GC	Gas chromatograph
GFU	Gas fractionation unit
HC	Hydrocarbons
HCO	Hydrocarbon oxidizing
HDPE	High density polyethylene
HPLC	High-performance liquid chromatography
HT	Heterotrophic bacteria
IEF	Iso-electric focusing
IPP	Isotactic polypropylene
IR	Infrared
IR MDCIR	Infrared microscopy of multiply disturbed complete inner reflection
LDPE	Low density polyethylene
MBTS	Dibenzothiazole disulfide
MFI	Melt flow index

MLR	Mass loss rates
MNPs	Magnetic nanoparticles
MNSs	Magnetically targeted nanosystems
MPN	Most probably number
NMR MAS	Nuclear magnetic resonance, magic-angle spinning
OM	Optical microscopy
OSR	Oxidized starch reagent
PE	Polyethylenes
PEG	Poly(ethy1ene glycol)
PIB	Polyisobutylene
PMPA	Polyelectrolyte-mediated protein adsorption
PMPAS	Polyelectrolyte mediated protein association
PP	Polypropylene
PVC	Polyvinyl chloride
RHR	Rate of heat release
SA	Sodium alginate
SALS	Small angle light scattering
SBR	Styrene-butadiene rubber
SBS	Styrene-butadiene-styrene
SC	Sodium caseinate
SA	sodium alginate
SEM	Scanning electron microscopy
SiH	Silicon hydride
SLS	Static light scattering
SP	Softening point
TEAT	Triethanolaminotitanate
TMTD	Tetramethylthiuram disulfide
TPEs	Thermoplastic elastomers
TPV	Thermoplastic vulcanizates
Trp	Tryptophan
WS	Water steam
XPS	X-Ray photoelectron spectroscopy

PREFACE

This book, *Key Engineering Materials: Volume II: Interdisciplinary Concepts and Research*, provides both a rigorous view and a more practical and understandable view of key engineering materials for graduate students and scientists in related fields. This book will satisfy readers with both direct and lateral interests in the discipline.

This volume is structured into different parts devoted to key engineering materials and their applications. Every section of this book has been expanded, where relevant, to take account of significant new discoveries and realizations of the importance of key concepts. Furthermore, emphases are placed on the underlying fundamentals and on acquisition of a broad and comprehensive grasp of the field as a whole.

This book contains innovative chapters on the growth of educational, scientific, and industrial research activities among chemists, biologists, and polymer and chemical engineers and provides a medium for mutual communication between international academia and the industry. This book presents significant research and reviews reporting new methodologies and important applications in the fields of industrial chemistry, industrial polymers, and biotechnology as well as includes the latest coverage of chemical databases and the development of new computational methods and efficient algorithms for chemical software and polymer engineering.

This book's aim is to provide comprehensive coverage on the latest developments of research in the ever-expanding area of polymers and advanced materials and their applications to broad scientific fields spanning physics, chemistry, biology, materials, and so on.

This new book:

- provides physical principles in explaining and rationalizing polymeric phenomena
- features classical topics that are conventionally considered part of chemical technology
- covers the chemical principles from a modern point of view
- analyzes theories to formulate and prove the polymer principles
- presents future outlook on application of bioscience in chemical concepts
- focuses on topics with more advanced methods

**— François Kajzar, PhD, Eli M. Pearce, PhD,
Nikolai A. Turovskij, PhD, and Omari Vasilii Mukbaniani, DSc**

CHAPTER 1

CASE STUDIES ON KEY ENGINEERING MATERIALS IN NANOSCALE: KEY CONCEPTS AND CRITERIA

A. K. HAGHI and G. E. ZAIKOV

CONTENTS

1.1 INTRODUCTION

Technical and technological development demands the creation of new materials, which are stronger, more reliable, and more durable that is materials with new properties. Up-to-date projects in creation of new materials go along the way of nanotechnology.

Nanotechnology can be referred to as a qualitatively new round in human progress. This is a wide enough concept, which can concern any area like information technologies, medicine, military equipment, robotics and so on. We narrow the concept of nanotechnology and consider it with the reference to polymeric materials as well as composites on their basis.

The prefix nano- means that in the context of these concepts the technologies based on the materials, elements of constructions and objects are considered, whose sizes make 10^{-9} meters. As fantastically as it may sound but the science reached the nanolevel long ago. Unfortunately, everything connected with such developments and technologies for now is impossible to apply to mass production because of their low productivity and high cost.

That means nanotechnology and nanomaterials are only accessible in research laboratories for now but it is only a matter of time. What sort of benefits and advantages will the manufacturers have after the implementation of nanotechnologies in their manufactures and starting to use nanomaterials? Nanoparticles of any material have absolutely different properties rather than micro-or macroparticles. This results from the fact that alongside with the reduction of particles' sizes of the materials to nanometric sizes, physical properties of a substance change too. For example, the transition of palladium to nanocrystals leads to the increase in its thermal capacity in more than 1.5 times, it causes the increase of solubility of bismuth in copper in 4000 times and the increase of self-diffusion coefficient of copper at a room temperature on 21 orders.

Such changes in properties of substances are explained by the quantitative change of surface and volume atoms' ratio of individual particles that is by the high-surface area. Insertion of such nanoparticles in a polymeric matrix while using the apparently old and known materials gives a chance of receiving the qualitatively and quantitatively new possibilities in their use.

Nanocomposites based on thermoplastic matrix and containing natural, laminated inorganic structures are referred to as laminated nanocomposites. Such materials are produced on the basis of ceramics and polymers however, with the use of natural laminated inorganic structures such as montmorillonite or vermiculite which are present, for example, in clays. A layer of filler ~1 nm thick is saturated with monomer solution and later polymerized. The laminated nanocomposites in comparison with initial polymeric matrix possess much smaller permeability for liquids and gases. These properties allow applying them to medical and food processing industry. Such materials can be used in manufacture of pipes and containers for the carbonated beverages.

These composite materials are eco-friendly, absolutely harmless to the person and possess fire-resistant properties. The derived thermoplastic laboratory samples have been tested and really confirmed those statements.

It should be noted that manufacturing technique of thermoplastic materials causes difficulties for today, notably dispersion of silicate nanoparticles in monomer solution. To solve this problem it is necessary to develop the dispersion technique, which could be transferred from laboratory conditions into the industrial ones.

What advantages the manufacturers can have, if they decide to reorganize their manufacture for the use of such materials, can be predicted even today. As these materials possess more mechanical and gas barrier potential in comparison with the initial thermoplastic materials, then their application in manufacture of plastic containers or pipes will lead to raw materials' saving by means of reduction of product thickness.

On the other hand, the improvement of physical and mechanical properties allows applying nanocomposite products under higher pressures and temperatures. For example, the problem of thermal treatment of plastic containers can be solved. Another example of the application of the valuable properties of laminated nanocomposites concerns the motor industry.

As mentioned earlier, another group of materials is metal containing nanocomposites. Thanks to the ability of metal particles to create the ordered structures (clusters), metal containing nanocomposites can possess a complex of valuable properties. The typical sizes of metal clusters from 1 to 10 nm correspond to their huge specific surface area. Such nanocomposites demonstrate the superparamagnetism and catalytic properties; therefore, they can be used while manufacturing semiconductors, catalysts, optical and luminescent devices, and so on.

Such valuable materials can be produced in several ways, for example, by means of chemical or electrochemical reactions of isolation of metal particles from solutions. In this case, the major problem is not so much the problem of metal restoration but the preservation of its particles that is the prevention of agglutination and formation of large metal pieces.

For example under laboratory conditions metal is deposited in such a way on the thin polymeric films capable to catch nano-sized particles. Metal can be evaporated by means of high energy and nano-sized particles can be produced, which then should be preserved. Metal can be evaporated while using explosive energy and high voltage electric discharge or simply high temperatures in special furnaces.

The practical application of metal-containing nanocomposites (not going into details about high technologies) can involve the creation of polymers possessing some valuable properties of metals. For example, the polyethylene plate with the tenth fractions of palladium possesses the catalytically properties similar to the plate made of pure palladium.

An example of applying metallic composite is the production of packing materials containing silver and possessing bactericidal properties. By the way, some countries have already been applying the paints and the polymeric coverings with silver nanoparticles. Owing to their bactericidal properties, they are applied in public facilities (painting of walls, coating of handrails, and so on).

The technology of polymeric nanocomposites manufacture forges ahead, its development is directed to simplification and cheapening the production processes of composite materials with nanoparticles in their structure. However, the nanotechnologies develop high rates; what seemed impossible yesterday, will be accessible to the introduction on a commercial scale tomorrow.

The prospects in the field of polymeric composite materials upgrading are retained by nanotechnologies. Ever increasing manufacturers' demand for new and superior materials stimulates the scientists to find new ways of solving tasks on the qualitatively new nanolevel.

The desired event of fast implementation of nanomaterials in mass production depends on the efficiency of cooperation between the scientists and the manufacturers in many respects. Today's high technology problems of applied character are successfully solved in close consolidation of scientific and business worlds.

Nanocomposites are polymers containing nanofillers. The microstructure of nanocomposites has inhomogeneities in the scale range of nanometers. Nanocomposite materials cover the range between inorganic glasses and organic polymers. Fillers of polymers have been used for a long time with the goal of enhanced performance of polymers and especially of rubber.

Nanofillers lists increased within years as well as the matrix in which they are used and interactions with traditional fillers. Nowadays, the development of polymer nanocomposites is one of the most active areas of development of nanomaterials. The properties imparted by the nanoparticles are various and focus particularly on strengthening the electrical conduction and barrier properties to temperature, gases, and liquids as well as the possible improvement of fire behavior. As a method which consists of reinforcing polymer chains at the molecular scale in the same way than the fibers at the macroscopic scale, nanocomposites represent the new generation of two-phased materials, associating a basic matrix to nanofillers inserted between polymer chains. Nanofillers can significantly improve or adjust the different properties of the materials into which they are incorporated. The properties of composite materials can be significantly impacted by the mixture ratio between the organic matrix and the nanofillers.

Fillers play important roles in modifying the desirable properties of polymers and reducing the cost of their composites. In conventional polymer composites, many inorganic filers with dimensions in the micrometer range, for example calcium carbonate, glass beads, and talc have been used extensively to enhance the mechanical properties of polymers. Such properties can indeed be tailored by changing the volume fraction, shape, and size of the filler particles. A further improvement of the mechanical properties can be achieved by using filler materials with a larger aspect ratio such as short glass fibers. It is logical to anticipate that the dispersion of fillers with dimensions in the nanometer level having very large aspect ratio and stiffness in a polymer matrix could lead to even higher mechanical performances. These fillers include layered silicates and carbon nanotubes. Rigid inorganic nanoparticles with a smaller aspect ratio are also promising reinforcing and/or toughening materials for the polymers. The dispersion of nanofillers in the polymers is rather poor due to their incompatibility with polymers and large surface-to-volume ratio. Therefore, organic surfactant and compatibility additions are needed in order to improve the dispersion of these nanofillers in polymeric matrices. For example, layered silicate surfaces are hydrophilic and proper modification of the clay surfaces through the use of organic surfactants is needed. The obtained product is known as 'organoclay'. In this context, organoclays can be readily delaminated into nanoscale platelets by the polymer molecules, leading to the formation of polymer clay nanocomposites. These nanocomposites belong to an emerging class of organic–inorganic hybrid materials that exhibit improved mechanical properties at very low loading levels compared with conventional microcomposites.

The behavior of polymer–nanofiller composites is directly related to their hierarchical microstructures.

Therefore, the mechanical properties of polymer–nanofiller composites are controlled by several microstructural parameters such as properties of the matrix, properties and distribution of the filler as well as interfacial bonding, and by the synthetic or processing methods. The interfaces may affect the effectiveness of load transfer from the polymer matrix to nanofillers. Thus, surface modification of nanofillers is needed to promote better dispersion of fillers and to enhance the interfacial adhesion between the matrix and fillers. Fabrication of homogeneous polymer nanocomposites and advanced computational techniques remains a major scientific challenge for materials scientists.

1.2 CASE STUDY-I

In the present case study, the formation of chemical bond between the atoms of polymer coating components and carbon metal-containing nanostructures with additions of silver or zinc is investigated for improving electrical conduction in coatings.

The analysis has been conducted by the X-ray photoelectron spectroscopy (XPS) method. The XPS allows to investigate an electronic structure, chemical bond, and nearest surrounding of an atom.

The investigation was conducted on a unique automated X-ray electron magnetic spectrometer with double focusing allowing the investigation of samples in both solid and liquid state, which has the following performance specifications: the resolution is 10^{-4} and luminosity is 0.185% [1-10].

The electron magnetic spectrometer has a number of advantages in comparison with electrostatic spectrometers, which are the constancy of luminosity and resolution independent of the electron energy, and high contrast of spectra. The XPS method is a non-destructive method of investigation which is especially important for studying metastable systems.

Two samples of polymer coatings were studied:

- Silver-containing coating modified with carbon metal-containing nanostructures (70% Ag and 1% nanostructures) and
- Zinc-containing coating modified with carbon metal-containing nanostructures (60% Zn, and 1% nanostructures).

The carbon metal-containing nanostructure samples were prepared by low temperature synthesis. Carbon copper-containing nanostructures were mixed with polyethylene-polyamine $H_2N[CH_2\text{-}CH_2\text{-}NH]_mH$, where m = 1–8, through mechanic activation.

For improving the interaction of the polymer coating and carbon metal-containing nanostructures, the interaction of nanostructures with d metals, that is Ag and Zn, was used. The spectra of the C1s, Ag3d, Ag3p, and Zn2p core levels and the spectra of valence bands of the composites prepared were studied. The composites were carbon metal-containing nanostructures containing silver, a polymer coating with addition of carbon copper-containing nanostructures functionalized with silver. The spectra of reference samples, that is Ag, Zn, and carbon copper-containing nanostructures, were studied as well. The cleanness of the sample surface was controlled by the O1s-spectrum. At heating to 300°C, oxygen on the surface is absent.

The decomposition of the spectra into components was performed with the help of a program based on the least squares method. For the spectrum decomposition, energy position, width of spectrum components and components' intensity based on the data obtained from the reference samples' spectra were entered into the program. The accuracy in the determination of the peak positions was 0.1 eV. The error in the determination of the contrast of the electronic spectra was less than 5%.

The C1s-spectrum (Figure 1 (a)) of carbon copper-containing nanostructures consists of three components: C-Cu(283 eV), C-H (285 eV), and C-O (287 eV).

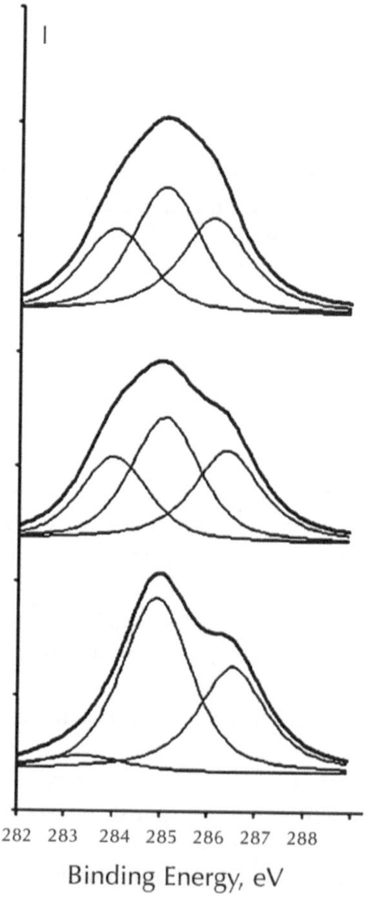

FIGURE 1 C1s spectrum ((a) C1s spectrum of the copper/carbon nanocomposite, (b) C1s spectrum of the copper/carbon nanocomposite with silver, and (c) C1s spectrum coating) filled by silver and modified by copper/carbon nanocomposite.

In the C1s-spectrum (Figure 1 (b)) of the studied composite consisting of carbon copper-containing nanostructures functionalized with silver, the intensity of the first component C-Me significantly increases and the binding energy changes (283.8 eV), which corresponds to the data from experimental works and is associated with a larger localization of d electrons in silver than that in copper.

There are models, which describes the connection of the parameters of multiplet splitting of the XPS 3s spectra with a spin state, namely, the relative intensity of the maxima of the multiplets in the 3s spectra correlates with the value of the magnetic moment of the atoms in the d-metal systems; the distance between the maxima of the multiplets (Δ) provide the information about the exchange interaction of the 3s–3d shells. The change of their overlapping is associated with changes in the distance between the atoms and average atomic volume. The presence of changes in the shape of the 3s spectra provides the information about the changes in the chemical bond of the atoms in the nearest surrounding in the composite.

The analysis of the parameters of the multiplet splitting in the Cu3s and Ag4s spectra shows the presence of the atomic magnetic moments on the copper atoms (1.6 MB, $\Delta = 2.5$ eV) and the argentum atoms (2.2 MB, $\Delta = 2.0$ eV). In contrast to carbon copper-containing nanostructures, a decrease in the distance between the multiplets indicates the enhancement of the chemical bond between the copper d electrons and carbon p electrons and the formation of a strong covalent bond. The appearance of the magnetic moment on the silver atoms is associated with the formation of non-compensated d electrons on the silver atoms and the involvement of the d electrons of the silver atoms into the covalent bond with the carbon atoms.

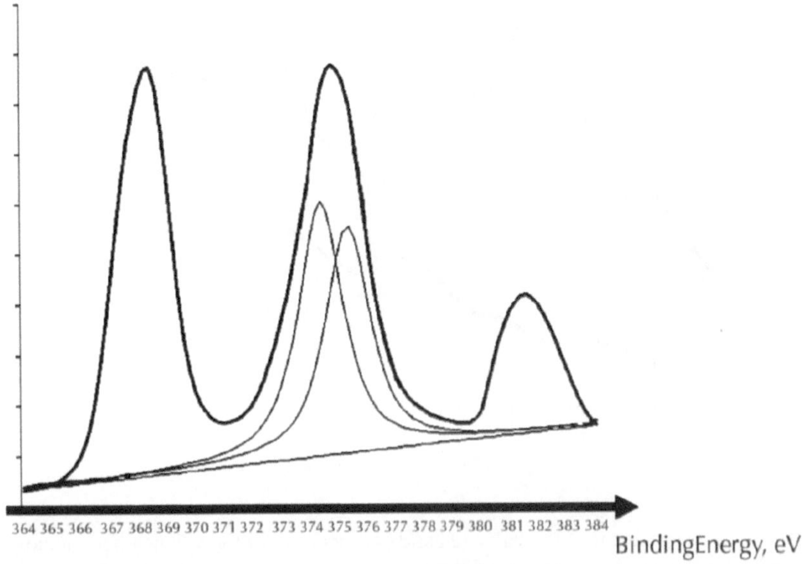

FIGURE 2 Ag3d spectrum ((a) Ag3d spectrum of the carbon copper-containing nanostructures at interaction with silvera and (b) Ag3d spectrum of coating).

The pure silver $Ag3d_{5/2,\ 3/2}$ spectra are formed by two components of the spin-orbital splitting (Figure 2 (a)). Similar situation is observed in carbon copper-containing nanostructures functionalized with silver (Figure 2 (b)).

The complex composite Ag3d spectrum (Figure 2 (c)) consists of the following maxima: the first—Ag or Ag-C-Cu, and at the distance of 7 eV from it there is a less intensive maximum corresponding to an ionic component of the chemical bond between silver and carbon.

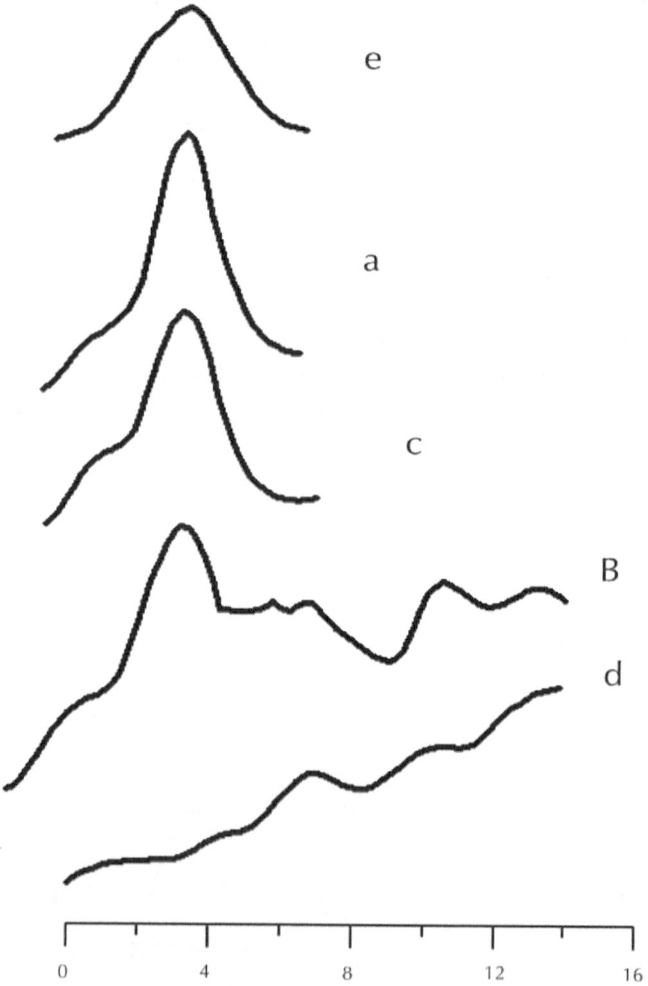

FIGURE 3 The spectrum of the valence band (a) spectrum of the valence band nanostructures interacted with silver, (b) spectrum of the valence band coating, (c) spectrum of the valence band Ag_2O, (d) spectrum of the valence band C, and (e) spectrum of the valence band pure Ag.

To clarify the nature of the first Ag or Ag-C component, the spectrum of the valence band of the studied composite has been taken (Figure 3). In contrast to pure silver, the growth of an additional maximum at E_f and a similar maximum in the Ag_2O valence band indicates the hybridization of Ag4d5s-electrons with 2p electrons of the second component —carbon. The shape and intensity of the valence band at the distance of 7 eV is associated with the presence of the Ag-C-O component, which is confirmed by the coincidence of them with the maxima in the valence band of pure graphite [4] and the calculations of the oxygen density of states.

The increase of the density of states at E_f leads to the growth of the electron density of the polymer composite, and, correspondingly, of the electrical conduction in it, similar to that observed for Ag_2O which is used for increasing conduction and is an obligatory element in fabricating conductive glass [15-30].

The data on the measurement of electrical resistance show that when 1% carbon metal-containing nanostructures is added in polymer, the electrical resistance decreases from 10^{-4} down to 10^{-5} Ω cm.

1.2.1 CONCLUSION

It is shown that nanomodification of the argentum-containing polymer coating improves the electrical conduction in it by one order of magnitude.

The analysis of the X-ray photoelectron spectra of the nanomodified composite shows the presence of the covalent bond of the Ag-C-Cu atoms and the growth of the electron density at E_f due to the hybridization of the C2p and Ag4d5sp electrons.

1.3 CASE STUDY II

The modification of functional groups in the composition of a protein macromolecule is one of the approaches in the development of biotechnology for pharmaceutics. The virus protection of pharmaceutical preparations of plasma is an acute international task. It is necessary to find albumin modifiers which do not negatively influence human organism.

The main task of the present investigation is the establishment of the regularities of the formation of the energy spectra of electrons and the determination of the chemical bond between the atoms of protein and a modifier by X-ray photoelectron spectroscopy, which allows to define the direction in the investigation of an increase in the protein stability and to choose an optimal modifier for albumin.

The XPS method has been chosen for the above purpose because in contrast to methods using ion and electron beams it is a non-destructive method. The choice of an X-ray electron magnetic spectrometer is conditioned by a number of advantages over electrostatic spectrometers that is constant luminosity and resolution independent of the electron energy and high contrast of spectra. In addition, the constructional separation of the magnetic type energy analyzer from the spectrometer vacuum chamber allows differently effect a sample in vacuum directly during spectra taking [30-37].

The present investigation tasks are as follows:

- The development of the method for the decomposition of X-ray photoelectron spectra into components for finding the spectra parameters indicating the transition of protein atoms into the state of stabilization.
- The development of the method for the determination of the temperature of the protein change and the establishment of a criterion for structural transitions.
- The investigation of the spectra of ordinary and compound amino acids with the purpose of the interpretation of the protein C1s, O1s, and N1s spectra.
- The study of the formation of the chemical bond between the atoms of protein and modifiers, namely, copolymers, super-dispersed particles of d metals and carbon metal-containing nanoforms.
- The study of the influence of the degree of the modification of protein with polymer (vinyl pyrrolidone–acrolein diacetal copolymer) on the protein thermal stability.
- The study of the influence of the protein modification on the protein thermal stability
- The selection of an optimal modifier for protein, which would provide the best protein thermal stability.

The objects of investigation are the native form of protein and modified protein. Protein was modified with carbon metal-containing nanostructures and additions of functional sp groups for increasing the activity of the interaction of nanostructures with the environment, NiO super-dispersed powder and vinyl pyrrolidone–acrolein diacetal copolymer at the temperature changing from room temperature to 623K.

Carbon metal-containing nanostructures are multilayer nanotubes growing on a metal particle due to the penetration of carbon atoms and carbon adsorption on the particle surface. Samples were prepared by the low energy synthesis method from polymers in the presence of metal systems. 3d metals (Cu and Ni) were used in the form of oxides and super-dispersed particles [12-20]. To increase the activity of the synthesis, the functional sp elements of ammonium polyphosphate were used [13-16].

The investigation of the formation of bionanostructures of a certain form and their properties is based on the concept of studying the interatomic interaction of initial components, the formation of the hybridized chemical bond of d electrons of metal atoms with p electrons of atoms of sp elements.

The C1s, O1s, and N1s spectra of the core levels of the samples of native albumin and modified albumin were studied at the temperatures from room temperature to 573. For investigating the state of atoms of carbon, oxygen, and nitrogen the investigation of reference samples of amino acids (glycine and histidine), copolymer, carbon metal-containing nanostructures, and super-dispersed d metal particles were studied and the data on the electronic structure of graphite and hydrocarbons were used [4].

Figure 4 shows the X-ray photoelectron 1s spectra of carbon and nitrogen taken from glycine and albumin samples at room temperature.

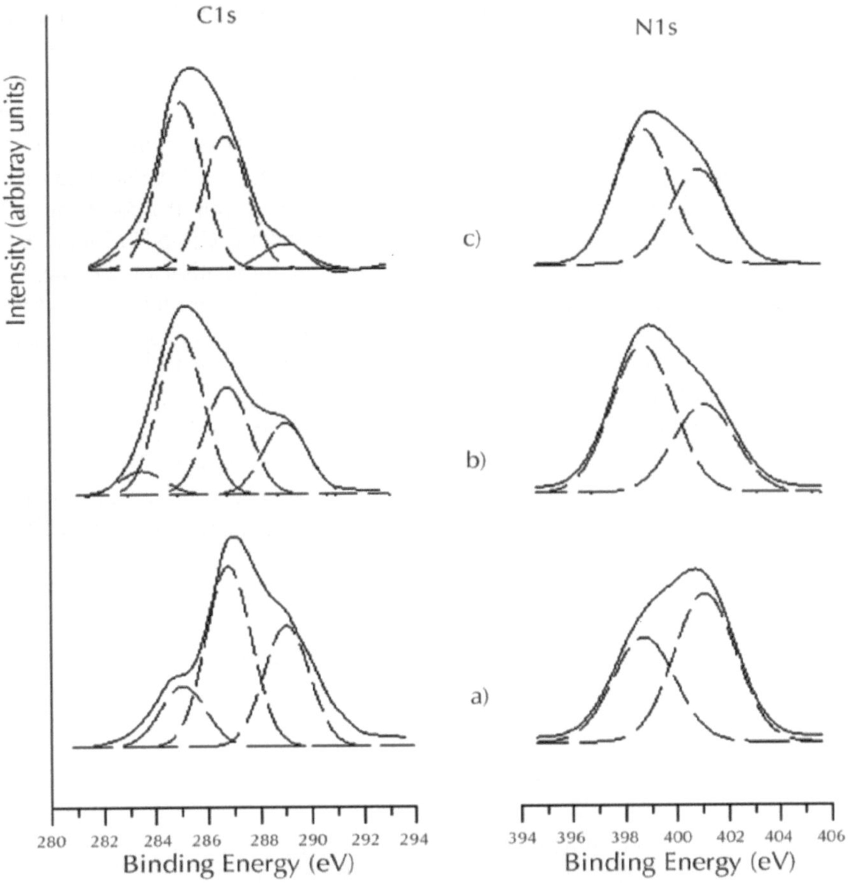

FIGURE 4 The X-ray photoelectron C1s and N1s spectra of glycine (a), histidine (b), and albumin (c) at 300K.

At room temperature the C1s spectrum of glycine (Figure 4 (a)) consists of three components, bound with different surroundings of carbon atoms, namely, C-H (285 eV), CH-NH (286.5 eV), and COOH (290.1 eV). At room temperature the N1s spectrum of glycine (Figure 4 (b)) consists of two components bound with different surroundings of nitrogen atoms, namely, CH-NH (398.5 eV), and N-O (401 eV).

The histidine C1s spectrum (Figure 4 (b)) consists of four components bound with different surrounding of carbon atoms at room temperature: C-C (283.5 eV), C-H (285 eV), CH-NH (286.5 eV), and COOH (289.1 eV). The histidine N1s spectrum (Figure 4 (b)) consists of two components bound with different surrounding of nitrogen atoms: CH-NH (398.5 eV) and N-O (401 eV). At room temperature, the albumin C1s-spectrum (Figure 4 (c)) also consists of four components bound with different surrounding of carbon atoms: C-C (283.5 eV), C-H (285 eV), CH-NH (286.5 eV), and COOH (289.1 eV). The presence of the C-C bond is also indicated by a satellite at 306 eV [4]. The albumin N1s spectrum (Figure 3 (a)) consists of two components reflecting the bonds of nitrogen with hydrogen (N-H) and oxygen (N-O). The appearance of oxidized nitrogen can be explained by the oxidation of protein on the sample surface (several tens of Angstroms) or by the formation of the N-O bonds in the protein structure. At temperature growing above 350 K, the spectra shapes of the studied samples change significantly. In the C1s spectrum, the contributions from the components C-C (283.5 eV) and C=O (287.1 eV) appear; and the components CH-NH (286.5 eV) and COOH (290.1 eV) disappear, which indicates the decomposition of the samples at heating, and the spectrum consists of three components C-C, C-H and C=O. In the N1s spectrum the contribution from the CH-NH component disappears at heating, and the spectrum consists of one component reflecting the bond N-O (401 eV).

Thus, the oxidation of protein leads to its destruction. The appearance of the carbonyl group C=O indicates the protein destruction (Figure 4 (a), (b), and (c)). The concentration of the carbonyl groups shows the degree of the protein destruction [5]. The growth of the C-C bonds indicates the partial breakage of the C-H bonds. Thus, we have determined the parameters of the X-ray photoelectron spectra characterizing the state of protein. The NH components in the N1s spectrum and COOH in the C1s spectrum indicate the presence of protein. The absence of these components in the N1s and C1s spectra and the growth of N-O and C=O bonds indicate the oxidative destruction of protein. With the change of temperature the amino groups of glycine, histidine, and albumin have similar behavior.

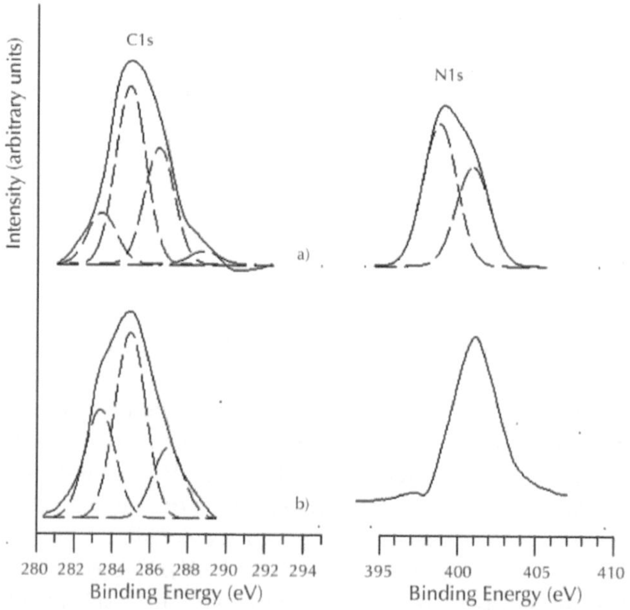

FIGURE 5 The X-ray photoelectron C1s and N1s spectra of albumin—(a) T = 300 K and (b) T = 450 K.

The samples of vinyl pyrrolidone–acrolein diacetal copolymer have been studied. The copolymer contains 84% of vinyl pyrrolidone and 16 mol% of acrolein diacetal.

The X-ray photoelectron C1s and N1s spectra of vinyl pyrrolidone–acrolein diacetal copolymer were taken at room temperature and at heating to 473K. At room temperature, the C1s spectrum of vinyl pyrrolidone–acrolein diacetal copolymer consists of three components bound with different surrounding of carbon atoms: C-H (285 eV), N-C (N-CH) (287.3 eV), and COOH (290.1 eV). At the increase of temperature to 373 K, the contribution of the COOH component decreases in the carbon spectra and the components C-H (285 eV) and N-C (N-CH) (287.3 eV) remain. At the temperature above 473, the component C=O (287.1 eV) appears in the C1s spectrum. At room temperature and at the temperatures up to 373K the N1s spectrum consists of two components N-C (397 eV) and N-O (401 eV) and at the temperature above 373K there is one component N-O (401 eV) in the N1s spectrum.

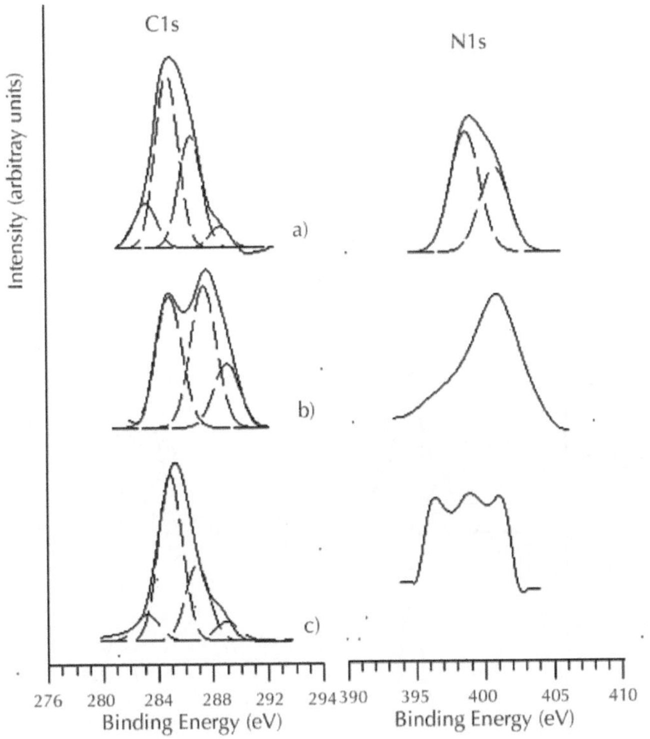

FIGURE 6 The X-ray photoelectron C1s and N1s spectra of (a) albumin, (b) vinyl pyrrolidone–acrolein diacetal copolymer, and (c) conjugate.

The comparison of the C1s spectra (Figure 6) of albumin, vinyl pyrrolidone–acrolein diacetal copolymer and albumin modified with vinyl pyrrolidone–acrolein diacetal copolymer (conjugate) taken at room temperature shows that in the conjugate prepared from albumin and vinyl pyrrolidone–acrolein diacetal copolymer in the ratio

1:5, there are three components characteristic of albumin, that is C-C (283.5 eV), C-H (285 eV) and COOH (290.1 eV), and the fourth component has the binding energy 286.8 eV characteristic of the CH-NH bond and vinyl pyrrolidone–acrolein diacetal copolymer which has a component with the binding energy 287.3 eV (CH-N). Figure 6 shows the N1s spectra of albumin, vinyl pyrrolidone–acrolein diacetal copolymer and conjugate. In the conjugate N1s spectrum there are components characteristic of vinyl pyrrolidone–acrolein diacetal copolymer and albumin. In contrast to albumin, an increase in the thermal stability of conjugate can be explained by the formation of stronger bonds C-N in it. The dependence of the conjugate thermal stability on the degree of the albumin modification with vinyl pyrrolidone–acrolein diacetal copolymer (1:1, 1:3, and 1:5) has been studied. The samples of albumin modified with vinyl pyrrolidone–acrolein diacetal copolymer in the ratios 1:1, 1:3, and 1:5 were investigated by the XPS method. The C1s and N1sspectra of all the samples (Figure 7) were studied.

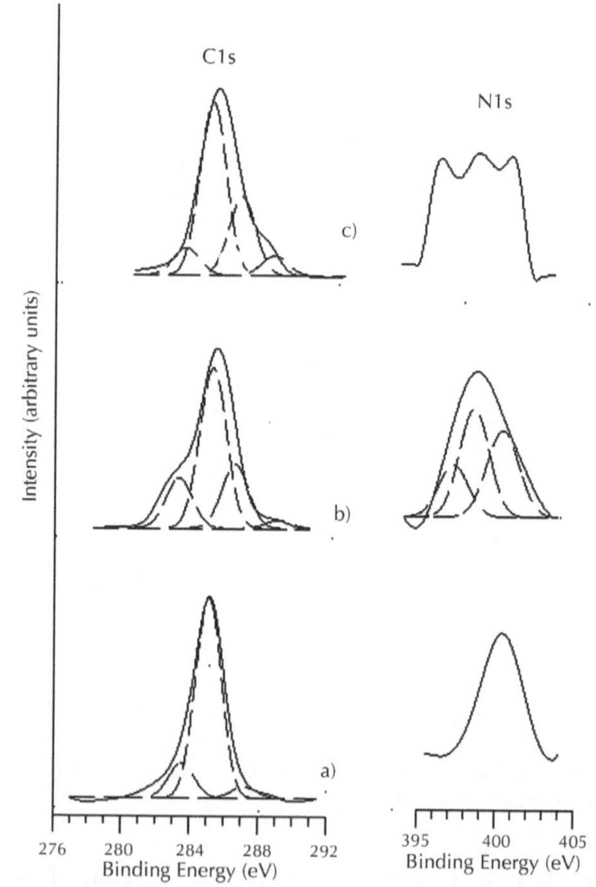

FIGURE 7 The X-ray photoelectron C1s spectra of protein modified with vinyl pyrrolidone–acrolein diacetal copolymer in the ratios—(a) 1:1, (b) 1:3, and (c) 1:5.

In the C1s spectrum (Figure 7 (a)) of the sample with equal content (1:1) of the modifier (vinyl pyrrolidone–acrolein diacetal copolymer) and albumin there are three components C-C, C-H, and C=O. In the N1s spectrum of the sample there is only one component N-O and the contributions from N-H and COOH characteristic of protein are absent at room temperature.

For the sample containing protein and vinyl pyrrolidone–acrolein diacetal copolymer in the ratio 1:3, the C1s spectrum (Figure 7 (b)) shows maxima corresponding to the bonds C-C, C-H, C-NH, and COOH and in the N1s spectrum, maxima characteristic of the C-NH, NH, and N-O bonds are observed. Consequently, there are components N-H and COOH characteristic of protein.

The amino acid group NH is bound with carbon (C-NH) which stabilizes the N-H group and prevents its decomposition. With further growth of the vinyl pyrrolidone–acrolein diacetal copolymer content in protein (5:1), the C-NH, and COOH components' contributions grow relative to that of C-H (Figure 7 (c)). The thermal stability grows with the increase of the relative content of vinyl pyrrolidone–acrolein diacetal copolymer in the conjugate and it reaches 473 K for the 5:1 modification.

For studying the consequences of the interaction of nanobiosystems with artificial nanostructures the modification of protein with nanostructures was conducted. At the protein modification with multilayer carbon copper-containing nanostructures, the position and relative content of the components in the C1s and N1s spectra (Figure 8 (a)) differ from those of the components observed for unmodified protein. For modified protein, in the C1s and N1s spectra, in addition to C-C, CH, and COOH, the $C\text{-}N(H_2)$ components are observed. Heating up to 523K changes insignificantly the shape of spectra (Figure 8 (a) and (b)). Apparently, the bond between protein atoms (N-H) and carbon atoms becomes stronger due to the formation of double or triple bond of C with NH_2 or, which is more likely, the formation of the hybridized sp^3 bond on the $C\text{-}N(H_2)$ atoms. In contrast to the weak bond of the NH_2 groups, the presence of the strong covalent bond $C\text{-}N(H_2)$ leads to an increase in the thermal stability of the protein modified with carbon copper-containing nanostructures to 523 K. With the temperature increase above 523 K, the intensity of the $C\text{-}N(H_2)$ and COOH components decreases in the C1s spectrum and at T = 623 K these components completely disappear. In the C1s spectrum the components similar to destructed albumin that is C-C, C-H, and C=O, remain and in the N1s spectrum the component N-O remains which is characteristic of destructed protein. Sharp worsening of vacuum and the appearance of the Au4f spectrum from the substrate indicate the protein coagulation. Similar to the case of albumin, the growth of C-C bonds is observed.

Nanoparticles can influence protein molecules by penetrating into cells and often destructing them [6-18]. There is a minimal radius of a nanoparticle at which it can be captured inside the cell. Consequently, the biocompatibility of protein and nanoparticles depends on the nanoparticle size that is the surface area can be used as a measure of oxidation (toxicity).

Then, the protein modified by multilayer carbon nickel-containing nanotubes prepared by low temperature synthesis of polymers and nickel oxides was investigated. The multilayer carbon nickel-containing nanotubes have external dimensions much smaller (20–40 nm) than those of the above multilayer copper-containing nanotubes (from 50 nm and larger). The protein modified by multilayer carbon nickel-containing nanotubes destruction is observed at room temperature. In the C1s spectrum, the COOH carboxyl group is absent, and the C=O carbonyl group and C-C and C-H appear. In the N1s spectrum the N-H group disappears and the component N-O grows.

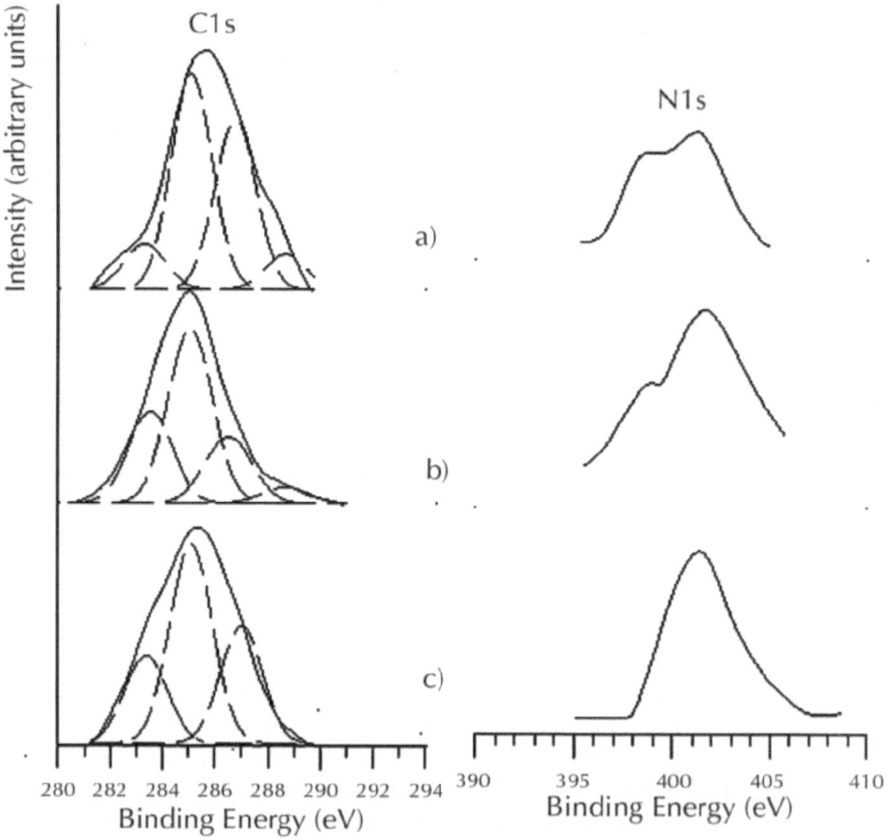

FIGURE 8 The X-ray photoelectron C1s and N1s spectra of albumin with multilayer carbon copper-containing nanotubes—(a) T = 373K, (b) T = 523K, and (c) T = 623K.

The same result is observed when protein is modified with NiO nanoparticles which are oxidation catalysts. The destruction of albumin modified with multi-layer carbon nickel-containing nanotubes is observed at room. The decomposition of the spectra into the components shows that in the C1s and N1s spectra the components COOH and N-H are absent and the components C-C, C-H, C=O, and N-O are present similar to the case of the modification with carbon nickel-containing nanostructures, that is the protein destruction takes place at room temperature. Thus, the growth of carbonyl groups, the breakage of the N-H bond and the formation of the N-O bond occur at an elevated temperature or in the presence of the interaction with a metal catalyst.

1.3.1 CONCLUSION

Based on the conducted fundamental investigations, the regularities are obtained, which can be used for the choice of the direction in the purposeful increase of protein stability. The investigations of the protein modification show that:

- Low thermal stability of protein (albumin) is due to the weak bond of the $NH_3(NH_2)$ group and the decrease of the content of NH (NH_3 and NH_2) groups in the protein composition and nitrogen oxidation at heating.
- One of the consequences of the formation of the strong covalent bond between the atoms of the NH_2 amino group and the carbon atoms (C-NH) at the protein modification is an increase in the thermal stability of the modified protein. The destruction of carboxyl groups (COOH) and the formation of the carbonyl groups (C=O) as it follows from the change of the C1s spectrum at heating indicate the destruction of protein. At the protein destruction, the bond between hydrogen and nitrogen is broken, and nitrogen forms the bond with oxygen.
- The presence of metal catalysts (Ni, NiO) in protein leads to the oxidation and destruction of protein due to the formation of carbonyl groups and the destruction of the N-H groups.

Also:

1. A method for the determination of the temperature of the protein destruction has been developed and the criteria of the protein destruction have been established.
2. At the modification of protein with carbon copper-containing nanotubes a strong covalent bond (C-NH) of the atoms of the N-H amino group with the carbon atoms is formed, which results in the increase of the modified protein thermal stability up to 523–573 K.
3. The compatibility of bio and nanostructures depends on the size of the nanostructures, which can serve as a measure of oxidation.
4. In addition to the influence of the temperature on the oxidative destruction of protein, the presence of metal catalysts (Ni and NiO) in protein leads to the oxidation and destruction of protein due to the formation of carbonyl groups and the decomposition of the N-H groups at room temperature.
5. In contrast to the modification of protein with carbon copper-containing nanotubes, the modification of protein with carbon nickel-containing nanotubes leads to the protein destruction at room temperature due to the presence of Ni atoms.
6. One of the consequences of the formation of the covalent bond C-N(-H) between nitrogen atoms and carbon atoms is an improvement in the thermal stability of the modified protein.
7. The dependence of the thermal stability of protein modified with vinyl pyrrolidone–acrolein diacetal copolymer on its content in the mixture is found. The

growth of the temperature of stabilization with the increasing content of vinyl pyrrolidone–acrolein diacetal copolymer is shown.

8. At the vinyl pyrrolidone–acrolein diacetal copolymer concentration three times smaller than that of protein, the protein destruction is observed at room temperature. With the threefold and more increase of the vinyl pyrrolidone–acrolein diacetal copolymer concentration, the temperature of the protein stability grows and is maximal at the ratio 1:5.

Based on the results obtained some recommendations on the modification of protein for increasing its stability are given, which can be used in pharmaceutical biotechnology.

A model for the protein stabilization is offered on the basis of the formation of the strong hybridized chemical bond of the atoms of protein and a modifier.

The results of the work on choosing an optimal modifier for protein for medico-biological and pharmaceutical technologies show that the highest thermal stability of protein (523–573 K) is achieved at the modification of protein with carbon copper-containing nanotubes having functional groups of sp elements in the ratio 1:0.01, which is more rational in comparison with the protein modification with vinyl pyrrolidone–acrolein diacetal copolymer because in this case the required amount of the modifier is significantly larger (in 3–5 times) than the amount of protein.

1.4 CASE STUDY III

For many years, the improvement of the mechanical properties of structural materials was mainly performed by the development of new alloys with new chemical and phase compositions. Lately, new ways have appeared for improving properties of structural materials, namely by the well-directed formation of micro- and nanocrystalline structure.

The set of the experimental methods that are used for studying the chemical structure of carbon cluster nanostructures is limited. Therefore, one of the main tasks is the development of diagnostic methods, which will allow controlling intermediate and final results in the creation of new materials.

At present, the analysis of numeral works shows that classical methods for determining shapes, sizes, and compositions of carbon nanostructures are transmission electron microscopy, methods based on electron diffraction and Raman spectroscopy. However, more and more publications appear referring to the investigation of nanostructures with the use of the XPS method. Further development of the XPS method and related methods for the surface (from 1 to 10 nm) investigation will lead to an increase in the number of methods for studying compositions, electronic properties, and structures of nanostructures.

The XPS has been used for the determination of the type of carbon structures. The XPS method allows investigating the electron structure, chemical bond, and nearest environment of atoms. One of the important specific features of the method is its non-destructive character of action since the X-ray radiation used for photoelectron excitation does not practically because any damages in most materials. This cannot be said about the surface analysis methods that involve ion or electron bombardment of a surface. In most cases, a sample can further be used for some other investigations after it has been studied by the XPS method.

In addition, the method provides the possibility to analyze thin layers and films, which are very important for the case of the formation of fullerenes, nanotubes and nanoparticles, and to obtain the information on the sample chemical composition based on spectra, which provides the control over chemical purity of materials.

The XPS method allows investigating electron structure, chemical bond, and the nearest environment of atoms with the use of an X-ray photoelectron magnetic spectrometer. The X-ray photoelectron magnetic spectrometers with automated control system [1] are not inferior to the best foreign spectrometers in their main parameters. The preference is given to the X-ray electron magnetic spectrometer because of a number of advantages compared with electrostatic spectrometers [2], which are the high spectrum contrast, the permanency of optical efficiency and resolution capacity that are not influenced by electron energy. Moreover, the XPS method is a non-destructive investigation method.

In this work, the samples prepared from iron powder modified with fullerenes or graphite was studied with the use of the XPS method. The samples were prepared in two ways—fusion or pressing. The modification was carried out for obtaining nanocarbon structures in metal matrices in order to improve strength properties of a material. The samples were verified with the use of X-ray diffraction, which showed that the sample structures were mainly fcc iron. The description of the samples studied is given in Table 1.

TABLE 1 The investigated samples

Sample No	Sample composition	Sample form
1	Fe + 0.5% $C_{60/70}$	Ingot (T = 1410°C)
2	Fe + 0.5% graphite	Ingot (T = 1410°C)
3	Fe + 2% $C_{60/70}$	Powder
4	Fe + 2% graphite	Powder
5	Fe + 2% $C_{60/70}$	Pellet (P = 800 MPa)
6	Fe + 2% graphite	Pellet (P = 800 MPa)

The X-ray photoelectron investigations were carried out for studying the changes in the nearest environment of the carbon atoms in the samples prepared in different ways.

The investigations were carried out using the X-ray photoelectron magnetic spectrometer with double focusing and instrumental resolution of 0.1 eV at the excitation of AlKα-lines (1486.6 eV).

For the XPS investigations of carbon-metal cluster nanomaterials, the method of the C1s spectra identification by the satellite structure was employed. To do this, reference samples were studied, the carbon components of which could give the C1s spectrum——C-H–hydrocarbons [1-25], C-C (sp^2)–graphite [4], and C-C (sp^3)–diamond [25-37]. The spectra parameters are presented in Table 2, where E_b is binding energy, E_{sat} is the energy characterizing the satellite position, I_{sat} is the satellite intensity, and I_0 is the intensity of the main maximum.

TABLE 2 The C1s spectra parameters for reference samples

	E_b, eV	E_{sat}, eV	$\Delta E = E_{sat} - E_b$, eV	FWHM, eV	I_{sat}/I_0, Δ = 10%
C-H [3] hydrocarbons	285.0 ± 0.1	292.0 ± 1.0	~ 6–7	2.0 ± 0.1	0.10
C-C (sp^2) [4] graphite	284.3 ± 0.1	306.0 ± 1.0	~ 22	1.8 ± 0.1	0.10
C-C (sp^3) [5] diamond	286.1 ± 0.1	313.0 ± 1.0	~ 27	1.8 ± 0.1	0.15

To identify the structures studied in [6], the C1s spectra of carbon nanostructures were studied. The nanostructures were obtained in the electric arc during graphite electrode sputtering. The carbon nanostructures obtained were fullerenes C_{60}, single-walled and multi-walled carbon nanotubes and amorphous carbon.

It is shown that in all the C1s spectra there is a satellite structure related to different effects (a shake-up process and characteristic losses—plasmons [17-24]), which allows to create a calibration technique and to determine not only the energy position of the components but their intensities as well. In the C1s spectrum of fullerenes C_{60}, there is a satellite with binding energy of 313 eV [8-14] and the relative intensity of 15% from the main peak. This satellite is characteristic of the sp^2hybridization of the valence electrons of the carbon atoms.

In the C1s spectrum of single-walled carbon nanotubes, in addition to a gradually rising spectrum in the high energy region, two satellites are observed characteristic of the C-C bonds with the sp^2 and sp^3 hybridization of the valance electrons of the carbon atoms. Consequently, in the C1s spectrum of the one-layer nanotubes, there are these two components with the binding energies of 284.3 and 286.1 eV and the intensities of 1:0.1 and 1:0.15 relative to their satellites and the width of 1.8 eV. The ratio between the C-C bonds with the sp^2 and those with sp^3 hybridization of valence electrons is 2. The similar situation is observed for the C1s spectrum of multi-walled carbon nanotubes.

In the region of 313 eV of the amorphous carbon spectrum, there is a satellite characteristic of sp^3 hybridization of the valence electrons with the relative intensity of 15%. Consequently, in the C1s spectrum there is a component characteristic of C-C bond with sp^3 hybridization of the valence electrons at the distance of 27 eV from the satellite. Thus, the amorphous carbon presents carbon inclusions that are like a globe-shaped form of graphite. The development of a calibration method for spectra in the X-ray photoelectron investigations of reference samples allows realizing the decomposition of the C1s spectrum to the components that determine the chemical bond, the hybridization type of s-p valence electrons of the carbon atoms and the nearest environment of the carbon atoms. Studying the nanoparticles with a known structure gives the possibility of the identification of studied carbon structures by examining the C1s spectra shapes.

The C1s spectra identification method developed by us was successfully used for the investigation of nanostructures in the iron matrix.

The XPS method was used to obtain the C1s, O1s, and Fe2p spectra. The O1s spectra show that large amounts of adsorbed oxygen and iron oxides are present on the sample surfaces. The Fe2p spectra are also indicative of the presence of the oxidized iron layer on the surfaces of the samples. During the experiment, the shift of the spectra was not observed. The experimental C1s line spectra for samples №1 and №2 are shown in Figure 9 (a) and (b).

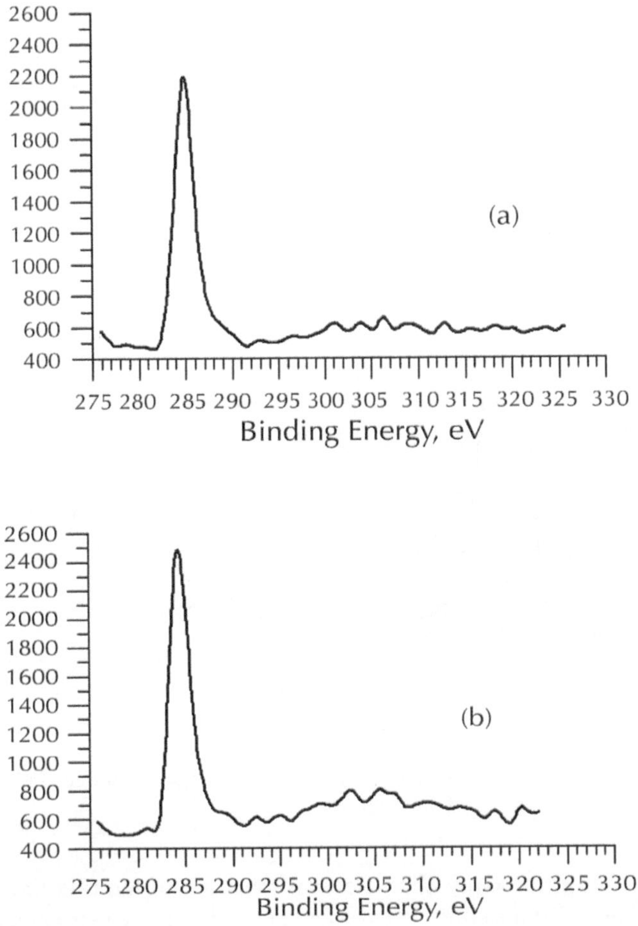

FIGURE 9 The experimental C1s spectra of samples №1 (a) and №2 (b).

The mathematical treatment of the C1s spectra was performed that is the background subtraction and the procedure of the spectra smoothing and decomposition. The results are given in Figure 10.

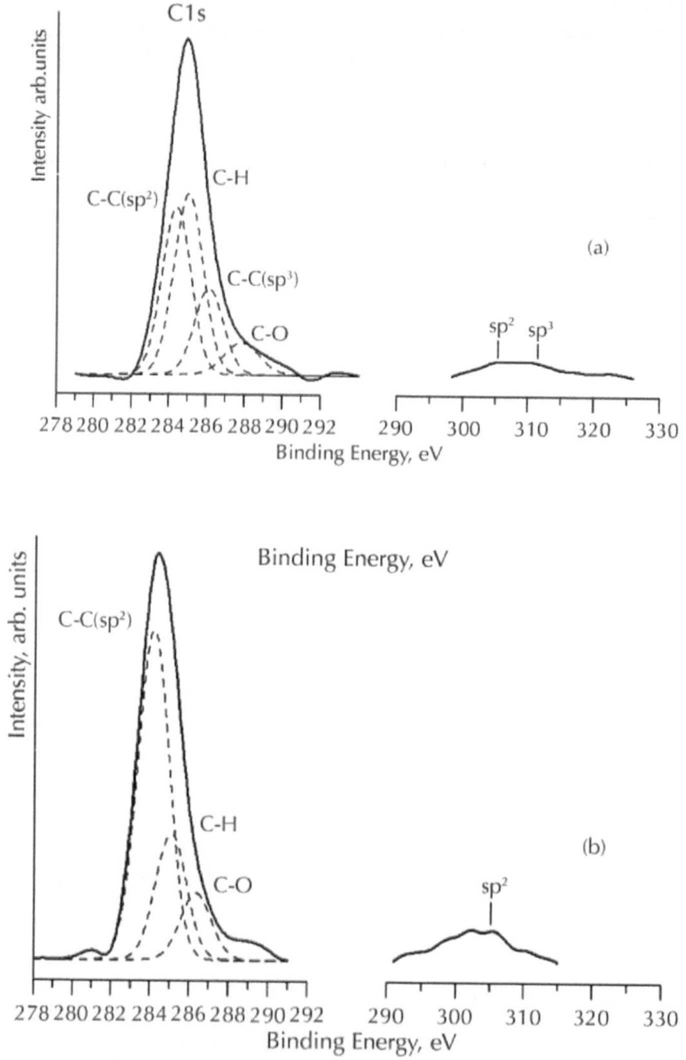

FIGURE 10 The X-ray photoelectron C1s spectra obtained from samples №1 (*a*) and №2 (*b*) after the spectra were decomposed into their components.

In Figure 10 (a), the X-ray photoelectron spectrum of the C1s line is displayed, which was obtained from sample №1 without heating in the spectrometer chamber. In the high energy region, two satellites are observed, which are characteristic of sp^2 and

sp^3 hybridization of the valence electrons of the carbon atoms judging by their binding energies (306.0 eV and 313.0 eV, respectively). According to the data in Table 2, in the C1s spectrum, there are components characteristic of the C-C (sp^2) and C-C (sp^3) bonds at distances of ~22 and ~27 eV. The relation of the C-C (sp^2) bond intensity to the C-C (sp^3) bond intensity is ~2, which is characteristic of carbon nanostructures. In the spectrum, there are also components of C-H and C-O bonds, which characterize surface contaminations.

When the sample is heated, the breakup of the C-C bonds with sp^3-type of hybridization of the valence electrons of the carbon atoms is taking place. At further heating of the sample, in the C1s spectrum there is a component characteristic of Fe-C bonds. In addition, at heating, the Fe2p spectra are observed on the sample surface, which is indicative of the fact that the sample surface is being cleaned during heating.

In Figure 10 (b), the C1s spectrum is displayed, which is obtained from sample №2 without heating in the spectrometer chamber. In the high energy region, a satellite with the binding energy of ~306.0 eV is observed, which is characteristic of the sp^2 hybridization of the valence electrons of the carbon atoms. Consequently, in the C1s spectrum, at the distance of ~22 there is a component characteristic of C-C bonds with sp^2 hybridization. There are also components characteristic of C-H and C-O bonds in the spectrum.

The X-ray photoelectron investigations of samples 3, 4, 5, and 6 show that the C1s spectra do not have a satellite and they have low intensity. Thus, only hydrocarbon contaminations are present on the surfaces of these samples.

1.4.1 CONCLUSION

The investigations conducted have demonstrated that when iron powder is mixed with fullerene mixture C_{60}/C_{70} or with graphite powder and then the mixtures obtained are treated differently (fusion and pressing), there are strong differences in the X-ray photoelectron spectra.

On the surfaces of the samples prepared from the mixture of the iron powder and the mix of fullerenes C_{60}/C_{70} and subjected to fusion, carbon nanostructures are present. On the surface of the samples prepared from the mixture of the iron powder and graphite powder, which were also subjected to fusion, graphite-like structures, are present.

1.5 CASE STUDY IV

The question of the mechanisms of modified cast irons and steels and the reasons for the formation of ordering phases in them is one of the most important in obtaining new materials with enhanced properties of use.

In spite of the large number of theoretical and experimental investigations devoted to the nanotypes, there is no single model that allows us to explain the structure and properties of steel and cast iron samples we obtain. The number of experimental techniques used to investigate the chemical structure of nanostructures at the atomic level is limited.

In this work, we examine industrial samples of nano modified cast irons and steels by means of XPS on an EMS-100 photoelectron magnetic spectrometer with double focusing [1]. The XPS allows us to investigate electronic structure, chemical bonds, and local atomic environments. We chose a photoelectron magnetic spectrometer because of its numerous advantages over electrostatic spectrometers.

These include constant light power and accuracy independent of electron energies and high spectral sharpness. In addition, XPS is a non-destructive technique, which is particularly important in investigations of metastable systems.

The objects of investigation were relegated cast iron samples obtained by means of rotary casting, modified aluminum doping agents, samples of 08X18N10T steel, and stainless nonmagnetic modified nanostructures of 08X21G11AN6 steel containing sp elements (carbon and nitrogen).

The chemical composition of the samples is shown in Table 3.

TABLE 3 Results of chemical analysis

Sample	Chemical composition, weight %							
	C	Mn	Si	Cr	Ni	Cu	Al	N
Cast iron	3.03	0.54	1.92	0.28	0.13	0.31	0.19	0.01
Modified cast iron	3.06	0.55	2.07	0.27	0.12	0.30	1.40	0.01
Stainless steel	0.08	1.20	0.50	18.10	9.85	0.25	0.10	0.01
Modified stainless steel	0.05	10.84	0.36	20.98	5.39	0.09	0.01	0.61

Experiments were performed on a unique EMS 100 automatic X-ray photoelectron magnetic spectrometer [1] with a resolution of 10^{-4} and a luminosity of 0.185%.

The spectra of inner energy levels Fe2p, Al2p, C1s, and O1s were investigated. The time of signal accumulation at each point of a spectrum was from 3 to 30 sec. The purity of the surface was monitored using the C1s and O1s spectra. Spectral decomposition was performed using software based on the least square method. Energy status, spectral width, and intensity data were considered in the software and compared to the spectra of reference samples. Decomposition was performed using a Gaussian function with maximum approximation of the envelope curve to the experimental curve. The accuracy of determining peak positions was 0.1 eV. The error in determining the sharpness of the electron spectra was no more than 5%.

The decomposition of the C1s spectrum was performed to examine chemical bonds, sp hybridization of valence electrons, and the local atomic environment of carbon atoms. To accomplish this, reference spectra (the carbon components of the C1s spectrum) were investigated—C–C (sp^2) graphite, C–C (sp^3) fullerene, and C–H hydrocarbons [2].

Figure 11 shows the C1s spectra of carbon in the ordinary and modified cast irons.

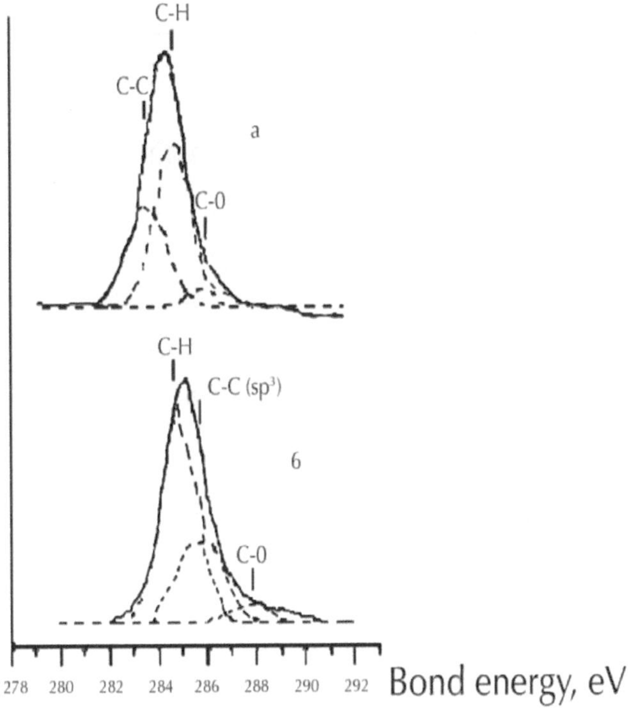

FIGURE 11 C1s spectra of carbon in (a) ordinary and (b) modified cast iron.

The spectra are shifted toward each other by 1.7 eV. The energy value of the C1s level in the ordinary cast iron is close to the one corresponding to the line in the Fe_3C spectrum and the C–C bond in graphite. Consequently, the bonds between carbon and iron in these samples are close in nature. Hydrocarbon and adsorbed oxygen impurities were present in all of the investigated samples, since all of the samples were placed into the chamber from the air. The energy values for the C1s level in modified cast iron coincide with the energy of the C1s level for diamond. Diamondlike structures with sp3 electron configuration thus, form in the carbon atoms of modified cast iron.

The C1s spectra of C_{60} carbon nanostructures obtained in an electric arc during the diffusion of the carbon electrode were examined to identify the objects under investigation. A satellite with bond energy 313 eV and an intensity of 15% relative to the main peak intensity is observed in the C1s fullerene spectrum. This particular satellite peak is characteristic of the sp^3 hybridization of valence electrons in a carbon atom. Satellite peaks are observed in graphite and diamond spectra at 22 eV and 27 eV, respectively, from the main maximum with relative intensities of 10% and 15%, respectively [3-12].

The values of the bond energies of the electrons of the Fe2p line of the cast iron samples are shown in Table 4.

TABLE 4 Values of electron bond energies of inner bonds

Substance	Energy level, eV	C1s energy level
Fe		
Fe_3C	708.7 ± 0.2	284.1 ± 0.2
Fe_3Al	708.1 ± 0.2	
Fe_3Si	708.5 ± 0.3	
Graphite	–	284.4 ± 0.2
Cast iron	708.8 ± 0,3	284.2 ± 0.2
Modified cast iron	708.1 ± 0.3	286.2 ± 0.2

In contrast to the spectra of pure iron and like Fe_3C in highly carbon phase, the Fe2p spectra have no clear features.

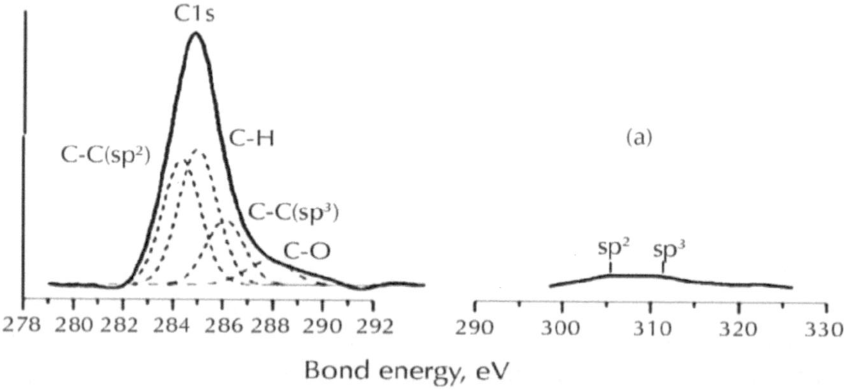

FIGURE 12 C1s spectra of carbon.

Figure 12 presents the XPS spectrum of a C1s line obtained from the surface of the modified steel sample.

Two satellites are observed in the high energy area that in terms of energy bonds are characteristic of the sp2 (306.0 eV) and sp^3 (313.0 eV) hybridizations of the valence electrons of carbon atoms, respectively. Components characteristic of C–C (sp^2) and C–C (sp^3) bonds are observed in the C1s spectrum at distances of around 22 and 27 eV. The ratio of the intensities of the main maximum of C–C (sp^2) and the main maximum of C–C (sp^3) bonds is ~2, a feature of carbon structures. The spectral components of C–H and C–O bonds also indicate surface contamination. The investigations reveal that the obtained steel is 1.5–3 times more durable, with identical or increased plasticity (Table 5).

TABLE 5 Mechanical properties of samples

Sample	Yield limit, $\sigma_{0.2}$, kgf/mm²	Breakdown limit, σ_t, kgf/mm²	Relative elongation, δ, %
08X18H10T	25	55	35
08X21Г11AH6	75	99	55

Iron and aluminum atoms in cast irons and steels have different properties. Aluminum, distinguished by its high chemical activity, is an energetic reductant in steels and cast irons, and partially interacts with oxygen, drawing the oxygen to itself.

The high carbon phase plays an important role in the formation of structures of cast iron and steel. In triple Fe–Al–C alloys, their physical nature under goes qualitative change. Aside from two known phases of the high carbon type (graphite and cement carbide) special types of carbon appear in alloys in the Fe–Al–C system (spherical graphite and Fe–Al phase). Aluminum atoms interact with carbon atoms and are adsorbed primarily at crystal faces where free bonds are present. As a result, the spread of graphite along the basis planes is suppressed, graphitization is slowed, and conditions are created for transverse crystal growth and the formation of compact (spherical) graphite inclusions. The conclusions are also confirmed by optical microscopy data. Figure 13 presents photographs of the microstructure of samples.

Direct Fe–Al and C–C bonds with sp^3 hybridization of valence electrons begin to form during the production of modified iron by means of rotary casting and aluminum alloying. This leads to the appearance of new structural elements in electron spectra, relative to ordinary cast irons in which Fe–Si and C–C bonds with sp^2 hybridization are present. The durability of aluminum cast irons is enhanced due to the stronger interatomic interaction of Fe–Al relative to Fe–Si, since hybrid 3d(Fe) and 3p(Al) bonds are formed [14-28].

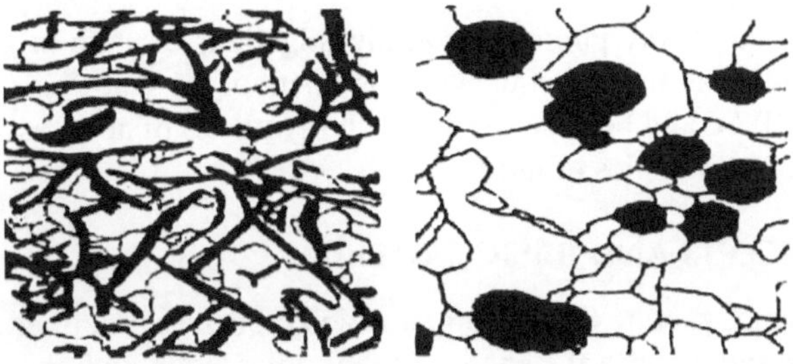

FIGURE 13 Graphite in modified and non modified cast iron.

The modification of cast iron with aluminum leads to the transfer of some aluminum valence electrons to graphite atoms, accompanied by the formation of a certain proportion of atoms with more energetically stable and diamond like sp3 hybridization of electrons.

The investigations of iron modified by fullerene has revealed the presence of carbon nanostructures on the surface of a sample, leading to its increased durability. Easily formed carbide nanostructures do not appear in steels alloyed with metals (V, Cr, and Mn). Such steels are alloyed with nitrogen. Alloying steel with nitrogen demonstrates that of all incorporated elements, nitrogen is the most effective hardener of solid alloys, and nitrides are more easily separated than carbides. Nitrogen's solubility in solid alloys of the Fe–Cr system is especially high due to the strong austenite forming properties of this element.

1.5.1 CONCLUSION

Using an X-ray photoelectron magnetic spectrometer, the XPS technique was applied for the first time to the investigation of metallic carbon nano modified cast irons obtained by means of rotary casting.

The results from XPS allowed us to determine the character of the ineratomic interaction of cast irons alloyed with aluminum and the mechanisms of the formation of spherical graphite in them.

The satellite structure of C1s spectra was used to identify C1s spectra and determine the sp hybridization type of valence electrons in the investigated samples.

Scientifically valuable results were thus obtained:

1. The role of the electronic structure in obtaining nano modified cast irons was identified.
2. The dependence of structure formation on the composition of initial substances was revealed.
3. The high durability of modified cast irons is explained by the formation of strong hybrid 3d(Fe) and 3p(Al) bonds, and by the presence of C–C complexes with diamond like sp3 hybridization of electrons.
4. Carbon nanostructures were present on the surface of the sample prepared from a mixture of iron powders and a C_{60}/C_{70} fullerene mixture and exposed to melting.
5. The high effectiveness of nitrogen's influence on the properties of steel use was demonstrated.

Results from our X-ray photoelectron investigations of nano modified cast irons and steels point the way for subsequent development of the synthesis of new cast irons and steels. The XPS could be used to control the synthesis of new cast irons and steels with enhanced characteristics of use.

1.6 KEY CONCEPTS AND BASIC CRITERIA

The activity of nanostructures in self-organization processes is defined by their surface energy thus corresponding to the energy of their interaction with the surroundings. It is known that when the size of particles decreases, their surface energy and particle activity increase. The following ratio is proposed to evaluate their activity:

$$a = \varepsilon_S/\varepsilon_V \quad (1)$$

Where, ε_S—nanoparticle surface energy and ε_V—nanoparticle volume energy.

Naturally in this case $\varepsilon_S >> \varepsilon_V$ conditioned by the greater surface "defectiveness" in comparison with nanoparticle volume. To reveal the dependence of activity upon the size and shape we take ε_S as $\varepsilon_S{}^{\circ}$—S, and $\varepsilon_V = \varepsilon_V{}^{\circ}$—V.
where, $\varepsilon_S{}^{\circ}$—average energy of surface unit, S—surface, $\varepsilon_V{}^{\circ}$—average energy of volume unit, and V—volume.

Then the Equation (1) is converted to:

$$a = d \cdot \varepsilon_S{}^{\circ}/\varepsilon_V{}^{\circ} \cdot S/V, \quad (2)$$

Substituting the values of S and V for different shapes of nanostructures, we see that in general form the ratio S/V is the ratio of the number whose value is defined by the nanostructure shape to the linear size connected with the nanostructure radius or thickness.

The Equation (3) can be given as:

$$a = d \cdot \varepsilon_S{}^{\circ}/\varepsilon_V{}^{\circ} \cdot N/r(h) = \varepsilon_S{}^{\circ}/\varepsilon_V{}^{\circ}\ 1/B \quad (3)$$

Where, B equals r(h)/N, r—radius of bodies of revolution including hollow ones, h—film thickness depending upon its "distortion from plane", and N—number varying depending upon the nanostructure shape.

Parameter d characterizes the nanostructure surface layer thickness, and corresponding energies of surface unit and volume unit are defined by the nanostructure composition. For the corresponding bodies of revolution the parameter B represents an effective value of the interval of nanostructure linear size influencing the activity at the given interval r from 1 to 1000 nm (Table 6). The table shows spherical and cylindrical bodies of revolution. For nanofilms the surface and volume are determined by the defectiveness and shape of changes in conformations of film nanostructures depending upon its crystallinity degree. However, the possibilities of changes in nanofilm shapes at the changes in the medium activity are higher in comparison with nanostructures already formed. At the same time, the sizes of nanofilms formed and their defectiveness (disruptions and cracks on the surface of nanofilms) are important.

TABLE 6 Changes in interval B depending upon the nanoparticle shape

Nanostructure shape	Internal radius as related to the external radius	Interval of changes B, nm
Solid sphere	–	0.33(3)–333.(3)

TABLE 6 *(Continued)*

Solid cylinder	–	0.5–500
Hollow sphere	8/9	0.099–99
Hollow sphere	9/10	0.091–91
Hollow cylinder	8/9	0.105–105
Hollow cylinder	9/10	0.095–95

The proposed parameters are called the nanosized interval (B) may be used to demonstrate the nanostructures activity. The nanostructures differing in activity are formed, depending on the structure and composition of nanoreactor internal walls, distance between them, shape and size of nanoreactor. The correlation between surface energy, taking into account the thickness of surface layer, and volume energy was proposed as a measure of the activity of nanostructures, nanoreactors, and nanosystems.

It is possible to evaluate the relative dimensionless activity value (A) of nanostructures and nanoreactors through relative values of difference between the modules of surface and volume energies to their sum:

$$A = (\varepsilon_S - \varepsilon_V)/(\varepsilon_S + \varepsilon_V) = [(\varepsilon_S°d)S - \varepsilon_V°V]/[(\varepsilon_S°d)S + \varepsilon_V°V] = \\ = [(\varepsilon_S°d/\varepsilon_V°)S - V]/[(\varepsilon_S°d/\varepsilon_V°)S + V] \quad (4)$$

If $\varepsilon_S >> \varepsilon_V$, A tends to 1.

If $\varepsilon_S°d/\varepsilon_V° \approx 1$, the equation for relative activity value is simplified as follows:

$$A \approx [(S - V)/(S + V)] = [(1 - B)/(1 + B)] \quad (5)$$

If we accept the same condition for *a*, the relative activity can expressed *via* the absolute activity:

$$A = (a - 1)/(a + 1) \quad (6)$$

At the same time, if $a >> 1$, the relative activity tends to 1.

Nanoreactors represent nanosized cavities, in some cases nanopores in different matrixes that can be used as nanoreactors to obtain desired nanoproducts. The main task for nanoreactors is to contribute to the formation of "transition state" of activated complex being transformed into a nanoproduct practically without any losses for activation energy. In such case the main influence on the process progress and direction is caused by the entropic member of Arrhenius equation connected either with the statistic sums or the activity of nanoreactor walls and components participating in the process.

The surface energy of nanostructures represents the sum of parts assigned forward motion (ε_{fm}), rotation (ε_{rot}), vibration (ε_{vib}), and electron motion (ε_{em}) in the nanostructure surface layer:

$$\varepsilon_S = \Sigma(\varepsilon_{fm} + \varepsilon_{rot} + \varepsilon_{vib} + \varepsilon_{em}) \quad (7)$$

The assignment of these parts on values depends on nature of nanostructure and medium. The decreasing of nanostructure sizes and their quantity usually leads to the increasing of the surface energy vibration part, if the medium viscosity is great. When the nanostructure size is small, the stabilization of electron motion takes place and the energy of electron motion is decreased. Also, the possibility of coordination reaction with medium molecules is decreased. In this case the vibration part of surface energy corresponds to the total surface energy.

The nanostructures formed in nanoreactors of polymeric matrixes can be presented as oscillators with rather high oscillation frequency. It should be pointed out that for nanostructures (fullerenes and nanotubes) the absorption in the range of wave numbers 1300–1450 cm^{-1} is indicative. These values of wave numbers correspond to the frequencies in the range $3.9–4.35 \cdot 10^{13}$ Hz, that is in the range of ultrasound frequencies.

If the medium into which the nanostructure is placed blocks its translational or rotational motion giving the possibility only for the oscillatory motion, the nanostructure surface energy can be identified with the vibration energy:

$$\varepsilon_S \approx \varepsilon_v = m\upsilon_v^2/2, \quad (8)$$

where, m—nanostructure mass and υ_v—velocity of nanostructure vibrations.

Knowing the nanostructure mass, its specific surface and having identified the surface energy, it is easy to find the velocity of nanostructure vibrations:

$$\upsilon_v = \sqrt{2\varepsilon_v/m} \quad (9)$$

If only the nanostructure vibrations are preserved, it can be logically assumed that the amplitude of nanostructure oscillations should not exceed its linear nanosize, that is $\lambda < r$. Then the frequency of nanostructure oscillations can be found as follows:

$$\nu_v = \upsilon_v / \lambda \quad (10)$$

Therefore, the wave number can be calculated and compared with the experimental results obtained from IR spectra.

However, with the increasing of nanostructures numbers in medium the action of nanostructures field on the medium is increased by the inductive effect.

The reactivity of nanostructure or the energy of coordination interaction may be represent as:

$$\Sigma\varepsilon_{coord} = \Sigma[(\mu_{ns} \cdot \mu_m)/r^3] \quad (11)$$

and thus, the activity of nanostructure:

$$a = \{[\varepsilon_{vib} + \Sigma[(\mu_{ns} \cdot \mu_{m})/r^3]\}/\varepsilon_V \quad (12)$$

When $\varepsilon_{vib} \rightarrow 0$, the activity of nanostructures is proportional product $\mu_{ns} \cdot \mu_{m}$, where, μ_{ns}—dipole moment of nanostructure and μ_{m}—dipole moment of medium molecule.

1.7 MATERIALS MODIFIED BY NANOSTRUCTURES

At present the precise definition of nanomaterials and nanocomposites creates certain difficulties. Prefix "nano" assumes small-dimensional characteristics of materials, and the notion "material" has a macroscopic meaning. Unlike substance, material has heterogeneous or heterophase composition.

What is nanomaterial? Formally it is a nanoparticle or aggregation of nanoparticles with properties necessary to produce several articles, for instance, in nanoelectronics or nanomachine–building. Let us assume that, we obtained rather active nanoparticles, which can be interconnected in certain succession with the help of UV-laser radiation to produce spatial articles. Naturally, the swarm of nanoparticles in certain surroundings should be compacted accordingly in pulse electrostatic or electromagnetic field to afterwards treat the "phantom" formed with several laser beams successively connecting nanoparticles in accordance with computer software. In this case, the corresponding nanoparticles can actually be called nanomaterial further used to produce articles. To tell the truth, the material formed from nanoparticles will not completely repeat the characteristics of nanoparticles constituting it. The material obtained from nanoparticles is also called nanomaterial or nanocomposite if nanoparticles differ by their nature. At the same time, the notion nanocomposite or composition material has a wider meaning. Under composites we understand any macroheterophase materials consisting of two or more heterogeneous components with different physical or mechanical properties. In general case, composition material represents multicomponent and multiphase material formed from the composition hardened when simultaneously obtaining material or article and containing mineral and organic, often polymeric materials, as a rule, with predominance of one of the components, for instance, a mineral one, thus, allowing to obtain unique properties of composite formed. The definition of nanocomposites represents similar complexity as the definition of nanomaterials. Scientists mentioned that fullerenes and tubules filled with various materials got to be called nanocomposites. At the same time, it was mentioned that effects for filling nanotubes with various substances are also interesting from the point of studying capillary properties of tubules.

Intercalated nanotubes or threads-like bunches of nanotubes, ordered layers of nanotubes in combination with different matrixes are referred to nanotubular composites. The notion "nanocomposite" in classical variant should also contain heterogeneous nanostructures or nanostructures with encapsulated nanoparticles or nanocrystals. At the same time, under nanoparticles we understand nanoformations of different character without strict internal order. For gigantic tubules and fullerenes consisting of several spheres or tubes filled with other microphases and with various shapes, the corresponding verbal designations or terms, for instance "onions" or "beads", can appear. In case of formation of extensive structures such as tubules or nanofibers their mutual interlacing with the formation of nets or braids is possible, between which particles of other components and other phases are located. Here, it should be mentioned that the mixture of phases with different shapes and structures can be referred to nanocomposites, though their composition remains the same. However, when the shape and ordering system change, the properties of nanoparticle change as well (that is its surface energy and, consequently, interaction potentials between the particles of different shapes will be different in comparison with corresponding interaction potentials of nanoparticles homogeneous in composition, dimensions, and shape). To some extent such conclusions follow from well-known definitions of regularities for polymeric, ceramic, and metallic composites. It is known that multiphase material with unique properties can be created under mechanical-chemical effect (monoaxial pulling from polymer melt), in which crystalline regions with different crystallization degree and amorphous regions will alternate. At the same time, in such material the formation of self-organizing reinforcing phases is possible. The availability of other component in the material, which can also form different nanophases, increases the abilities of nanocomposite formed.

What is nanophase? By analogy with common definition of the phase it can be said that nanophase is a homogeneous part of nanosystem, which is isolated by physical boundaries on nanometer level from other similar parts. However, the difference in properties of nanophases can be insignificant. This is conditioned by the tendency to decrease total energy of nanosystem.

A considerable number of recent researches are dedicated to nanocrystals. When we extend the notion "cluster" adding particles with nuclei, substances consisting of nonmetallic components and shells different in energy and composition to this class (the stability of clusters, nanoclusters to be more exact, is provided by their shape or surface on which these particles are located). Besides, in several cases nanoclusters identify with nanocrystals for which there are no corresponding protective shells, therefore nanoclusters and nanocrystals can be classified based on the type of crystalline structure, composition, sizes, and form of cluster structure, if a certain crystalline structure is preserved. At the same time, it should not be forgotten that cluster is translated as a group or bunch, and with nanoclusters this notion transforms into a group of chemical particles. Now, it is possible to produce clusters with a precisely determined number of atoms. Due to final sizes these small particles have structures and properties differing from "volumetric" characteristics of crystals of bigger sizes. At the same time, properties of nanoclusters change considerably even when one atom is removed from the cluster. It should be mentioned that a cluster can contain aggregates or groups of 2 up to 10^4 atoms. Therefore, aggregates of weakly bound condensed molecules can be presented as clusters. Thus, nanocrystals or quasicrystals can be classified based on their sizes. Is there any difference between nanoclusters and nanocrystals? From the aforesaid translation of the word "cluster" it can be seen that a group or bunch of atoms or molecules in this chemical formation differs from the corresponding nanocrystals or crystals by weaker bonds. At the same time, the availability of crystalline lattice in nanocrystals assumes stronger organization of the system and increase in the strength of chemical bonds between the particles constituting nanocrystals. By their composition nanoclusters with a certain order, that can be an element of crystalline structure, can consist of the element of one type, for instance, carbon, metallic, silicon, and so on particles. It is interesting to note that nanoclusters of "noble" metals have fewer atoms in comparison with the clusters formed from more active elements. In Figure 14, there are three nanostructures of elements differing by electronegativity and electrochemical potentials.

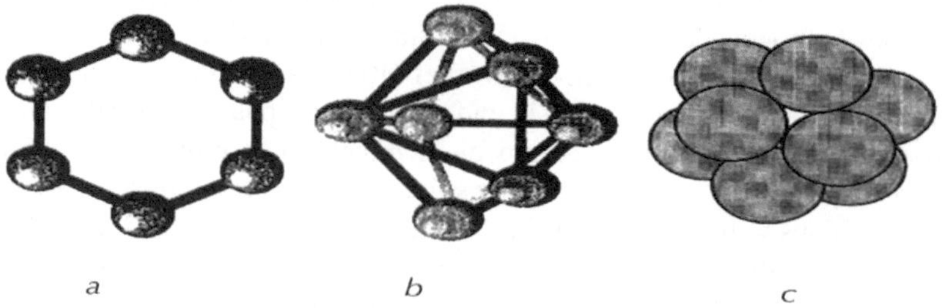

FIGURE 14 Nanoclusters of elements given according to the decrease of electronegativity of metals—(a) gold, (b) silicon, and (c) aluminum.

Based on the increase of their activity in redox processes these elements can be arranged as follows—Au< Si< Al. The least number of atoms is in nanocluster of gold—6 atoms (χ_{Au} = 2, 54, by Polling), in nanocluster of silicon—7 atoms (χ_{Si} = 1, 9), and in nanocluster of aluminum—13 atoms (χ_{Al}=1,61). Atoms of gold form six-term ring, atoms of silicon–pentagonal pyramid, and nanocluster of aluminum has a shape of icosahedrons. However, the least amount of atoms in a nanocluster is also determined, apparently, by the sizes of components and possibilities of their aggregation into certain initial geometrical formations. Let us point out that the structures given in the Figure refer to the third and sixth periods. The activity of elements of the second period is higher therefore the least number of particles in clusters of these elements can be expected to increase. For example, for carbon a six-term ring with π-electrons distributed between the atoms is initially the most stable, and the least number of atoms in carbon clusters is determined as 32 that was obtained. However, in fullerenes with the shape of icosahedrons five-term rings are also found. Silicon being an analog of carbon in chemical properties can form relatively stable nanostructures from three-term rings, and phosphorus atom can form four-term rings. Clusters are classified by the number of atoms or valence electrons participating in cluster formation. For instance, for atoms of group 1 in the periodic table it is pointed out that an even number of valence electrons always participate in their cluster formation, starting from 2 and further 8, 18, 20, and 40. The stability of clusters, on the one hand, is conditioned by the stability of their electron composition, and, on the other hand, by the corresponding external potential of the medium surrounding particles. To increase the stability neutral clusters can be transferred into ionic form. Therefore, clusters and nanocrystals can also be classified by their charge. Apart from neutral clusters, clusters anions, and clusters cations are known. Such clusters are investigated with the help of photoelectron spectroscopy. For metallic clusters anions stable electron shells appear at odd number of atoms in a cluster. For example, for Cu_7^- p orbital of the cluster is completely filled, similar situation of complete filling of d orbitals is noted for cluster Cu_{17}. For clusters cations the increase in stability at odd number of atoms in a cluster is seen. For instance, spherically asymmetrical particles of single-charged silver cations have the following numbers of atoms 9, 11, 15, and 21. The stability of nanoclusters and nanocrystals increases when the changes decreasing the cluster surface energy are introduced into the electron structure, this can be done when other atoms interacting with atoms "hosts" are introduced into the cluster, or the protective shell is formed on the cluster, or the cluster is precipitated on a certain substrate that stabilizes the cluster. Thus, clusters and nanocrystals can be classified by the stabilization method. Formation of clusters and nanocrystals on different surfaces is of great interest for researches. There is the phenomenon of epitaxy was mentioned, when the formation of or change in the structure of clusters proceed under the influence of active centers of surface layer located in certain order, thus contributing to the minimization of surface energy. The introduction of a certain structure of nanocrystals into the substrate surface layer results in the changes of nanoparticle electron structure and energy of corresponding active centers of surface layer providing the decrease of surface energy of the system being formed. Since nanocrytsals have a rather active surface, their application to the substrate should proceed under definite conditions that could not considerably change their energy and shape. However, depending upon the energy accumulated by a nanocrystal (nanocluster), even at identical element composition different results in morphology are achieved. The influence of substrate increases due to the dissipation energy brought by the cluster onto the substrate surface (Figure 15).

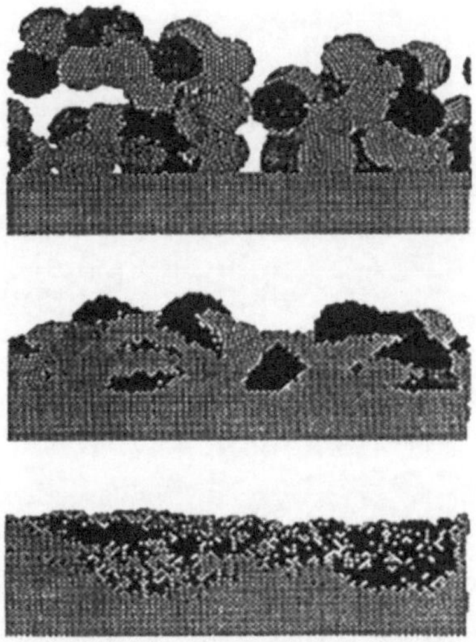

FIGURE 15 Morphology of cluster-assembled films created on Mo (100) surface as a function of the cluster energy—(a) 0.1 eV/atom, (b) 1 eV/atom, and (c) 10 eV/atom.

As seen from the Figure 15 the elevation of nanoparticle energy from 0.1 eV up to 10 eV results in considerable changes in thin films being formed. The elevation of energy up to 10 eV leads to finer mixing of structures of Mo cluster on molybdenum surface (100). The given type of morphology is observed on many other systems, such as manganese clusters on clean and covered with fullerenes C_{60} silicon surface (111), clusters of gold on gold surface (111), as well as clusters of cobalt and nickel on glass surface. In some cases, nanocrystals applied or formed on the surface are joined with the formation of relatively stable dendrite or star-like structures comprising hundreds of atoms. In Figure 16 there are pictures of such nanostructures of antimony on graphite surface.

FIGURE 16 The tunnel electron microscopy (TEM) images of islands formed by similar doses of Sb clusters on HOPG at 298K (a) and 373K (b).

The "Island" structure is produced. The clusters antimony formed contain 2300 atoms. Such nanostructures are less expressed on amorphous carbon and depend upon the substrate material and number of atoms in the cluster.

When the number of atoms in antimony cluster goes down, the nanostructures formed join to a lesser extent and represent "drop-like" formations.

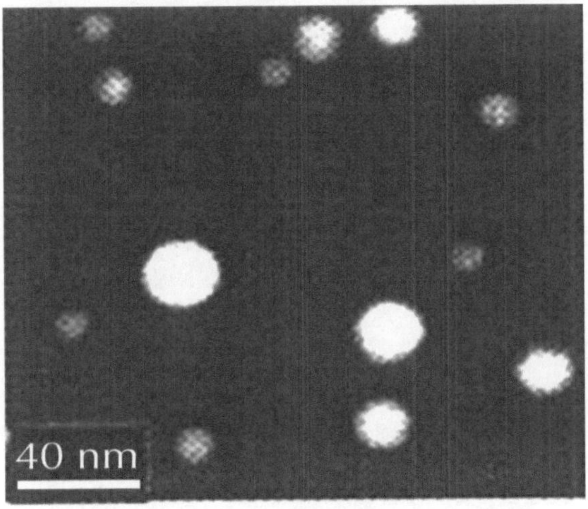

(a)

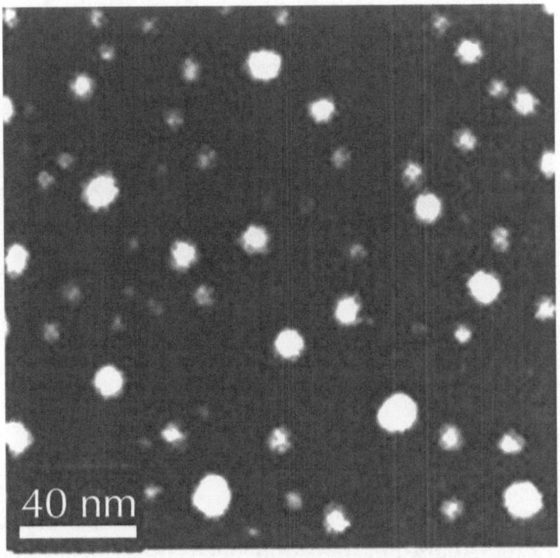

(b)

FIGURE 17 *(Continued)*

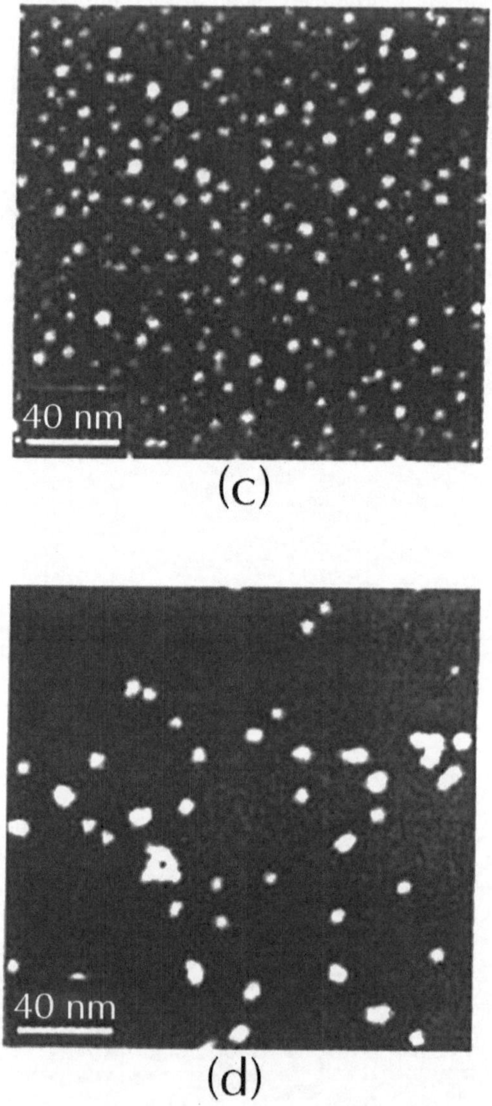

FIGURE 17 The scanning transmission electron microscopy (STEM) images of islands formed by the deposition of different-sized Sb clusters on amorphous carbon—(a) Sb_4, (b) Sb_{60}, (c) Sb_{150}, and (d) Sb_{2200}.

The "Island" morphology changes from spherical to the aforesaid shapes for the sizes from 4 up to 2200 atoms. For transition metals the critical size for the formation of branched nanostructures is much less than for alkaline metals. Thin films from clusters of silver containing less than 200 atoms also have certain joints.

The morphology of structures formed can keep on changing in comparison with the initial one. This is explained, by the analogy with terminology from polymeric chemistry, by Ostwald's maturation when nanoparticles "merge" into larger formations.

The investigations of metallic or shifted nanocrystals are closely connected with the studies on carbon and carbon-containing (metal carbides) nanoclusters (nanocrystals). Therefore, nanocrystals are split into one-, two-, and multi-element structures. The stability of nanocrystals and nanoclusters increases if they comprise two or more elements. At the same time, the combination of elements active and passive by redox potentials contributes to the stabilization of multi-element clusters being formed. Clusters containing Co/Cu, (Fe-Co)/Ag and Fe/Au, Co/Pt are known. Several clusters have a stabilizing shell in the form of oxides, carbides or hydrides, for example, Al_2O_3, Sm_2O_3, SiO_2, CO, Met_xC_y, and $B_{2n}H_{2n}$.

Thus, clusters and nanocrystals can be classified based on the composition of the shell and nucleus, alongside with their evaluation based on size and shape.

In turn, nanocrystals are classified by the type of "superlattice". Overview contains the classification of nanoclusters and nanosystems based on their production method.

The classification of carbon clusters (fullerenes and tubules) with metals or their compounds intercalated in them has a special position. The most complete idea was obtained during theoretical and experimental investigations of fullerene C_{60} filled with alkaline metals.

The shape of nanocluster or nanocrystal and their sizes are determined by the size and shape of the shell that often consists of ligands of different nature. Here, metal carbonyls are studied best of all, though some factors have to be explained. For instance, the nucleus of cluster Ni_4 is surrounded by the shell of ten ligands CO, and the nucleus of cluster Pt_4—only eight ligands CO. This is probably connected with the stabilization of electron shells and transition into a "neutral" cluster.

By their definition nanoreactors are energetically saturated nanosized regions (from 10 nm up to 1 m cm), intended for conducting directed chemical processes, in which nanostructures or chemical nanoproducts of certain application are formed. The term "nanoreactor" is applied when considering nanotechnological probe devices. In these devices a nanoreactor is the main unit and is formed between the probe apex and substrate under the directed stream of active chemical particles into the clearance "probe–substrate". Therefore, nanoreactors are classified by energy and geometrical characteristics, by shape and origin of the matrix in which nanoreactors are formed, possibilities to apply for definite reactions. One of the types of reactions in nanoreactors are topochemical ones proceeding with the participation of solids and localized on the surface of splitting of solid phases of reagent and product. Gaseous and liquid phases can participate in these reactions. Here, belong topotaxic polymerization that consists in obtaining macromolecules under low temperatures from monomers located in crystal channels under the radiation and thermal action.

Thus, the reactions for producing nanostructures–macromolecules in the channels being nanoreactors were proposed long time ago. The classical example is the topotaxic polymerization of butadiene and its derivatives in the crystals of thiourea and urea. Researchers studied the polymerization of 2,3-dimethylbutadiene and 2,3-dichlorbutadiene in clathrate complexes of thiourea. The molecules of these substances are arranges in the complexes as shown in Figure 18.

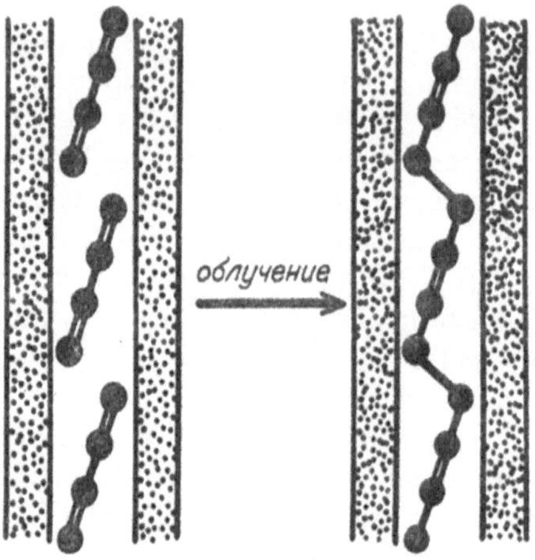

FIGURE 18 Polymerization scheme of dienes in the channel of thiourea Molecules of included butadiene (a) and Macromolecule of polybutadiene (b).

Molecules of butadiene—flat and fill the rectangle 0.76 × 0.5 nm (Figure 18). In a complex crystal the distance between the centers of molecules of butadiene derivatives mentioned equals 0.625 mn. Molecules are arranged obliquely that allows optimizing the interaction. At the same time, inner diameters of the channel (nanoreactor) in thiourea crystals (or clathrate complexes of thiourea and urea) are a little bit bigger than transverse size of butadiene molecule and equal 0.7 nm. The temperature of monomer polymerization reaction under radioactive impact equals approximately 80°C. Since the monomer polymerization process flows in a condensed state, these reactions can be referred to topochemical reactions. However, stereoregulator macromolecules or nanostructures of certain regularity are rarely obtained, in other words, there are only a few topotaxic processes among topochemical reactions. The corresponding orientation of nanoparticles obtained in nanoreactors is determined, based on the results of topotaxic polymerization, by the interaction level of reacting chemical particles with the walls of nanoreactor. Depending upon the value of the interaction energy of chemical particles located inside the nanoreactors with the walls of corresponding containers, it is advisable to separate nanoadsorbers and so-called nanocontainers from nanoreactors. Nanoadsorbers are applied to adsorb differesnt substances, mainly toxic but nanocontainers to store certain products. However, the same nanostructures can be nanoadsorbers, nanocontainers, and nanoreactors. For instance, carbon nanotubes are applied to remove dioxins. These nanostructures are used to clean industrial wastes from sulfurous gas. Nanotubes are also applied to adsorb hydrogen as nanocontainers or hydrogen storage. All these containers should have active atoms in their walls to absorb the corresponding chemical particles.

Nanoadsorbers and nanoreactors are evaluated and classified by the adsorption value that, in turn, means the number of particles (N) or volume (v*) of adsorbed particles absorbed by the unit of adsorption space (w) of nanoadsorber.

Nanoreactors can have cavities dimensionally corresponding to the pores from 0.7 nm up to 100 nm. The pores are classified by their radii—micropores (under 0.7 nm), supermicropores (from 0.7 up to 1.5 nm), mesopores (from 1.5 up to 100 nm), and macropores (over 100 nm). If we limit the size of the pores by the radius of 1 m cm, all the pores in porous absorbers can be considered potential nanoadsorbers and nanoreactors.

Mono- and polymolecular layers are formed in nanoadsorbers with equivalent radius over 1.5 nm during the sorption process, thus making the conditions to conduct condensation or chemical transformations. If the equivalent radius is less and when nanoadsorbers are filled completely, the Equation (13) written in accordance with the theory of volumetric filling of micropores is proposed:

$$a = (w_0/v^*)\exp[-(A/\beta E_0)^n] \quad (13)$$

Where, a—value of equilibrium adsorption, w_0—ultimate volume of adsorption space, E_0—characteristic adsorption energy (or interaction energy), v*—volume of adsorbed particles, β—similarity coefficient approximated by the ratio of parachors of the substance adsorbed to the substance taken as a standard (for example, benzene), N—equation parameter (for the majority of coals equals 2), and A—differential work

Work of adsorption equals:

$$A = RT\ln(p_s/p) \quad (14)$$

Where, T—adsorption temperature, p_s and p—pressure of the substance saturated steam, approximately, this ratio equals n_s/n, in which:

n_s—number of particles participating in the interaction and n—equilibrium number of adsorbed particles.

The sense of "a" for nanoadsorbers means an ultimate number of chemical particles adsorbed in nanoadsorber till the equilibrium is reached under the given conditions of the process. The dimensionalities of corresponding parameters will differ from classical Dubinin's equation but the external form of the equation will be preserved.

As mentioned earlier than, materials modified with introduced nanoparticles or substances in which due to the formation conditions self-organizing processes with the formation of certain nanostructures are initiated can be referred to nanocomposites. In these variants for the formation of nanostructures the appearance of nanoparticles will be in the form of separate fragments of bodies of rotation. Images of bodies of rotation and their separate fragments realized in nanostructures as cones, cylinders, spheres, ellipsoids, slit ellipsoids, slit spheres, hyperboloids, paraboloids, sphere and elliptical segments and sectors, and barrel-like nanostructures. These nanostructures can be described by the formulas of analytical geometry using the corresponding terms.

The diversity of shapes of nanoparticles of one component results in the response reaction of another. If corresponding capillaries are available and caps closing the ends of nanotubes are absent, it is possible to fill them with atoms and ions of another component. An increased activity of one component leads to the dispersion and activity growth of another one with the formation of new nanophases.

Since, a considerable number of interface layers can be found in nanocomposites, the production of composites to be destructed with the formation of corresponding nanophases in certain media at given local pulse impacts (for instance, laser beam can be perspective). Such materials are of interest for their further utilization after the exploitation resource or functional depreciation of articles produced from these composites. The production of nanocomposites of such level is perspective and contributes to the ecological purity of the environment. However, to solve the tasks connected with the problem of environmental pollution by the products of human vital activity it is necessary to change engineering psychology and education of manufacturers of materials, articles, various structures, machines and mechanisms. At present any engineer after getting the task to produce the corresponding article does not think as for the destiny of this product after its life span. This already resulted in littering the Earth and space around it. To clean the environment from the garbage has been produced by the people for many years the task of present and further generations of mankind.

In nanocomposite and nanomaterial terminology the terms taken from adjacent disciplines, such as coordination chemistry, play an important role. Since nanoparticles and nanoclusters are stabilized in particular surroundings, their transition into different medium results in decreasing their stability and changing their "architecture". This increases interactions in the composition consisting of nanostructures (nanoparticles) and can be equated with "re-coordination" reactions. The possibility of such processes can be characterized by corresponding constants of instability or stability of nanoparticles in the medium or composite. If certain nanoparticles obtained in different medium are introduced into the composition, the change in their surface energy results in the particle shape distortion or its destruction with the formation of new bonds and shapes.

When nanoparticles (clusters) are stabilized in the surroundings, such notions as "conformation" and "conformation energy" can be applied. In this case conformation reflects spatial location of separate components of nanosystem. Since, any nanoparticles in nanosystems or nanocomposites are highly effective "traps" of energy coming from their habitat, this energy can be conditionally divided into energy consumption connected with reactions of polycoordination or re-coordination (self-organizing energy) and consumption conditioned by orientation processes in the habitat (conformation energy). The increase in the number of nanoparticles can result in the destruction of their shells, changes in the nanoparticle shapes and sizes with the transition to micron level, thus finally leading to coagulation and separation of phases. Let us mention that coagulation means the adhesion of particles of disperse phase in colloid systems and it is conditioned by the system tendency to decrease its free energy. Mainly the process of phase separation in nanocompositions or nanosystems composed of a mixture of particles of different components (nanocomponets) depends upon the "architecture" of nanostructures and physical structure of nanocomposites. In turn, the decrease in concentration of nanoparticles in nanosystem can result in destruction (depending upon the medium activity) or decrease of nanoparticle dimensions, that is dissolution of nanoparticles in the medium. Since, the stability of nanosystems in different media is connected with dissolution processes of physical-chemical nature, such parameters as the degree of nanocomposite filling with nanopartciles, nanocomposite density in general and interface layers in comparison with nanophase density, nanocomposition effective viscosity require great importance. The determination of dissolution phase diagrams is of interest for investigating nanosystems and nanocomposites. In this regard the terminology widely used when studying polymeric systems and compositions seems appropriate.

Why is it possible to use this terminology and notions of polymer chemistry but not chemistry of metals with more advanced study of phase diagrams? It can be explained that nanocomponents are more active due to increasing their surface energy. This, in turn, contributes to the increase of interactions between nanocomponents and their mutual solubility with the formation of new strengthening nanophases.

Such notions as compatibility and operational stability of systems are applied. Interaction parameters (critical on nanophase boundaries, enthalpy, and entropy) are calculated based on experimental data. Upper and lower critical mixing points are determined and phase diagrams with nodes, coexistence curves, and spinodals are drawn. Such notions as spinodal decay and critical embryo are used in phase separation. Such notion as "compatibility" defines the ability to form in certain conditions a stable system comprising dispersed (fine) components or phases. In case of nanomaterials or nanocomposites the compatibility is determined by the interaction energy of nanophases and, consequently, the density of interface layer being formed between them. The time period during which the changes in material characteristics are within the values permissible by operational conditions is defined as operational stability. When this time period is exceeded the changes in phase energy are possible, this can result in their destruction, coagulation, and, finally, phase separation and material destruction. The nanocomposite stability and compatibility of nanophases are conditioned by the values of interaction parameters or forces (energies) of interactions.

Many notions and terms in the field under consideration are taken from chemistry and physics of surface.

1.8 DEFINITIONS

In physics and chemistry of surface of materials the basic notions still raising debates are the following: "surface", "interface layers", "surface", "boundary" layers. Round-table discussions, workshops and conferences are dedicated to estimating these notions and assessing the rationality of their use.

First, when studying the surface, what surface do we mean? There is purely mathematical notion of the surface as geometrical space of points dividing the phases. A geometrical surface is simplistically shown as a line. However, everyone understands that this is an abstract conception and a surface cannot be presented as a line. Irregularities are always observed on the boundary of gas and liquid separation, liquid and solid separation, and gas and solid separation caused by energy fluctuations. Physics of surface indicates that chemical particles on the surface are, on the one hand, in the action field of the particles of solid or liquid, and, on the other hand, in the action field of molecules of gas or liquid being in the contact with this body. Therefore, it is better to estimate properties, structure, and composition of surface layers, the dimensions of which change in thickness from 1 nm up to 10 nm depending upon the nature of material being investigated (conductor or dielectric) and depth of surface influence on material inner layers. The relief of surface layer or morphology of material surface is determined by particular features of its formation and nature (composition and structure) of material. Energy characteristics of surface of solids are usually estimated by curved angles of wetting. In turn, molecules from the layer about 1 nm thick (from the side of solid or liquid phase volume) contribute to the surface energy, and then the influence of deeper layers decreases. Changes in the surface layer energy are determined by its chemical composition and structure, as well as the aggregate of chemical particles surrounding this surface layer and belonging to material and medium in which this material is placed. We are going to consider a surface layer of any body as a boundary separating the body and surroundings. At the same time, surface energy always tends to get balanced with the energy of surroundings (liquid or gas medium) [1-28].

The notion "interface layer" has a wider interpretation and can refer to "surface layer" and "boundary layer". The notion "surface layer" is used to determine the boundaries—gas—solid, liquid—gas, and liquid—solid. The notion "boundary layer" is used when phase boundaries in a solid, suspension and emulsion are considered.

However, interface layers are more often considered in complex compositions or composites containing a big number of components. The notion "boundary layer" is more often used for multilayer materials with a clearly lamellar structure, or for coatings, materials of solids and liquids located on a certain substrate. At the same time "boundary layers" can define the boundaries between phases in solid materials. Naturally, it is difficult to imagine quick jumps when transferring from one phase to another. As was already mentioned, surface energy tends to get balanced with chemical particles surrounding it. Here, the notion "surface energy" will be used both for surface and interface and boundary layers of materials [29-40].

When investigating polymer films on different substrates (metal substrates) it is mentioned that intermolecular interactions on the boundary of phase separation results in the appearance of structural heterogeneity on molecular and per molecular levels in polymer films. At the same time, the structure defective and heterogeneous by its film thickness is formed. The influence of nature and structure of substrate spreads to over 400 m cm. The more intermolecular interactions in a polymer or on a substrate boundary are revealed, the greater distance the substrate influences to film structure changes [41-71].

The surface characteristics determine many practicable macroscopic properties of materials. Such characteristics are revealed in the surface chemical and physical structure and morphology that is immediately connected with surface chemical functional composition, crystalline degree, shape, and roughness. Chemical composition of material surface represents complex characteristics determining their properties and reactivity expressed *via* the rate of chemical interaction of the surface of material being considered with other adjoining materials or media. Under the material surface chemical composition we understand its chemical composition and availability of molecule fragments or separate atoms on the surface that enhance the surface activity or, on the contrary, passivity the surface. The surface of solids always contains the layers of gases physically and chemically sorbed. The layer of adsorbed gases or impurities (usually hydrocarbonaceous) is about several monolayers thick. The layer of chemisorbed molecules can be found under the layer of molecules physically adsorbed [50-71]. The thickness of such layer depends upon chemical activity of centers on the material surface. The material surface layer can be represented as adsorbed particles and the layer of surface atoms and molecules of material itself. Not only chemical composition but also surface geometry influence gas and impurity adsorption. It is possible to present the models reflecting the influence of surface geometry and various roughnesses of the surface in a simple way, when the active center or heteroatom is surrounded by different numbers of surface layer atoms (Figure 19).

FIGURE 19 Model of various surface roughness (a–c) (heteroatom × is surrounded by different numbers of other atoms).

At the same time, on ledges the activity of heteroatoms on roughnesses increases in comparison with other surrounding atoms. The activity of neighboring atoms with chemical particles of environment increases. The surface peculiarities result in local elevations of surface energy or surface potential and influence the formation of adsorbed layers and their thickness on the surface.

It is known that on the surface of monocrystals active centers are located on crystallographic steps or at the points where dislocations intersect the surface. Then an adsorbed particle can interact with several lattice atoms, as a result, the total interaction increases. For instance, the "adhesion" of oxygen molecule to silicon stepped surface is 500 times more probable than to the smooth surface.

Surface layers can also be found in the volume of material with defective regions such as micro- and macro-cracks, pores. Here, it is possible that surface layers in one and the same material have different chemical composition. For instance, for cellular plastics chemical composition of such layers is practically the same (surfaces of material and pore), since pores are channels connected with material surface. For foam plastics, in which there are basically closed pores not connected with material surface, the surfaces of material and pore differ in chemical composition since gaseous medium in foam plastics bubble is considerably different from the gaseous medium surrounding the material.

As the majority of per molecular structures in polymers can be referred to nanostructures being investigated and polymeric materials to nanocomposites, in opinion, several notions from the science of polymers can be referred to chemical physics of nanostructures. The main stage of polymer and nanoparticle formation is the formation of closed shape or their surface determining the existence area of polymeric molecule or nanoparticle. The particle surface formed in non-equilibrium conditions tends to transfer into the state of thermodynamic equilibrium. The rates of relaxation processes mainly depend on the surroundings ("habitat"), temperature and nature of the particle or polymeric material. However, we can distinguish two most important factors influencing the formation of properties and structure (chemical and physical) of the surface of polymers and nanoparticles—mobility of constituent groups of atoms and molecules (taking into account the conformation energy) and surface energy. In equilibrium conditions the surface energy is usually minimal and this minimum is provided by the mobility of chemical particles migrating from the volume to the surface and *vice versa* or adsorption and hemisorption of chemical particles from the environment onto the surface. This takes place even when "artificial pairs" are formed, for instance, polystyrene covered with poly(ethylene oxide). The surface energy of polystyrene is lower the corresponding energy of poly(ethylene oxide). Therefore, when this material is kept under the temperature above the vitrification temperature of both polymers, the outer layer is enriched with polystyrene. Polystyrene nanoformations in poly(ethylene oxide) film are 5 nm thick. The depending upon the substrate surface energy that can be connected with its polarity, the chemical composition and structure of surface and boundary layers of static and graft copolymers change as well. For example, depending upon substrate material the polarity of boundary layers of vinylchloride–vinylacetate copolymers changes due to the decreased number of polar groups in boundary region copolymer substrate. The changes in surface energies γ of copolymer and some materials of substrates are given:

Substrate material	Au	Ni	Al	PTFE *
Surface energy, γ, mN/m of				
substrate material	43	37	33	19
copolymer	51	48	46	38

*PTFE—polytetrafluorine ethylene film (sheet)

Based on XPS the greatest number of polar acetate groups is present in the layer boundary to substrate of gold that is characterized by the highest surface energy. Depending upon the formation conditions and substrate types the surface energy of polymers changes as well. The change in the surface energy (γ, mN/m) of some copolymers in comparison with polyethylene depending upon their formation conditions are given in Table 7.

TABLE 7 Values of surface energy for sheets and films obtained on the boundary of different media

Polymer	Surface energy, γ, mN/m		
	Sheets of polyester film obtained by hot pressing	**Sheets of PTFE obtained by die casting**	**Film of PTFE obtained by bulge formation**
Non-polar polymers:			
Low-density polyethylene	32	31(32)	32

TABLE 7 *(Continued)*

Copolymer of vinylacetate ethylene (86:14)	33	32(33)	33
Polar polymers:			
Statistic copolymer of ethylene and methacrylic acid (85:15)	44	37	38
Graft copolymer of maleic anhydride (2.1%) to high-density polyethylene	50	33 (33)	-
Graft copolymer of vinyltrime toxisilane (1.1%) to copolymer of ethylene and vinylacetate (72:28)	39	36	33

Note: Values of surface energy during the mold water-cooling are given in brackets

The results given were obtained with the help of XPS methods and infrared microscopy of multiply disturbed complete inner reflection (IR MDCIR). The essence of investigation methods should be given together with currently widely known nanoscale methods for examining nanostructures, nanosystems, and nanocomposites. To determine the morphology of surface and shape of nanoparticles and nanosystems, various types of electron microscopy, such as scanning, transmission, tunnel, and atom force are widely applied. Diffractometric and spectroscopic methods are used to determine the structure and composition. Table 8 contains methods for investigating the surface of materials, boundary layers, and nanostructures.

TABLE 8 Methods for investigating surface and boundary layers and nanostructures

Method	Information	Depth	Sensitivity, profiling, nm %
X-Ray photoelectron spectroscopy (XPS)	Element composition Chemical surroundings Conformation analysis	< 10	10^{-1}
Auger-electron spectroscopy (AES)	Element composition	< 5	10^{-1}
Ultraviolet electron spectroscopy (UVES)	Chemical surroundings Conformation analysis	~ 1	10^{-2}

TABLE 8 *(Continued)*

Spectroscopy of ionic dissipation (SID)	Element analysis	< 1	10^{-2}
Secondary ionic mass spectrometry (SIMS)	Element analysis	< 1	10^{-3}
Laser microprobe mass analysis (LMMA)	Element analysis	10	3×10^{-7}
Infrared microscopy of multiply disturbed complete inner reflection (IR MDCIR)	Chemical surroundings Conformation analysis	>1000	monolayer portions
Spectroscopy of combination dissipation (SCD)	Chemical surroundings	~1000	monolayer portions
Raman spectroscopy (RS)	Conformation analysis Nanostructural analysis	~ 100	
Atom force microscopy (AFM)	Surface morphology and polarity	< 100	monolayer portions
Scanning transmission electron microscopy (STEM)	Surface morphology	< 10	10^{-1}
Transmission electron microscopy with electron microdiffraction (TEMEMD)	Chemical composition and structure Surface morphology	~ 1	10^{-1}
Tunnel electron microscopy (TEM)	Surface morphology	0.1	10^{-2}

KEYWORDS

- **Electron magnetic spectrometer**
- **Nanomaterials**
- **Nanoreactors**
- **Surface layer**
- **X-ray photoelectron spectroscopy**

REFERENCES

1. Kobayashi, S., Uyama, H., and Kimura, S. *Chemical Reviews*, **101**, 3793 (2001).
2. Gross, R. A., Kumar, A., and Kalra, B. *Chemical Reviews*, **101**, 2097 (2001).
3. Varma, I. K., Albertsson, A. C., Rajkhowa, R., and Srivastava, R. K. *Progress in Polymer Science*, **30**, 949 (2005).
4. Platel, R., Hodgson, L., and Williams, C. *Polymer Reviews*, **48**, 11 (2008).
5. Younes, H. and Cohn, D. *European Polymer Journal*, **24**, 765 (1988).
6. Moon, S. and Kimura, Y. *Polymer International*, **52**, 299 (2003).
7. Hiltunen, K., Seppala, J. V., and Harkonen, M. *Journal of Applied Polymer Science*, **63**, 1091 (1997).
8. Proikakis, C., Tarantili, P., and Andreopoulos, A. *Journal of Elastomers and Plastics*, **34**, 49 (2002).
9. Sodergard, A. and Stolt, M. *Progress in Polymer Science*, **27**, 1123 (2002).
10. Moon, S., Lee, C., Taniguchi, I., Miyamoto, M., and Kimura, Y. *Polymer*, **42**, 50591 (2001).
11. Fortunato, B., Pilati, F., and Manaresi, P. *Polymer*, **22**, 655 (1981).
12. Maharana, T., Mohanty, B., and Negia, Y. S. *Progress in Polymer Science*, **34**, 99 (2009).
13. Ovitt, T. M. and Coates, G. W. *Journal of the American Chemical Society*, **121**, 4072 (1999).
14. Malberg, S., Basalp, D., Finne-Wistrand, A., and Albertsson, A. C. *Journal of Polymer Science, Part A: Polymer Chemistry Edition*, **48**, 1214 (2010).
15. Dunsing, R. and Kricheldorf, H. R. *Polymer Bulletin*, **14**, 491 (1985).
16. Idage, B. B., Idage, S. B., Kasegaonkar, A. S., and Jadhav, R. V. *Materials Science and Engineering B*, **168**, 193 (2010).
17. Stassin, F. and Jerome, R. *Chemical Communications*, **2**, 232 (2003).
18. Mingotaud, A. F., Cansell, F., Gilbert, N., and Soum, A. *Polymer Journal*, **31**, 406 (1999).
19. Antonietti, M. and Landfester, K. *Progress in Polymer Science*, **689** (2002).
20. Ganapathy, H. S., Hwang, H. S., Jeong, Y. T., Lee, W. K., and Lim, K. T. *European Polymer Journal*, **43**, 119 (2007).
21. Varma, I. K., Albertsson, A. C., Rajkhowa, R., and Srivastava, R. K. *Progress in Polymer Science*, **30**, 949 (2005).
22. Zaks, A. and Klibanov, A. M. *Science*, **224**, 1249 (1984).
23. Cambou, B. and Klibanov, A. M. *Journal of the American Chemical Society*, **106**, 2687 (1984).
24. Gross, R. A., Kumar, A., and Kalra, B. *Chemical Reviews*, **101**, 2097 (2001).
25. Kobayashi, S., Uyama, H., and Kimura, S. *Chemical Reviews*, **101**, 3793 (2001).
26. Matsumura, S. *Macromolecular Bioscience*, **2**, 105 (2002).
27. Zaks, A. and Klibanov, A. M. *Science*, **224**, 1249 (1984).
28. Cambou, B. and Klibanov, A. M. *Journal of the American Chemical Society*, **106**, 2687 (1984).
29. Srivastava, R. K. and Albertsson, A. C. *Journal of Polymer Science, Part A: Polymer Chemistry Edition*, **43**, 4206 (2005).
30. Malberg, S., Finne-Wistrand, A., and Albertsson, A. C. *Polymer*, **51**, 5318 (2010).
31. Arvanitoyannis, I., Nakayama, A., Kawasak, N., and Yamamoto, N. *Polymer*, **36**, 2947 (1995).
32. Korhonen, H., Helminen, A., and Seppala, J. V. *Polymer*, **42**, 7541 (2001).

33. Stassin, F. and Jerome, R. *Journal of Polymer Science, Part A: Polymer Chemistry Edition*, **43**, 2777 (2005).
34. Namekawa, S., Uyama, H., and Kobayashi, S. *Polymer Journal*, **30**, 269 (1998).
35. Kobayashi, S. *Journal of Polymer Science, Part A: Polymer ChemistryEdition*, **37**, 3041 (1999).
36. Malberg, S., Finne-Wistrand, A., and Albertsson, A. C. *Polymer*, **51**, 5318 (2010).
37. Jacobsen, S., Fritz, H. G., Degee, P., Dubois, P., and Jerome, R. *Polymer*, **41**, 3395 (2000).
38. Xie, J. X. and Yang, R. J. *Polymeric Preprints (American Chemical Society, Division of Polymer Chemistry)*, **48**, 502 (2007).
39. Bhattacharyya, S. K. and Paul, N.B. *The Indian Textiles Journal*, **91**(3), 75 (1980).
40. Bisanda, E. and Anselm, M. *Journal of Material Science and Technology*, **27**, 1690 (1992).
41. Murherjee, P. S. and Satyanarayana, K. G. *Journal of Material Science*, **19**, 3925 (1984).
42. Torres, F. G., Ochoa, B., and Machicao, E. *International Polymer Processing*, **18**(1), 33 (2003).
43. Mohanty, A. K., Misra, M., and Hinrichsen, G. *Macromolecular Materials and Engineering*, **276–277**, 1 (2000).
44. Mwasha, A. and Petersen, A. *Journal of Materials and Design*, **31**, 2360 (2010).
45. Lu, B., Wang, Y., Liu, Y., Duan, H., Zhou, J., Zhang, Z., Wang, Y., Li, X., Wang, W., Lan, W., and Xie, E. *Small*, **6**(15), 1612 (2010).
46. Fang, D., Hsiao, B. S., and Chu, B. *Polymer Preprints*, **44**(2), 59 (2003).
47. Zhou, F. L., Gong, R. H., and Porat, I. *Journal of Applied Polymer Science*, **115**(5), 2591 (2010).
48. Theron, S. A., Yarin, A. L., Zussman, E., and Kroll, E. *Polymer*, **46**(9), 2889 (2005).
49. Varesano, A., Carletto, R. A., and Mazzuchetti, G. *Journal of Materials Processing Technology*, **209**(11), 5178 (2009).
50. Kim, G., Cho, Y. S., and Kim, W. D. *European Polymer Journal*, **42**(9), 2031 (2006).
51. Yamashita, Y., Ko, F., Miyake, H., and Higashiyama, A. Advanced Electrospinning Setups and Special Fibre and Mesh Morphologies. *Sen'i Gakkaishi*, **64**(1), 24–61 (2008).
52. Zhou, F. L., Gong, R. H., and Porat, I. *Polymer Engineering & Science*, **49**(12), 2475 (2009).
53. Dosunmu, O. O., Chase, G. G., Kataphinan, W., and Reneker, D. H. *Nanotechnology*, **17**(4), 1123 (2006).
54. Badrossamay, M. R., McIlwee, H. A., Goss, J. A., and Parker, K. K. *Nano Letters*, **10**(6), 2257 (2010).
55. Srivastava, Y., Marquez, M., and Thorsen, T. *Journal of Applied Polymer Science*, **106**(5), 3171 (2007).
56. Lukas, D., Sarkar, A., and Pokorny, P. *Journal of Applied Physics*, **103**(8), 084309 (2008).
57. Yarin, A. and Zussman, E. *Polymer*, **45**(9), 2977 (2004).
58. Thoppey, N. M., Bochinski, J. R., Clarke, L. I., and Gorga, R. E. *Polymer*, **51**(21), 4928 (2010).
59. Tang, S., Zeng, Y., and Wang, X. *Polymer Engineering & Science*, **50**(11), 2252 (2010).
60. Cengiz, F., Dao, T. A., and Jirsak, O. *Polymer Engineering & Science*, **50**(5), 936 (2010).
61. Jirsak, O., Sysel, P., Sanetrnik, F., Hruza, J., and Chaloupek, J. *Journal of Nanomaterials*, 1, (2010).
62. Wang, X., Niu, H., Lin, T., and Wang, X. *Polymer Engineering & Science*, **49**(8), 1582 (2009).
63. Salem, D. R. In *Nanofibers and Nanotechnology in Textiles*, P. J. Brown and K. Stevens (Eds.)., CRC Press, Boca Raton, FL, USA, p.1 (2007).
64. Sun, D., Chang, C., Li, S., and Lin, L. *Nano Letters*, **6**(4), 839 (2006).

65. Chang, C., Limkrailassiri, K., and Lin, L. *Applied Physics Letters*, **93**(12), 123111 (2008).
66. Levit, N. and Tepper, G. *The Journal of Supercritical Fluids*, **31**(3), 329 (2004).
67. Larrondo, L. and Manley, R. S. J. *Journal of Polymer Science. Part B. Polymer Physics*, **19**(6), 909 (1981).
68. Dalton, P. D., Grafahrend, D., Klinkhammer, K., Klee, D., and Möller, M. *Polymer*, **48**(23), 6823 (2007).
69. Rangkupan, R. and Reneker, D. H. *Journal of Metals, Materials, and Minerals*, **12**(2), 81 (2003).
70. Lin, T., Wang, H., and Wang, X. *Advanced Materials*, **17**(22), 2699 (2005).
71. Gupta, P. and Wilkes, G. L. *Polymer*, **44**(20), 6353 (2003).

CHAPTER 2

SMART DELIVERY OF DRUGS

A. V. BYCHKOVA and M. A. ROSENFELD

CONTENTS

2.1 INTRODUCTION

The problem of creation of magnetically targeted nanosystems for a smart delivery of drugs to target cells has been solved with the application of a fundamentally novel method that obtaining from stable protein coatings. The method is based on the ability of proteins to form interchain covalent bonds under the action of free radicals which are generated locally on nanoparticles surface. By the set of physical and biochemical methods it has been proved that the free radical cross-linking of proteins allows obtaining stable single layer protein coatings of several nanometers in thickness. The cross-linked coatings were formed on the surface of individual magnetic nanoparticles (MNPs) ($d \sim 17$ nm) by free radical processes taking place strictly in the adsorption layer. Spin labels technique has been applied for studying of macromolecules adsorption on nanoparticles. The free radical cross-linking of proteins on the surface of nanoparticles has been shown to keep native properties of the protein molecules (was demonstrated on thrombin coating). The method may be used to reach various biomedical goals concerning a smart delivery of drugs and biologically active substances revealing new possibilities of design of single layer multiprotein polyfunctional coatings on all the surfaces containing metals of variable valence (for example, Fe, Cu, Cr).

The MNPs have many applications in different areas of biology and medicine. The MNPs are used for hyperthermia, magnetic resonance imaging, immunoassay, purification of biologic fluids, cell and molecular separation, and tissue engineering [1-6]. The design of magnetically targeted nanosystems (MNSs) for a smart delivery of drugs to target cells is a promising direction of nanobiotechnology. They traditionally consist on one or more magnetic cores and biological or synthetic molecules which serve as a basis for polyfunctional coatings on MNPs surface. The coatings of MNSs should meet several important requirements [7]. They should be biocompatible, protect magnetic cores from influence of biological liquids, prevent MNSs agglomeration in dispersion, provide MNSs localization in biological targets and homogenity of MNSs sizes. The coatings must be fixed on MNPs surface and contain therapeutic products (drugs or genes) and biovectors for recognition by biological systems. The model which is often used when MNSs are developed that is presented in Figure 1.

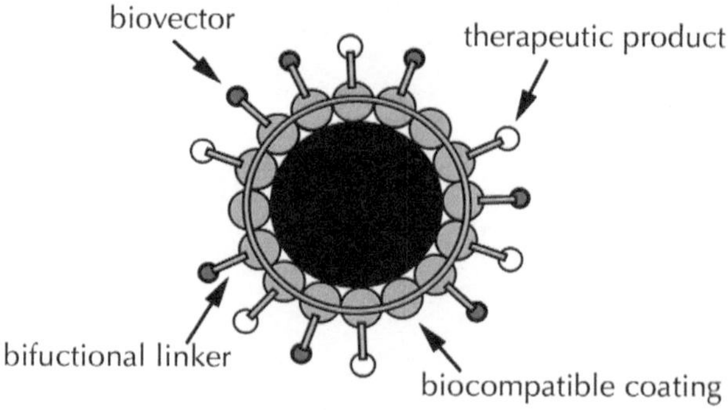

FIGURE 1 The classical scheme of magnetically targeted nanosystem for a smart delivery of therapeutic products.

Proteins are promising materials for creation of coatings on MNPs for biology and medicine. When proteins are used as components of coatings it is of the first importance that they keep their functional activity [8]. Protein binding on MNPs surface is a difficult scientific task. Traditionally bifunctional linkers (glutaraldehyde [9-10], carbodiimide [11-12]) are used for protein cross-linking on the surface of MNPs and modification of coatings by therapeutic products and biovectors. Authors of the study [9] modified MNPs surface with aminosilanes and performed protein molecules attachment using glutaraldehyde. The bovine serum albumin (BSA) was adsorbed on MNPs surface in the presence of carbodiimide [10]. These works revealed several disadvantages of this way of protein fixing which make it unpromising. Some of them are clusters formation as a result of linking of protein molecules adsorbed on different MNPs, desorption of proteins from MNSs surface as a result of incomplete linking, uncontrollable linking of proteins in solution (Figure 2). The creation of stable protein coatings with retention of native properties of molecules still is an important biomedical problem.

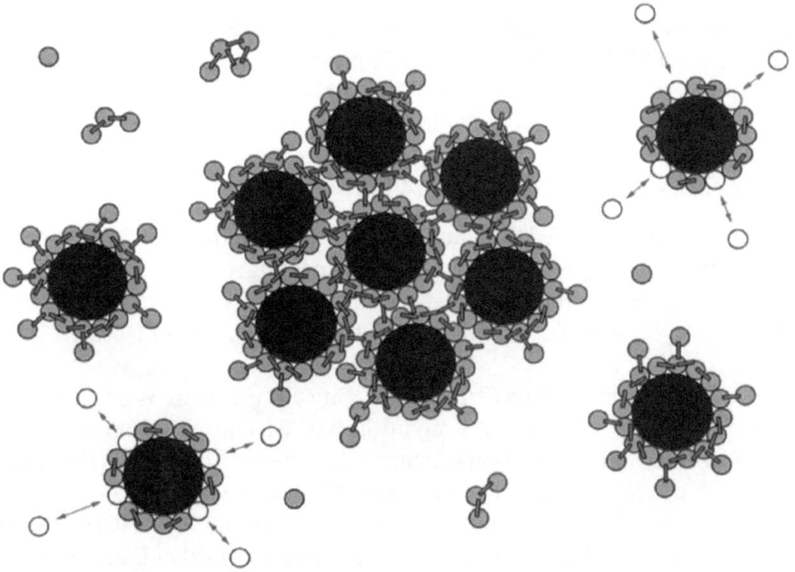

FIGURE 2 Nonselective linking of proteins on MNPs surface by bifunctional linkers leading to clusters formation and desorption of proteins from nanoparticles surface.

It is known that proteins can be chemically modified in the presence of free radicals with formation of cross-links [13]. The goals of the work were to create stable protein coating on the surface of individual MNPs using a fundamentally novel approach based on the ability of proteins to form interchain covalent bonds under the action of free radicals and estimate activity of proteins in the coating.

2.2 MATERIALS AND METHODS

2.2.1 MAGNETIC SORBENT SYNTHESIS

Nanoparticles of magnetite Fe_3O_4 were synthesised by co-precipitation of ferrous and ferric salts in water solution at 4°C and in the alkaline medium:

$$Fe^{2+} + 2Fe^{3+} + 8OH^{-} \rightarrow Fe_3O_4\downarrow + 4H_2O$$

The 1.4 g of $FeSO_4 \cdot 7H_2O$ and 2.4 g of $FeCl_3 \cdot 6H_2O$ ("Vekton", Russia) were dissolved in 50 ml of distilled water so that molar ratio of Fe^{2+}/Fe^{3+} was equal to 1:2. After filtration of the solution 10 ml of 25 mass% NH_4OH ("Chimmed", Russia) was added to it on a magnetic stirrer. 2.4 g of polyethylene glycol (PEG) 2 kDa ("Ferak Berlin GmbH", Germany) was added previously in order to reduce the growth of nanoparticles during the reaction. After the precipitate was formed the solution (with 150 ml of water) was placed on a magnet. Magnetic particles precipitated on it and supernatant liquid was deleted. The procedure of particles washing was repeated for 15 times until neutral pH was obtained. The MNPs were stabilized by double electric layer with the use of US-disperser ("MELFIZ", Russia). To create the double electric layer 30 ml of 0.1 M phosphate-citric buffer solution (0.05 M NaCl) with pH value of 4 was introduced. The MNPs concentration in hydrosol was equal to 37 mg/ml.

2.2.2 PROTEIN COATINGS FORMATION

The BSA ("Sigma-Aldrich", USA) and thrombin with activity of 92 units per 1 mg ("Sigma-Aldrich", USA) were used for protein coating formation. Several types of reaction mixtures were created: "A1-MNP-0", "A2-MNP-0", "A1-MNP-1", "A2-MNP-1", "A2-MNP-1-acid", "T1-MNP-0", "T1-MNP-0", and "T1-0-0".
All of them contained:

- 2.80 ml of protein solution ("A1" or "A2" means that there is BSA solution with concentration of 1 mg/ml or 2 mg/ml in 0.05 M phosphate buffer with pH 6.5 (0.15 M NaCl) in the reaction mixture, "T1" means that there is thrombin solution with concentration of 1 mg/ml in 0.15M NaCl with pH 7.3),
- 0.35 ml of 0.1 M phosphate-citric buffer solution (0.05 M NaCl) or MNPs hydrosol ("MNP" in the name of reaction mixture means that it contains MNPs),
- 0.05 ml of distilled water or 3 mass% H_2O_2 solution ("0" or "1" in the reaction mixture names correspondingly).

Hydrogen peroxide interacts with ferrous ion on MNPs surface with formation of hydroxyl-radicals by Fenton reaction:

$$Fe^{2+} + H_2O_2 \rightarrow Fe^{3+} + OH^{\cdot} + OH^{-}$$

"A2-MNP-1-acid" is a reaction mixture, containing 10 μl of ascorbic acid with concentration of 152 mg/ml. Ascorbic acid is known to form free radicals in reaction with H_2O_2 and generate free radicals in solution but not only on MNPs surface.

The sizes of MNPs, proteins and MNPs in adsorption layer were analyzed using dynamic light scattering (Zetasizer Nano S "Malvern", England) with detection angle of 173° at temperature 25°C.

2.2.3 STUDY OF PROTEINS ADSORPTION ON MNPS

The study of proteins adsorption on MNPs was performed using electron spin resonance (ESR)-spectroscopy of spin labels. The stable nitroxide radical used as spin label is presented in Figure 3. Spin labels technique allows studying adsorption of macromolecules on nano-sized magnetic particles in dispersion without complicated separation processes of solution components [14]. The principle of quantitative evaluation of adsorption is the following. Influence of local fields of MNPs on spectra of radicals in solution depends on the distance between MNPs and radicals [14-16]. If this distance is lower than 40 nm for magnetite nanoparticles with the average size of 17 nm [17] ESR spectra lines of the radicals broaden strongly and their intensity decreases to zero. The decreasing of the spectrum intensity is proportional to the part of radicals which are located inside the layer of 40 nm in thickness around MNP. The same happens with spin labels covalently bound to protein macromolecules. An intensity of spin labels spectra decreases as a result of adsorption of macromolecules on MNPs (Figure 4). We have shown that spin labels technique can be used for the study of adsorption value, adsorption kinetics, calculation of average number of molecules in adsorption layer and adsorption layer thickness, and concurrent adsorption of macromolecules [18-20].

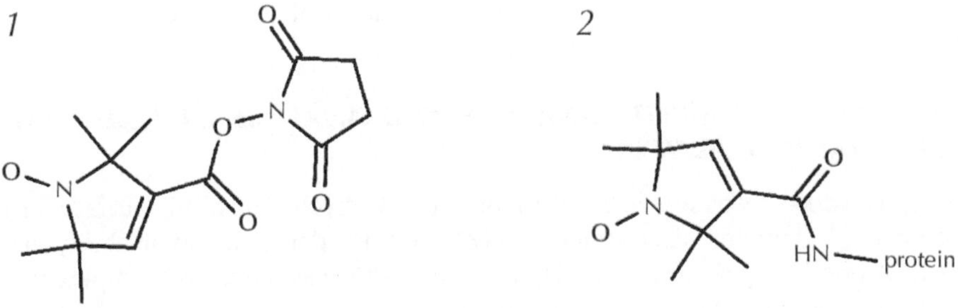

FIGURE 3 The stable nitroxide radical used for labelling of macromolecules containing aminogroups (1) and spin label attached to protein macromolecule (2).

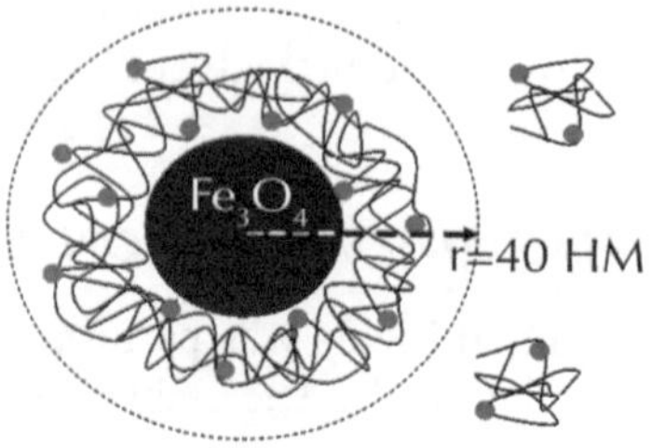

FIGURE 4 Magnetic nanoparticles and spin-labelled macromolecules in solution.

The reaction between the radical and protein macromolecules was conducted at room temperature. The 25 μl of radical solution in 96% ethanol with concentration of 2.57 mg/ml was added to 1 ml of protein solution. The solution was incubated for 6 hrs and dialyzed. The portion of adsorbed protein was calculated from intensity of the low-field line of nitroxide radical triplet I_{+1}.

The method of ferromagnetic resonance (FMR) was also used to study adsorption layer formation. The spectra of the radicals and MNPs were recorded at room temperature using Bruker EMX 8/2.7 X-band spectrometer at a microwave power of 5 mW, modulation frequency 100 kHz and amplitude 1 G. The first derivative of the resonance absorption curve was detected. The samples were placed into the cavity of the spectrometer in a quartz flat cell. Magnesium oxide powder containing Mn^{2+} ions was used as an external standard in ESR experiments. Average amount of spin labels on protein macromolecules reached 1 per 4–5 albumin macromolecules and 1 per 2–3 thrombin macromolecules. Rotational correlation times of labels were evaluated as well as a fraction of labels with slow motion (τ >1 ns).

2.2.4 COATING STABILITY ANALYSIS AND ANALYSIS OF SELECTIVITY OF FREE RADICAL PROCESS

In previous works, it was shown that fibrinogen (FG) adsorbed on MNPs surface forms thick coating and micron-sized structures [18]. Also, FG demonstrates an ability to replace BSA previously adsorbed on MNPs surface. This was proved by complex study of systems containing MNPs, spin-labelled BSA and FG with spin labels technique and FMR [20]. The property of FG to replace BSA from MNPs surface was used in this work for estimating BSA coating stability. 0.25 ml of FG ("Sigma-Aldrich", USA) solution with concentration of 4 mg/ml in 0.05 M phosphate buffer with pH 6.5 was added to 1 ml of the samples "A1-MNP-0", "A2-MNP-0", "A1-MNP-1", and "A2-MNP-1". The clusters formation was observed by dynamic light scattering.

The samples "A2-MNP-0", "A2-MNP-1", "T1-MNP-0", and "T1-MNP-1" were centrifugated at 1,20,000 g during 1 hr on «Beckman Coulter» (Austria). On these conditions MNPs precipitate, but macromolecules physically adsorbed on MNPs remain in supernatant liquid. The precipitates containing MNPs and protein fixed on MNPs surface were dissolved in buffer solution with subsequent evaluation of the amount of protein by Bradford colorimetric method [21]. Spectrophotometer CF-2000

(OKB "Spectr", Russia) was used. Free radical modification of proteins in supernatant liquids of "A2-MNP-0", "A2-MNP-1" and the additional sample "A2-MNP-1-acid" were analyzed by infrared (IR)-spectroscopy using fourier transform infrared (FTIR)-spectrometer Tenzor 27 ("Bruker", Germany) with deuterated triglycine sulfate (DTGS)-detector with 2 cm^{-1} resolution. Comparison of "A2-MNP-0", "A2-MNP-1", and "A2-MNP-1-acid" helps to reveal the selectivity of free radical process in "A2-MNP-1".

2.2.5 ENZYME ACTIVITY ESTIMATION

The estimation of enzyme activity of protein fixed on MNPs surface was performed on the example of thrombin. This protein is a key enzyme of blood clotting system which catalyzes the process of conversion of FG to fibrin. Thrombin may lose its activity as a result of free radical modification and the rate of the enzyme reaction may decrease. So, estimation of enzyme activity of thrombin cross-linked on MNPs surface during free radical modification was performed by comparison of the rates of conversion of FG to fibrin under the influence of thrombin contained in reaction mixtures. 0.15 ml of the samples "T1-MNP-0", "T1-MNP-1", and "T1-0-0" was added to 1.4 ml of FG solution with concentration of 4 mg/ml. Kinetics of fibrin formation was studied by Rayleigh light scattering on spectrometer 4400 ("Malvern", England) with multibit 64-channel correlator.

2.3 RESULTS AND DISCUSSION

The ESR spectra of spin labels covalently bound to BSA and thrombin macromolecules (Figure 5) allow obtaining information about their microenvironment. The spectrum of spin labels bound to BSA is a superposition of narrow and wide lines characterized by rotational correlation times of 10^{-9} sec and 2×10^{-8} sec respectively. This is an evidence of existence of two main regions of spin labels localisation on BSA macromolecules [22]. The portion of labels with slow motion is about 70%. So, a considerable part of labels are situated in internal areas of macromolecules with high microviscosity. The labels covalently bound to thrombin macromolecules are characterized by one rotational correlation time of 0.26 nsec. These labels are situated in areas with equal microviscosity.

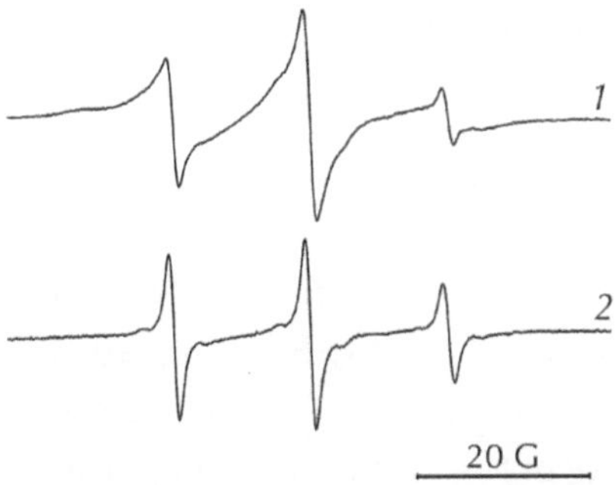

FIGURE 5 ESR spectra of spin labels on BSA (1) and thrombin (2) at 25°C.

The signal intensity of spin-labelled macromolecules decreased after introduction of MNPs into the solution that testifies to the protein adsorption on MNPs (Figure 6). Spectra of the samples "A1-MNP-0" and "T1-MNP-0" consist of nitroxide radical triplet, the third line of sextet of Mn^{2+} (the external standard) and FMR spectrum of MNPs. Rotational correlation time of spin labels does not change after MNPs addition. The dependences of spectra lines intensity for spin-labelled BSA and thrombin in the presence of MNPs on incubation time are shown in Table 1. Signal intensity of spin-labelled BSA changes insignificantly. These changes correspond to adsorption of approximately 12% of BSA after the sample incubation for 100 min. The study of adsorption kinetics allows establishing that adsorption equilibrium in "T1-MNP-0" takes place when the incubation time equals to 80 min and ~ 41% of thrombin is adsorbed. The value of adsorption 'A' may be estimated using the data on the portion of macromolecules adsorbed and specific surface area calculated from MNPs density (5200 mg/m^3), concentration and size. Hence BSA adsorption equals to 0.35 mg/m^2 after 100 min incubation. The dependence of thrombin adsorption value on incubation time is shown in Figure 7. Thrombin adsorption equals to 1.20 mg/m^2 after 80 min incubation.

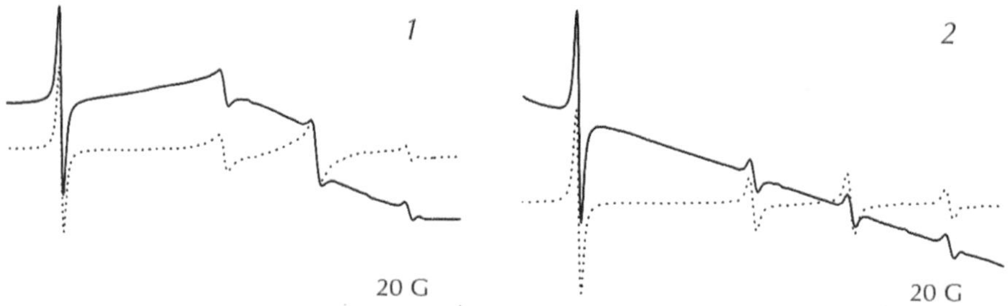

FIGURE 6 The ESR spectra of spin labels on BSA (*1*) and thrombin (2) macromolecules before (dotted line) and 75 min after (solid line) MNPs addition to protein solution at 25°C. External standard – MgO powder containing Mn^{2+}.

TABLE 1 The dependence of relative intensity of low-field line of triplet I_{+1} of nitroxide radical covalently bound to BSA and thrombin macromolecules, and the portion *N* of the protein adsorbed on incubation time *t* of the samples "A1-MNP-0" and "T1-MNP-0"

	Spin-labelled BSA		Spin-labelled thrombin	
t, min.	I_{+1}, rel. units	N, %	I_{+1}, rel. units	N, %
0	0.230 ± 0.012	0 ± 5	0.25 ± 0.01	0 ± 4
15	–	–	0.17 ± 0.01	32 ± 4
35	0.205 ± 0.012	9 ± 5	0.16 ± 0.01	36 ± 4
75	0.207 ± 0.012	10 ± 5	0.15 ± 0.01	40 ± 4
95	–	–	0.15 ± 0.01	40 ± 4
120	0.200 ± 0.012	13 ± 5	0.14 ± 0.01	44 ± 4

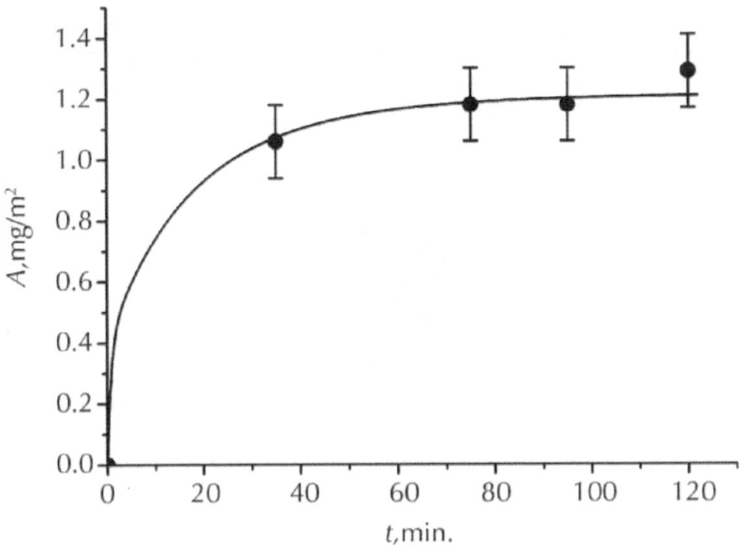

FIGURE 7 Kinetics of thrombin adsorption on magnetite nanoparticles at 25°C. Concentration of thrombin in the sample is 0.9 mg/ml, MNPs – 4.0 mg/ml.

The FMR spectra of the samples "A1-MNP-0", "T1-MNP-0", and MNPs are characterized by different position in magnetic field (Figure 8). The centre of the spectrum of MNPs is 3254 G, while the centre of "A1-MNP-0" and "T1-MNP-0" spectra is 3253 G and 3449 G respectively. Resonance conditions for MNPs in magnetic field of spectrometer include a parameter of the shift of FMR spectrum,

$$|M_1| = \frac{3}{2}|H_1|,$$

where H_1 is a local field created by MNPs in linear aggregates which form in spectrometer field.

$$H_1 = 2\sum_{1}^{\infty}\frac{2\mu}{(nD)^3},$$

Where D is a distance between MNPs in linear aggregates, μ is MNPs magnetic moment, n is a number of MNPs in aggregate [23]. Coating formation and the thickness of adsorption layer influence on the distance between nanoparticles decrease dipole interactions and particles ability to aggregate. As a result the centre of FMR spectrum moves to higher fields. This phenomenon of FMR spectrum centre shift we observed in the system "A1-MNP-0" after FG addition [20]. The spectrum of MNPs with thick coating becomes similar to FMR spectra of isolated MNPs. So, the similar centre positions of FMR spectra of MNPs without coating and MNPs in BSA coating point to a very thin coating and low adsorption of protein in this case. In contrast according to FMR centre position the thrombin coating on MNPs is thicker than albumin coating. This result is consistent with the data obtained by ESR spectroscopy.

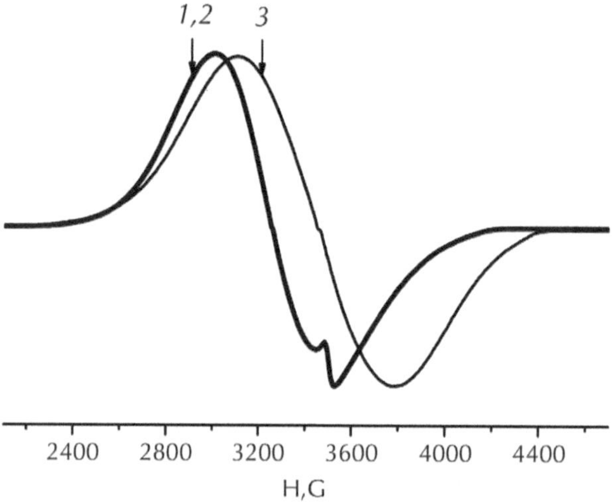

FIGURE 8 The FMR spectra of MNPs (1), MNPs in the mixture with BSA (the sample "A1-MNP-0") after incubation time of 120 min (2) and MNPs in the mixture with thrombin (the sample "T1-MNP-0") after incubation time of 120 min (3).

The FG ability to replace BSA in adsorption layer on MNPs surface is demonstrated in Figure 9. Initially, there is bimodal volume distribution of particles over sizes in the sample "A2-MNP-0" that can be explained by existence of free (unadsorbed) BSA and MNPs in BSA coating. After FG addition the distribution changes. Micron-sized clusters form in the sample that proves FG adsorption on MNPs [18]. In the case of "A2-MNP-1" volume distribution is also bimodal. The peak of MNPs in BSA coating is characterized by particle size of maximal contribution to the distribution of ~ 23 nm. This size is identical to MNPs in BSA coating in the sample "A2-MNP-0". It proves that H_2O_2 addition does not lead to uncontrollable linking of protein macromolecules in solution or cluster formation. Since, MNPs size is 17 nm, the thickness of adsorption layer on MNPs is approximately 3 nm.

After FG addition to "A2-MNP-1" micron-sized clusters do not form. So, adsorption BSA layer formed in the presence of H_2O_2 keeps stability. This stability can be explained by formation of covalent bonds between protein macromolecules [13] in adsorption layer as a result of free radicals generation on MNPs surface. Stability of BSA coating on MNPs was demonstrated for the samples "A1-MNP-1" and "A2-MNP-1" incubated for more than 100 min before FG addition. The clusters are shown to appear if the incubation time is insufficient.

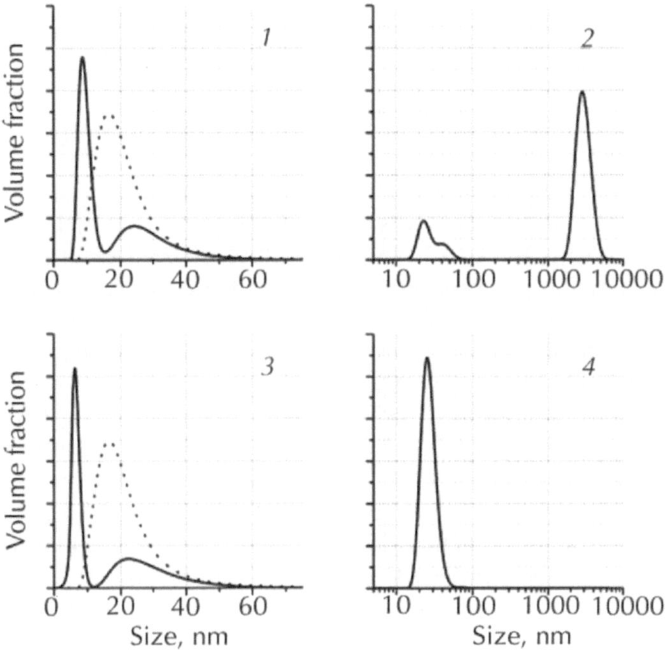

FIGURE 9 Volume distributions of particles in sizes in systems without (*1*, *2*) and with (*3*, *4*) H_2O_2 ("A2-MNP-0", "A2-MNP-1") incubated for 2 h before (*1*, *3*) and 20 min after (*2*, *4*) FG addition. Dotted line is the volume distribution of nanoparticles in sizes in dispersion.

The precipitates obtained by ultracentrifugation of "A2-MNP-0", "A2-MNP-1", "T1-MNP-0", and "T1-MNP-1" was dissolved in buffer solution. The amount of protein in precipitates was evaluated by Bradford colorimetric method (Table 2). The results showed that precipitates of systems with H_2O_2 contained more protein than the same systems without H_2O_2. Therefore, in the samples containing H_2O_2 the significant part of protein molecules does not leave MNPs surface when centrifuged while in the samples "A2-MNP-0" and "T1-MNP-0" the most of protein molecules leaves the surface. This indicates the stability of adsorption layer formed in the presence of free radical generation initiator and proves cross-links formation.

TABLE 2 The amount of protein in precipitates after centrifugation of the samples "A2-MNP-0", "A2-MNP-1", "T1-MNP-0" and "T1-MNP-1" of 3.2 ml in volume

Sample name	Amount of protein in precipitates, mg
"A2-MNP-0"	0.05
"A2-MNP-1"	0.45

TABLE 2 *(Continued)*

"T1-MNP-0"	0.15
"T1-MNP-1"	1.05

The analysis of content of supernatant liquids obtained after ultracentrifugation of reaction systems containing MNPs and BSA which differed by H_2O_2 and ascorbic acid presence ("A2-MNP-0", "A2-MNP-1" and "A2-MNP-1-acid") allows evaluating the scale of free radical processes in the presence of H_2O_2. As it was mentioned above in the presence of ascorbic acid free radicals generate not only on MNPs surface but also in solution. So, both molecules on the surface and free molecules in solution can undergo free radical modification in this case. From Figure 10 we can see that the IR-spectrum of "A2-MNP-1-acid" differs from the spectra of "A2-MNP-0" and "A2-MNP-1", while the spectra of "A2-MNP-0" and "A2-MNP-1" almost have no differences. The IR-spectra differ in the region of 1200–800 cm^{-1}. The changes in this area are explained by free radical oxidation of amino acid residues of methionine, tryptophane, histidine, cysteine, and phenylalanine. These residues are sulfur-containing and cyclic ones which are the most sensitive to free radical oxidation [13, 24]. The absence of differences in "A2-MNP-0" and "A2-MNP-1" proves that cross-linking of protein molecules in the presence of H_2O_2 is selective and takes place only on MNPs surfaces.

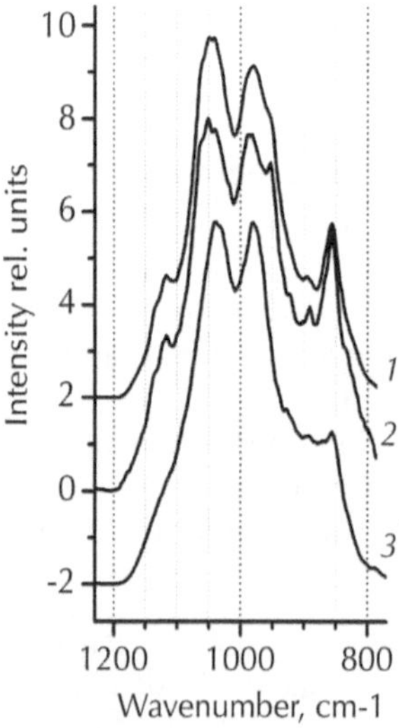

FIGURE 10 The IR-spectra of supernatant solutions obtained after centrifugation of the samples "A2-MNP-0" (*1*), "A2-MNP-1" (*2*), and "A2-MNP-1-acid" (*3*).

When proteins are used as components of coating on MNPs for biology and medicine their functional activity retaining is very important. Proteins fixed on MNPs can lose their activity as a result of adsorption on MNPs or free radical modification which is cross-linking and oxidation but it was shown that they do not lose it. The estimation of enzyme activity of thrombin cross-linked on MNPs surface was performed by comparison of the rates of conversion of FG to fibrin under the influence of thrombin contained in reaction mixtures "T1-MNP-0", "T1-MNP-1", and "T1-0-0" (Figure 11). The curves for the samples containing thrombin and MNPs which differ by the presence of H_2O_2 had no fundamental difference that illustrates preservation of enzyme activity of thrombin during free radical cross-linking on MNPs surface. Fibrin gel was formed during ~15 min in both cases. Rayleigh light scattering intensity was low when "T1-0-0" was used and small fibrin particles were formed in this case. The reason of this phenomenon is autolysis (self-digestion) of thrombin. Enzyme activity of thrombin, one of serine proteinases, decreases spontaneously in solution [25]. So, the proteins can keep their activity longer when adsorbed on MNPs. This way, the method of free radical cross-linking of proteins seems promising for enzyme immobilization.

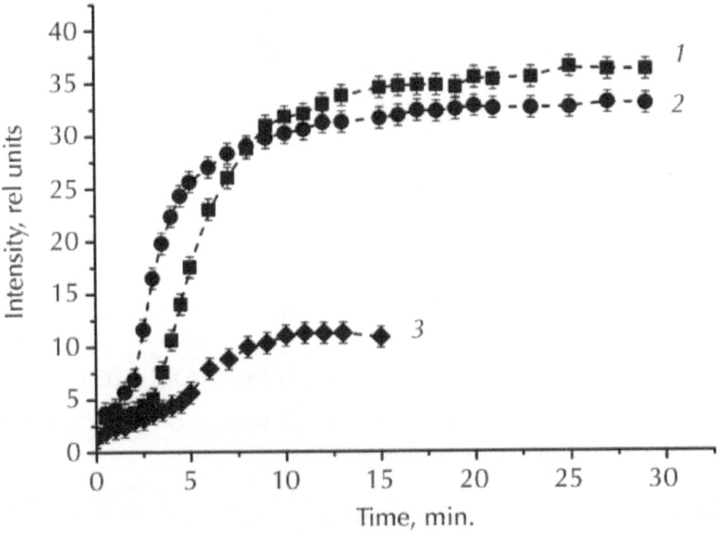

FIGURE 11 Kinetics curves of growth of Rayleigh light scattering intensity in the process of fibrin gel formation in the presence of "T1-MNP-0" (*1*), "T1-MNP-1" (*2*) and "T1-0-0" (*3*).

2.4 CONCLUSION

The novel method of fixation of proteins on MNPs proposed in the work was successfully realized on the example of albumin and thrombin. The blood plasma proteins are characterized by a high biocompatibility and allow decreasing toxicity of nanoparticles administered into organism. The method is based on the ability of proteins to form interchain covalent bonds under the action of free radicals. The reaction mixture for stable coatings obtaining should consist on protein solution, nanoparticles containing metals of variable valence (for example, Fe, Cu, Cr) and water-soluble initiator of free radicals generation. In this work, albumin and thrombin were used for coating being formed on magnetite nanoparticles. Hydrogen peroxide served as initiator. By the set of physical (ESR-spectroscopy, FMR, dynamic and Rayleigh light scattering, IR-spectroscopy) and biochemical methods it was proved that the coatings obtained are stable and formed on individual nanoparticles because free radical processes are localized strictly in the adsorption layer. The free radical linking of thrombin on the surface of nanoparticles has been shown to almost completely keep native properties of the protein molecules. Since, the method provides enzyme activity and formation of thin stable protein layers on individual nanopaticles it can be successfully used for various biomedical goals concerning a smart delivery of therapeutic products and biologically active substances (including enzymes). It reveals principally novel technologies of one-step creation of biocompatible magnetically targeted nanosystems with multiprotein polyfunctional coatings which meet all the requirements and contain both biovectors and therapeutic products (Figure 12).

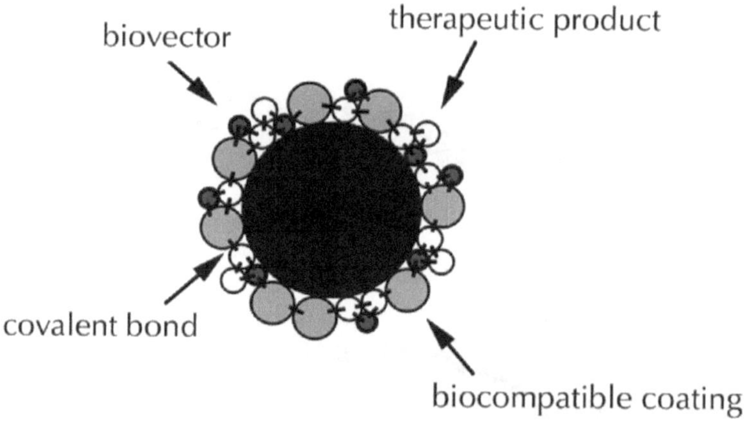

FIGURE 12 The scheme of magnetically targeted nanosystem for a smart delivery of therapeutic products based on the free radical protein cross-linking.

KEYWORDS

- **Activity**
- **Adsorption**
- **Albumin**
- **Coating**
- **Magnetite**
- **Nanoparticles**
- **Thrombin**

REFERENCES

1. Gupta, A. K. and Gupta, M., Synthesis and surface engineering of iron oxide nanoparticles for biomedical applications. *Biomaterials,* **26**, 3995-4021 (2005).
2. Vatta, L. L., Sanderson, D. R., and Koch, K. R. Magnetic nanoparticles: Properties and potential applications. *Pure Appl. Chem.*, **78**, 1793-1801 (2006).
3. Lu, A. H., Salabas, E. L., and Schűth, F. Magnetic Nanoparticles: Synthesis, Protection, Functionalization, and Application. *Angew. Chem. Int. Ed.*, **46**, 1222-1244 (2007).
4. Laurent, S., Forge, D., Port, M., Roch, A., Robic, C., Elst, L. V., and Muller, R. N. Magnetic Iron Oxide Nanoparticles: Synthesis, Stabilization, Vectorization, Physicochemical Characterizations, and Biological Applications. *Chem. Rev.*, **108**, 2064-2110 (2008).
5. Pershina A. G., Sazonov A. E., and Milto I. V., Application of magnetic nanoparticles in biomedicine. *Bulletin of Siberian Medicine* [in Russian], **2**, 70-78 (2008).
6. Trahms, L. Biomedical Applications of Magnetic Nanoparticles. *Lect. Notes Phys.*, **763**, 327-358 (2009).
7. Bychkova, A. V., Sorokina, O. N., Rosenfeld, M. A., and Kovarski, A. L. Multifunctional biocompatible coatings on magnetic nanoparticles. *Uspekhi Khimii (Russian journal)*, in press (2012).
8. Koneracka, M., Kopcansky, P., Antalik, M., Timko, M., Ramchand, C. N., Lobo, D., Mehta, R. V., and Upadhyay, R. V Immobilization of proteins and enzymes to fine magnetic particles. *J. Magn. Magn. Mater.*, **201**, 427-430 (1999).
9. Xu, L., Kim, M. J., Kim, K. D., Choa, Y. H., and Kim, H. T. Surface modified Fe_3O_4 nanoparticles as a protein delivery vehicle. *Colloids and Surfaces A: Physicochem. Eng. Aspects*, **350**, 8-12 (2009).
10. Peng, Z. G., Hidajat, K., and Uddin, M. S. Adsorption of bovine serum albumin on nanosized magnetic particles. *Journal of Colloid and Interface Science*, **271**, 277–283 (2004).
11. Šafăřík, I., Ptáčková, L., Koneracká, M., Šafăříková, M., Timko, M., and Kopčanský, P. Determination of selected xenobiotics with ferrofluid-modified trypsin. *Biotechnology Letters*, **24**, 355–358 (2002).
12. Li, F. Q., Su, H., Wang, J., Liu, J. Y., Zhu, Q. G., Fei, Y. B., Pan, Y. H., and Hu, J. H. Preparation and characterization of sodium ferulate entrapped bovine serum albumin nanoparticles for liver targeting. *International Journal of Pharmaceutics*, **349**, 274–282 (2008).
13. Stadtman, E. R. and Levine, R. L. Free radical-mediated oxidation of free amino acids and amino acid residues in protein. *Amino Acids*, **25**, 207–218 (2003).
14. Bychkova, A. V., Sorokina, O. N., Shapiro, A. B., Tikhonov, A. P., and Kovarski, A. L. Spin Labels in the Investigation of Macromolecules Adsorption on Magnetic Nanoparticles. *The Open Colloid Science Journal*, **2**, 15–19 (2009).
15. Abragam, A. *The Principles of Nuclear Magnetism*, Oxford University Press, New York (1961).
16. Noginova, N., Chen, F.,Weaver, T., Giannelis, E. P. Bourlinos, A. B., and Atsarkin, V. A. Magnetic resonance in nanoparticles: between ferro- and paramagnetism. *J. Phys.: Cond. Matter*, **19**, 246208–246222 (2007).
17. Sorokina, O. N., Kovarski, A. L., and Bychkova, A. V. Application of paramagnetic sensors technique for the investigation of the systems containing magnetic particles. In *Progress in Nanoparticles research*; Frisiras, C.T., Ed.; Nova Science Publishers, New York, 91–102 (2008).
18. Bychkova, A. V., Sorokina, O. N., Kovarskii, A. L., Shapiro, A. B., Leonova, V. B., and Rozenfel'd, M. A., Interaction of fibrinogen with magnetite nanoparticles. *Biophysics (Russian journal)*, **55**, 4, 544–549 (2010).

19. Bychkova, A. V., Sorokina, O. N., Kovarski, A. L., Shapiro, A. B., and Rosenfeld, M. A. The Investigation of Polyethyleneimine Adsorption on Magnetite Nanoparticles by Spin Labels Technique. *Nanosci. Nanotechnol. Lett. (ESR in Small Systems)*, **3**, 591–593 (2011).
20. Bychkova, A. V., Rosenfeld, M. A., Leonova, V. B., Lomakin, S. M., Sorokina, O. N., and Kovarski, A. L. Surface modification of magnetite nanoparticles with serum albumin in dispersions by free radical cross-linking method. *Russian colloid journal*, in print (2012).
21. Bradford, M. M., A rapid and sensitive method for the quantitation of microgram quantities of *protein utilizing* the *principle* of *protein-dye binding. Anal. Biochem.*, **72**, 248–254 (1976).
22. Antsiferova, L. I., Vasserman, A. M., Ivanova, A. N., Lifshits, V. A., and Nazemets, N. S. Atlas of Electron Paramagnetic Resonance Spectra of Spin Labels and Probes [in Russian], Nauka, Moscow (1977).
23. Dolotov, S. V. and Roldughin, V. I. Simulation of ESR spectra of metal nanoparticle aggregates. *Russian colloid journal*, **69**, 9–12 (2007).
24. Smith, C. E., Stack, M. S., and Johnson, D. A. Ozone effects on inhibitors of human neutrophil proteinases. *Arch. Biochem. Biophys.*, **253**, 146–155 (1987).
25. Blomback, B. Fibrinogen and fibrin – proteins with complex roles in hemostasis and thrombosis. *Thromb. Res.*, **83**, 1–75 (1996).

CHAPTER 3

MODERN BASALT FIBERS AND EPOXY BASALTOPLASTICS: PROPERTIES AND APPLICATIONS

A. V. SOUKHANOV, A. A. DALINKEVICH, K. Z. GUMARGALIEVA, AND S. S. MARAKHOVSKY

CONTENTS

3.1 INTRODUCTION

The results of investigations presented in review characterize modern continuous basalt fibers as a novel type of silicate fibers.

Basalt roving's have the improved elasticity modulus typical of high modulus and high strength magnesium aluminosilicate S-fibers and strength level close to the well-known E-fibers, and improved acid and alkali resistance. The mechanical properties of winded epoxy basaltoplastics are much similar to the properties of S-glass fibers reinforced plastics, and are over the same properties of E-glass fibers reinforced plastics. The alkali resistance and heat humidity resistance of epoxy basaltoplastics are much higher than appropriate characteristics of epoxycomposites derived from E- and S-glass fibers which is associated with the specificity of adhesive properties of basalt roving's.

Strong interest to basalt fibers as a perspective reinforcement for composites with organic and inorganic matrices stimulate intensive studies of their mechanical and physicochemical properties, technological and application aspects [1-7]. This review is dedicated to analysis of properties of modern continuous basalt fibers and basalt fiber reinforced epoxypolymer composites (basaltoplastics) and to their comparison with the properties of various glass fibers and glass fiber reinforced plastics.

3.2 BASALT SPECIFICATION AS NATURAL RAW SILICATE MATERIAL FOR MAKING CONTINUOUS FIBERS

Basalt is a volcanic rock (chilled magma) being silicate by its chemical origin. Basalt occurrences are spread extremely wide, easily accessible, and virtually unlimited. As shown in Table 1, chemical composition of basalt resembles widely spread industrial glasses used for fiber manufacture [1, 7-11].

TABLE 1 Chemical composition of basalt and several glasses for continuous fiberproduction [1, 8 – 11].

Oxide	Basalt	Alkali-free aluminoboro-silicate glass (E-glass)	Magnesium alu-minosilicate glass (S-glass)	Chemically resistant zirconium glass (AR-glass)
		Oxide content in fibers, wt.%		
SiO_2	47.5–52.5	53.4	58–73	61.5
Al_2O_3	13.3–18	14.3	15–25	5.19
Fe_2O_3	7.0–14	0.28	¾	0.14
CaO	8–11	19	¾	12
MgO	3.5–5	3.3	4–15	3.2

TABLE 1 *(Continued)*

B_2O_3	¾	10.3	¾	0.0
TiO_2	0.2–3.5	0.14	0.3–2.8	0.14
$Na_2O + K_2O$	2.5–6.0	0.29	¾	10.5
ZrO_2	0.0	¾	¾	5.8
MnO	0.17–0.22		¾	¾

Iron oxides in the basalt structure color its fibers grey-brown. Basalt rock capable of being processed to a fiber from melt has polycrystalline structure and contains sufficient amount of the glassy-like phase. Such types of basalt may be processed to fiber by a technology fundamentally similar to the technology for manufacture of the known glass fibers.

3.2.1 MECHANICAL PROPERTIES OF BASALT FIBERS

Depending on the occurrence, chemical composition of basalt suitable for processing into continuous fiber varies in a certain low range. Mechanical and physicochemical properties of basalt fibrous materials (roving's etc.) may also slightly vary thereof. By now, it has been sufficiently reliably stated that low variations in chemical composition have a minor effect on the level of mechanical properties of continuous basalt fibers [1, 10, 12]. Of the greatest effect on mechanical properties of continuous basalt fibers are direct molding conditions of the fibers (drawing temperature and the period of melt homogenization, and the fiber diameter). For example, for the same basalt composition, fiber drawing temperature rise by 160°C (from 1,220°C to 1,380°C) increased their strength from 1.3 to 2.23 GPa and modulus of elasticity from 78 to 90.3 GPa [10]. As the filament diameter increases from $d = 1 - 4$ μm (thin fibers) to $d = 7 - 10$ μm (long length and continuous fibers) the strength value decreases from 2.8 to 1.8 GPa [10]. Table 2 shows, according to data from references. [13 – 19], mean mechanical parameters of basalt roving's produced from raw basalt from different occurrences at different time, and by different manufacturers.

TABLE 2 Comparable mechanical properties of basalt roving's and some glass roving's [1, 16 - 18, 22].

No.	Roving type	Linear density, tex	Strength, GPa	Elasticity modulus, GPa	Filament diameter, μm
1	Basalt	1165	1.568	91	9
2	Basalt	1350	1.256	92	9

TABLE 2 *(Continued)*

3	Basalt	1250	1.770	91	9
4	S-type glass	1200	1.910	89.5	10
5	E-type glass	1260	1.600	71.4	13
6	ZrO_2-containing glass	2640	1.180	70	12.2
7	Basalt	1220	1.640	92.4	13
8	Basalt	2520	1.148	82	13
9	Basalt	2600	1.350	80	13
10	Basalt	1200	1.530	87.5	13
11	Basalt	2360	1.075	78.2	16
12	Basalt	3600	1.300	78.7	17
13	Basalt	4650	1.060	77.6	22
14	Basalt	4600	1.400	80	22
15	Basalt	4600	1.250	80	22

As may be inferred from Table 2, basalt roving's having close values of linear density and filament diameter have strength by 16-20% lower, as compared with high modulus and high strength roving from magnesium aluminosilicate glass (so-called S-glass [11]). Basalt roving strength is closer to the strength of roving's made of alkali-free aluminoborosilicate glass (so-called E-glass).

On the other hand, the modulus of elasticity of basalt roving's exceeds that of E-roving and, as is shown in Table 2, very closely approximates the modulus of elasticity of high modulus and high strength roving made of magnesium aluminosilicate glass (S-roving) [9-11, 16, 18, 19-22].

Thus, the data shown characterize modern continuous basalt fibers (roving's) as a new type of silicate fibers having the strength level close to that of the well-known E-fibers, but with rather high modulus of elasticity typical of high modulus and high strength magnesium aluminosilicate S-fibers (roving's).

3.2.2 ADHESIVE PROPERTIES

Adhesive interaction of basalt fibers with various polymeric matrices (epoxy [16-18], phenolic, and imide [10, 23]) is stronger than of glass fibers. This is bound to rather high content (up to 15%) of iron oxides and Fe^{+2}/Fe^{+3} iron ions, respectively, which are of high coordination ability, in the basalt composition. Catalytic effect of iron ions on epoxy oligomer curing in a thin layer (≈25 nm) directly at the fiber surface is also possible. In the event that basalt filaments are processed with a process silane coupling agent, the effort required for pulling basalt filaments from a polymeric matrix is also high [10, 24].

The higher level of adhesive interaction in basaltoplasicts makes also itself evident in the higher level of residual stresses in interface layers of polymeric matrices. Thus, according to estimates in references [10, 24], these stresses are up to 24–50 MPa in respect to the matrix type. The residual stresses in basalt fiber reinforced plastics were effectively reduced by applying a surface modification agent to fiber surface of organosilicon comb-shaped block copolymer KEP (having polyorganosiloxane in the backbone and poly(ethylene oxide) in lateral chains [10, 24-26]), which has pronounced surface activity. Because of the pronounced surface activity, when introduced into the binder composition this block copolymer reduces surface tension of the binder, makes wetting of fibers easier, and is absorbed on the fiber surface. Polysilicane chain is effectively absorbed on the fiber surface thereat, and ethylene oxide chain is combined with the binder [24-26]. As a copolymer with molecular weight of 7,000–8,000 is applied to the fiber (basalt and glass) surface, an elastic and rather strong shock absorbing adsorption layer 20–25 nm thick is formed on it. This is the reason for distinguished reduction of residual stresses and substantial increase of mechanical properties of the reinforced plastic. For instance, for phenolic basalt fiber reinforced materials with a block copolymer in the interface zone 2-fold shear strength increase, tensile strength increase along reinforcing fibers by 70% and transversal to reinforcing fibers by 35% is observed [24]. Such results were also obtained for epoxy and epoxy novolac basaltoplastics [25, 26].

3.2.3 CHEMICAL RESISTANCE OF CONTINUOUS BASALT FIBERS

As a rule, basalt fiber resistance to aggressive media (acid and alkaline) is determined by the strength change after some time of exposure (aging) in these media. Chemical resistance of basalt fibers depends upon their chemical composition, nature of aggressive medium and temperature - time conditions of influence. The ratio of silicon, aluminum, calcium, magnesium, and iron oxides in the basalt composition is of high importance [1, 12, 13, 27]. Exactly the presence of iron oxides in the silicate carcass of basalt fibers imparts to them higher chemical and heat resistance, as compared with glass fibers [1, 13, 27]. Beside chemical composition, in case of surface-active media (alkali, some salt solutions, and so on) the fiber surface layer condition (the level of defects presented) [14-18, 22, 27-30] example thermal (technological) prehistory of the fiber, is also valuable.

3.2.4 ACID RESISTANCE OF CONTINUOUS BASALT FIBERS

Basalt fibers have high acid inertness, which is far beyond inertness of widespread glass fibers from alkali-free aluminoborosilicate glass (E-type) and magnesium aluminosilicate glass (S-type), but is somewhat behind inertness of specific chemically resistant zirconium glasses. At short-term exposure in strong mineral acid solutions, no fiber strength was observed [10]. Long-term (over 100 hrs) impact of hydrochloric acid solutions caused strength reduction by 15–20% [10]. Acid resistance of basalt fibers was considered in more detail in references. [17, 22]. Kinetic curves in Figure

1 show that the breaking load of two different (basalt and E-glass) roving's after exposure to hydrochloric acid reduces much slower for basalt roving's rather than for E-glass [17, 22]. Hence, this reduction proceeds also slower for basalt roving's with smaller filament diameter than for the glass one.

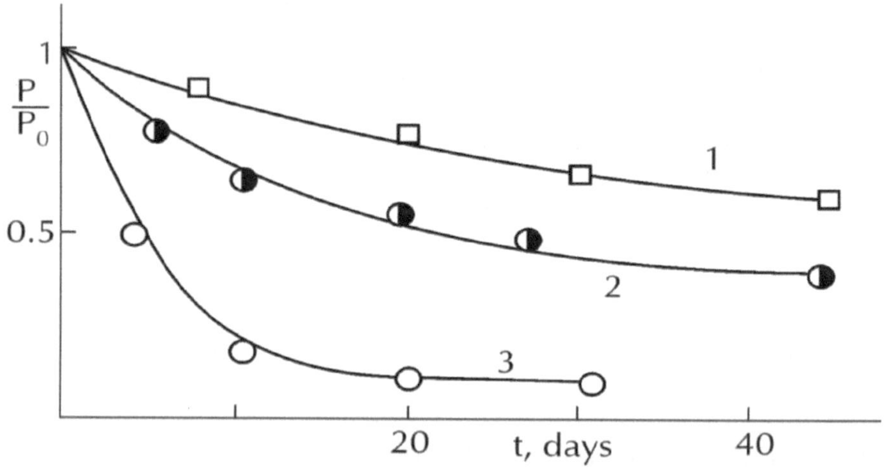

FIGURE 1 Relative breaking load reduction for roving's at their aging in 20% HCl. 1--basalt roving Number 13; 2--basalt roving Number 1; 3--roving Number 5 from E-type alkali-free aluminoborosilicate glass; T = 20°C; numbers of roving's represent their positions in Table 2.

Above all, the differences observed relate to differences in chemical compositions of these materials [1, 27], and with slightly different proceeding of E-glass and basalt components in acid, as is specified in reference [22].

It has been known [27] that as strong acids interact with E-glass fibers, no protective film of silicon dioxide is formed on the fiber surface, and therefore, strong acids are capable of rapidly transforming these fibers into high porosity silica. In the event basalt fibers aging in acid shown in Figure 1, as the initial stage of breaking load decrease, which is relatively fast and related to fast adsorption of aggressive medium and partial dissolution of the basalt components ends, such film is apparently formed with time, and the fiber etching and strength reduction decelerate as a consequence of the diffusion barrier formation.

Good mechanical properties of basalt fibers, increased chemical resistance, heat resistance, [1, 8, 13] and high filtering properties allowed [31] for their effective use in structures of various industrial filters for gaseous and liquid aggressive media.

For broader picture of basalt fiber properties, it should be noted that there are types of basalt having non-optimal (from positions of chemical resistance) content of silicon, alkaline earth metals, and aluminum oxides. Roving's made from such basalts have satisfactory mechanical properties, but their acid and alkali resistances are not far beyond E-glass roving's. Apparently, such basalt fibers may be of interest for both fabrication of composite materials and silica production.

3.3 ALKALI RESISTANCE

Basalt fibers resistance to the influence of various alkali media has been considered in different times and at different levels in a considerable number of works [10, 32, 33]. Historically, this is associated with the fact that since their occurrence basalt fibers (short first, and continuous later on) demonstrated higher alkali resistance compared with the majority of glass fibers. This feature of basalt fibers has specified the known attempts of using them as reinforcement of Portland cement concrete, which has durably preserved alkali medium [12, 32]. Data from the literature on this subject may be conditionally divided into two unequal groups.

In the first, the largest group of works [10, 32, 33] the main aspects of interaction between various alkali media (NaOH, KOH, $Ca(OH)_2$, cement extract etc.) and basalt and various glass fibers causing fiber strength decrease have been considered qualitatively. As stated in the works from this group united in the monograph [32], the model alkali media studied (at equal normality of solutions) can be arranged in appropriate sequences in accordance with aggressiveness to basalt and glass fibers [32]:

$$NaOH > KOH > LiOH > NH_4OH; \quad (1)$$

and

$$Ba(OH)_2 > Sr(OH)_2 > Ca(OH)_2 \quad (2)$$

In this group of works alkali resistances of basalt fibers and various glass fibers including specific, chemically resistant ZrO_2-containing ones, such as Cem-Fil [32] and analogous native glass fibers Shch-15-ZhT and STs-6 [34], were compared qualitatively. Among zirconium-free glass fibers, a perspective boron-free alkali resistant glass fiber should be intimated. In these works, the main features of the aging mechanism for glass fibers and basalt fibers in alkali media were determined [32].

As is demonstrated, aging example strength decrease of glass fibers having different chemical compositions (alkali, alkali-free, quartz, and zirconium) and basalt fibers after exposure to alkali media during some time, is the result of complex multistage physicochemical interactions between glass components and surrounding aqueous alkali medium. The first process among them, which is the most rapid (and sometimes the main one), is adsorption of water and other components of aggressive media on the fiber surface which, as a rule is reasonably developed [27]. Adsorptional filling of defects on the fiber surface (pores and microcracks) by liquid surface-active (aggressive) medium is accompanied by a substantial decrease of the fiber strength (the Rehbinder effect). Slower reaction causing dissolution of all or a part of basalt or glass oxide components, mainly, the silica carcass [27, 32], proceed in parallel.

Analysis of the strength decrease of glass fibers having different chemical composition in various model alkali media enables the authors [32] to range glass and basalt fibers studied in the following descending sequence by alkali resistance:

$$zirconium > basalt > quartz > alkali > alkali\text{-}free. \quad (3)$$

As may be inferred from the sequence (3), expensive zirconium-containing glass fibers are followed right away by relatively cheap basalt fibers having herewith higher physical and mechanical properties. This circumstance has attracted and attracts still attention of the researchers and manufacturers to these fibers.

In the later works from the second, the shorter group, attempts were made to describe qualitatively kinetics of the interaction of new improved continuous basalt fibers and some glass fibers with alkali medium [14-18, 22, 28-30]. In this group, a kinetic model for alkali aging (strength change) of basalt (and glass) fibers has been designed and evaluated. This model allows for separately estimate contributions of both adsorptional strength decrease (the Rehbinder effect) and chemical dissolution of basalt or glass components to the general picture of the fiber strength change. Analysis of the experimental data from references. [8, 14-18, 22, 28-30, 33] shows that kinetic curves of breaking load changes for fibrous materials under study may be divided into two types: monotonously decreasing [14-17, 22, 28-30] and ones having a local extreme [8, 17, 33].

For a simple event of the breaking load monotonous decrease during alkali aging of fibers (Figure 2) a kinetic model [14-17, 22, 28-30] yielding from the known conceptions [27, 32, 36] of the glass fiber strength decrease mechanism in aggressive adsorption active media has been developed.

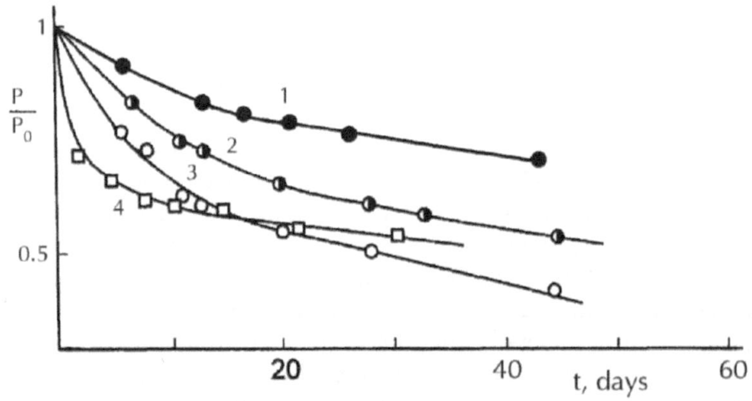

FIGURE 2 Kinetic curves for breaking load decrease for glass (1, 3) and basalt (2, 4) roving's at their aging in 1.25N NaOH. 1--glass roving Number 6 (zirconium glass); 2--basalt roving Number 1; 3--glass roving Number 5 (E-type glass); 4--basalt roving Number 13; T = 20°C; numbers of roving's represent their positions in Table 2.

According to this model, at the initial stage of roving's aging in the alkali medium rapid decrease of the breaking load is substantially associated with removal (washing off) of an oiling agent and prompt proceeding of aggressive medium adsorption on the fiber surface defects (in pores and microcracks), which reduces their strength (breaking load). Adsorptional processes on the fiber surface and relevant decrease of the breaking load obey the first order kinetic law [14-17, 22, 28-30, 36]. At this prompt stage of aging mass losses and fiber diameter decrease due to chemical dissolution of specifically glass or basalt are low, and their contribution into strength decrease is negligible [14-17, 22, 28-30].

The subsequent slow stage of breaking load decrease is associated with chemical dissolution of the fibers proceeding from the surface under diffusion-controlled conditions that is, in the external diffusion-kinetic zone [37, 38].

As shown by electron microscopic studies (Figure 3), at this slow stage of alkali aging on the basalt fiber surface a spongy porous cover consisting of insoluble products of basalt alkali hydrolysis and very weakly bound to undamaged internal part of the fiber is formed with the time.

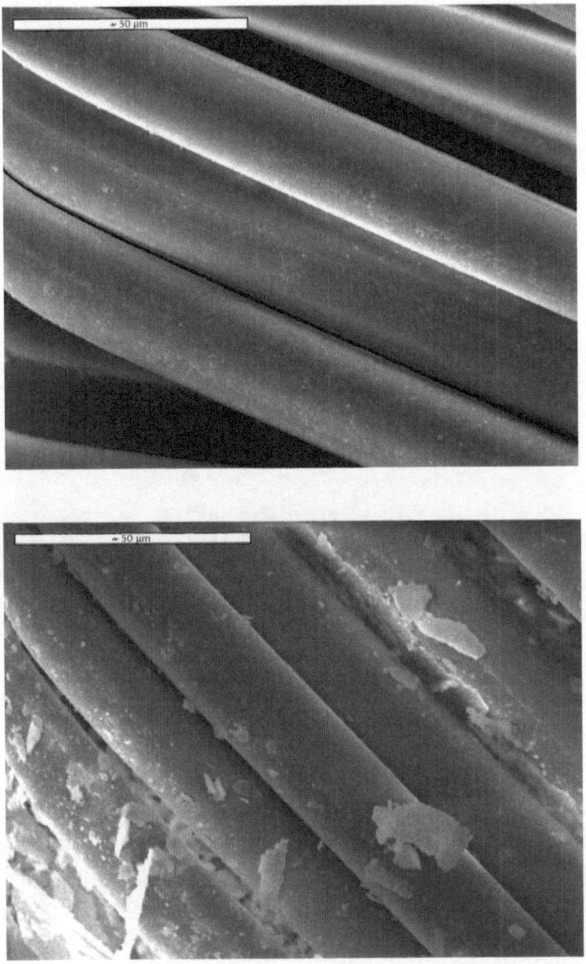

FIGURE 3 *(Continued)*

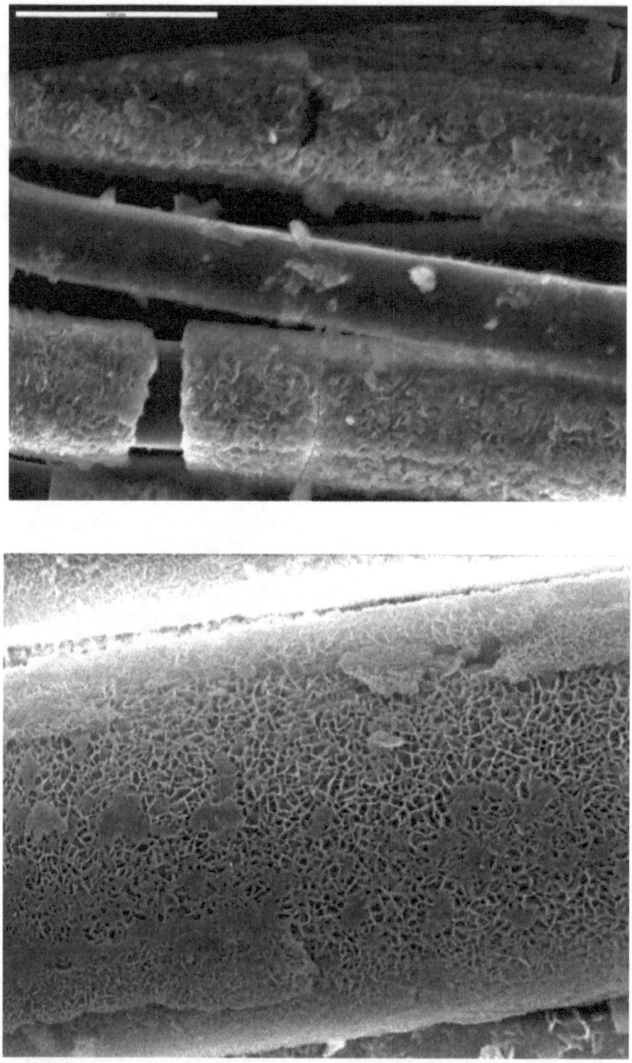

FIGURE 3 Microphotographs of basalt fibers.a--basalt fibers in the initial state; b, c,--after aging in NaOH solution (5%) at 20°C during 36, and 90 days, respectively; d--microporous structure of the surface aged layer after 90 days of aging in alkali.

As time of exposure to alkali solution increases, dissolution and removal of a part (as for basalt fibers) or all (as for E-glass fibers) oxide components from the fiber surface induce a gradual decrease of the fiber diameter part not affected by etching and respectively, the breaking load decreases. Basing on these conceptions, the following kinetic equation taking account of the breaking load decrease at simultaneous proceeding of adsorption and chemical dissolution of the fiber has been deduced [14-17, 22, 28-30]:

$$\frac{P}{P_0} = \frac{P_S(0)}{P_0} exp(-k_S t) + \frac{P_V(0)}{P_0}(1 - k_V \times t)^2 \quad (1)$$

Where:

P_0 and P are initial and current values of the breaking load;

t is time of exposure in alkali solution;

$P_S(0)/P_0$ is a part of breaking load, which decrease is associated with adsorption processes (the so-called "surface" component of the breaking load for the initial ($t = 0$) roving depending on defectiveness degree of the fiber surface);

k_S is effective constant of the fiber strength adsorptional decrease; characterizes the measure of surface defects "activity", that is depends on micropore and microcrack geometry on the surface layer;

$P_V(0)/P_0$ is a part of breaking load, which decrease is associated with chemical dissolution of the fiber components (the so-called "voluminous" component);

k_V is the effective constant of the fiber chemical dissolution depending on chemical composition of the fiber, its diameter, alkali solution concentration, and k_R constant of specifically chemical interaction between basalt (or glass) and alkali [37]:

$$k_V = k_R[NaOH](0.5d)^{-1} \quad (2)$$

k_R is a constant of specifically chemical interaction at basalt (glass) - alkali solution interface.

As observed from the Equation (1), in the context of this kinetic model the breaking load reduction is assumed as the sum of exponentially decreasing surface component and parabolically decreasing voluminous component. At $t = 0$ the following equality is true:

$$P/P_0 = 1 = P_S(0)/P_0 + P_V(0)/P_0. \quad (3)$$

As follows from the Equation (1), starting from some specifically high t, when it is reputed that $\exp(-k_S t) \rightarrow 0$, reduction of the breaking load measured in the experiment is only associated with the second summand in the Equation (1). In this case, the Equation (1) is reduced to:

$$\left(\frac{P}{P_o}\right)^{0,5} \cong \left(\frac{P_V(0)}{P_0}\right)^{0,5} (1 - k_V \cdot t) \quad (4)$$

As we depict experimental results on alkali aging of roving's in the Equation (4) coordinates, the effective constant k_V is determined from the straight line slope, and $(P_V(0)/P_0)^{0.5}$ value is determined from a line segment cut off the axis of ordinates and, further on, parameters $P_S(0)/P_0$ and k_S are calculated with respect to the Equation (3).

In the framework of the model under consideration, parameters from the Equation (1) (Table 3) were calculated via processing kinetic curves of various basalt and glass fibers alkali aging [14-17, 22, 28-30]. As is shown [14-17, 22, 28-30], the kinetic method considered allows for prediction of basalt and glass fiber materials aging at long-term exposure to alkali medium.

Figure 2 shows well the influence of the filament diameter and respectively the fiber surface condition on the strength decrease of studied roving's at the initial, adsorptional stage of alkali aging. In case of basalt roving aging with the greatest filament diameter ($d = 22$ μm), the breaking load decreases most rapidly (characterized by $k_S = 0.69$ day^{-1}, Table 3) at the fast initial (adsorptional) part of the kinetic curve, and for aging of roving's with the smallest filament diameter ($d = 9$ μm) it decreases most slowly ($k_S = 0.125$ day^{-1}, Table 3).

This is connected with the known fact of glass and basalt fibers surface defectiveness increase with their thickness [1, 27] and, as a consequence, with more rapid adsorptional decrease of the fiber strength (the Rehbinder effect). As is outlined, parameter k_S represents a characteristic (measure) of "activity" of the surface defects during adsorptional decrease of the fiber strength, and $P_S(0)$ value is defined by concentration of these defects.

TABLE 3 Kinetic parameters of basalt and glass roving aging in 1.25 N aqueous NaOH, T ≈ 20°C

Roving type[1]	$\frac{P_S(0)}{P_0}$	$\frac{P_V(0)}{P_0}$	**k$_S$, day^{-1}**	**k$_V$, day^{-1}**	**k$_R$·10^7** $\frac{cm}{(MO\pi b/\pi)\cdot cym}$	**Filament diameter, μm**
ZrO_2-containing glass, №6	0.15	0.85	0.15	$2.0\cdot10^{-3}$	9.8	12.2
Basalt, №1	0.28	0.72	0.125	$3.0\cdot10^{-3}$	10.8	9
E-type glass, №5	0.33	0.67	0.15	$4.15\cdot10^{-3}$	21.6	13
Basalt, №13	0.34	0.66	0.69	$3.25\cdot10^{-3}$	28.6	22
Basalt, №15	—	—	—	2.7510^{-3}	24.2	22
Basalt, №3	0.32 [2]	—		$2.20\cdot10^{-3}$	8.0	9
S-type glass, №4	0.22 [2]	—		$8.10\cdot10^{-3}$	32.4	10
Basalt, №9	0.40 [2]	—		$1.90\cdot10^{-3}$	7.0	13
Basalt, №15	0.47 [2]	—		$2.20\cdot10^{-3}$	19.4	22

1 Numbers of roving's represent their positions in Table 2.
2 Breaking load decrease from the initial level in the minimum of alkali aging curve (P_{min}/P_0).

The lower concentration of defects in the surface layer is, the shorter the breaking load decrease at the fast adsorptional stage is, that is the lower $P_S(0)$ is and the higher the voluminous component level of the breaking load $P_V(0) = P_0 - P_S(0)$ is thereby, from which slow chemical dissolution of glass or basalt components proceeds. It ought to be noted that the surface layer defectiveness depends on temperature-time fiber molding conditions. Somewhat further increase of alkali resistance of basalt fiber is therefore possible still due to reduction of its surface layer defectiveness *via* improvement of technological parameters of molding.

The aging constant of the fiber, k_V, represents an effective parameter of chemical resistance of fixed diameter fibers in alkali solution of fixed concentration. Intrinsically, k_V is a technical aging parameter of specific fibrous materials under fixed conditions. For more general characterization of the reactivity measure of basalt in relation to alkali solutions, specific constant k_R [37] can be used. Numerically, it equals k_V for the fiber of elementary diameter and elementary concentration of solution, as follows from the Equation (2). Table 2 shows that for different basalt fibers k_R values may change by more than twice. This apparently is associated with oscillations of $[SiO_2]/([Al_2O_3] + [FeO + Fe_2O_3])$ ratio [27, 39]. It ought to be noted that for various basalt fibers, k_R value may be close to or even exceed its values for unstable aluminoborosilicate glass, but it may also approach k_R value for specific, chemically resistant zirconium glass (Table 2) [11, 27, 32, 34].

The kinetic model considered clearly explains and allows for description of alkali aging of basalt and glass fibers in the simplest cases that is, in absence of local extremes on kinetic curves. Note also that for continuous fibers the case, when kinetic curves of the fiber alkali aging have extremes, is most often observed and, apparently, more general among all [8, 18, 33]. In this case, some limitations appear for the kinetic model considered.

The presence (Figure 4) or the absence (Figure 2)of extremes on alkali aging curves for various fibers was associated with unequal conditions of their surfaces [18], that is with the differences in types (size and shape) and concentrations of defects present on the surface of these fibers.

As is already mentioned, the surface condition (the defectiveness degree) of the fiber depends on fiber production technology and usually, the surface defectiveness of the fiber increases with the fiber diameter [27, 39]. Therefore, the kinetic curve shape for alkali aging of basalt and glass roving's somehow reflects thermal prehistory of the fiber. This is indicated by the presence, intensity or the absence of extreme on the fiber alkali aging curve, and also *via* the value and rate of the breaking load reduction at the initial, fast (adsorptional) part of this kinetic curve.

The absence and the presence of local extreme and its different intensity were primarily associated [17] with the type of defects and their concentration on the fiber surface. Surface defect (a spiny pore of a crack [27]) may be considered as an effective stress concentrator, which negative effect on the fiber strength intensifies with adsorption proceeding and reaches its maximum after adsorption processes end, that is in the minimum of alkali aging curves (Figure 4).

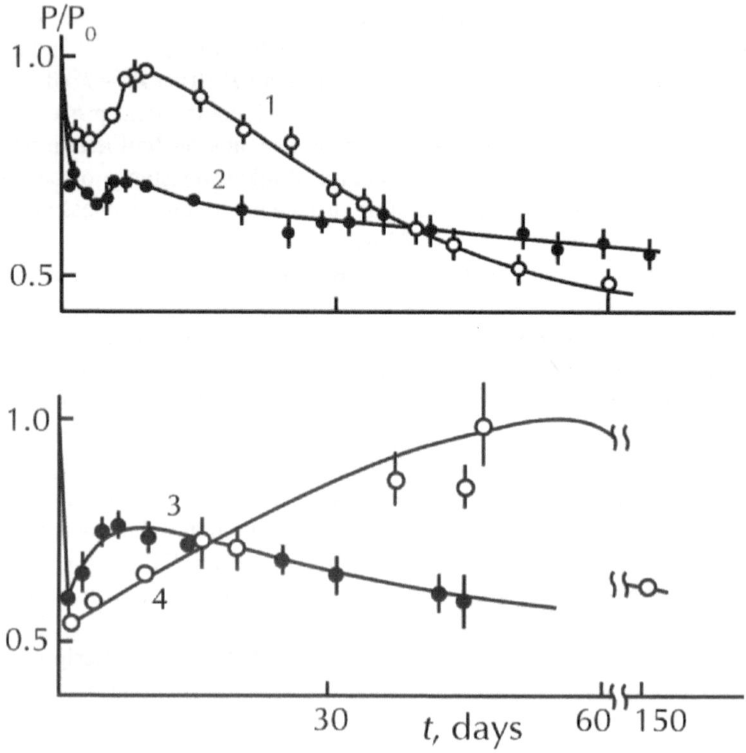

FIGURE 4 Kinetic curves of breaking load decrease for high strength and high modulus S-glass roving (1) and basalt roving's (2 – 4) as aged in 1.25N NaOH. 1--S-glass roving Number 4; 2--basalt roving Number 3; 3--basalt roving Number 9; 4--basalt roving Number 14; T = 20°C; numbers of roving's represent their positions in Table 2.

As time of exposure to alkali increases, the spiny pore or the crack transforms to a defect of high curvature due to chemical dissolution (etching) of its walls (Figure 5).

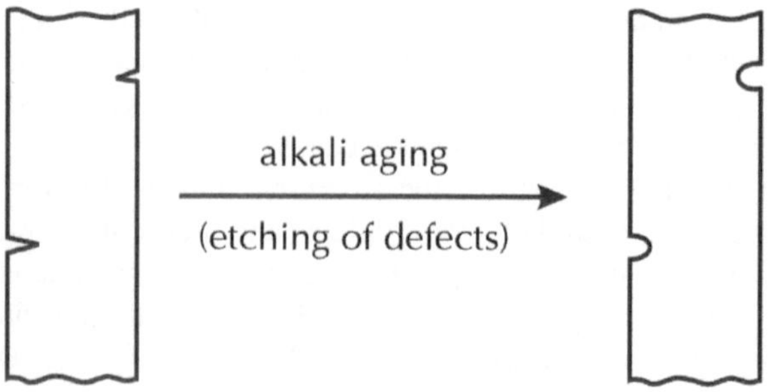

FIGURE 5 The scheme of surface defects etching in basalt (or glass) fibers.

The radius curvature increase in the defect (crack) mouth induces the loss of defect feature as an effective stress concentrator [27]. As a result, at this stage of alkali aging the breaking load will be determined by effective diameter of the fiber and, without influence of the surface defects stress concentrators, may exceed the value conformed to the adsorptional process end, that is to the minimum on alkali aging curves. In the framework of such structural kinetic approach, the absence of a maximum on the alkali aging curves of basalt fibers (Figure 2) or its low intensity (Figure 4, curve 2) means that in these cases, the defects existing on the fiber surface are not effective stress concentrators, and etching of these defects and "removal" either causes no increase of the breaking load at all, or results in insignificant increase.

The main limitation for the model approach considered is the fact that in case of kinetic curves with extremes, k_S value may only be determined at a particular degree of approximation. However, the breaking load value in the minimum of the aging curve (Figure4) example at the end of the adsorption stage, is determined rather accurately. This reduction degree may also characterize the surface defectiveness, which increases with the filament diameter [1, 27, 39]. For instance, Figure 4 shows that as the filament diameter increases, for basalt roving's, the breaking load in the minimum of the aging curve decreases more intensively. For basalt roving with the filament diameter $d = 9$ µm, the breaking load decreased by ≈ 32%, and for roving's with $d = 13$ µm, it equaled ≈ 40%, whereas for roving with the filament diameter $d = 22$ µm, it equaled ≈ 46%, respectively. Data in Figure 4 [18] also show that the defectiveness of high strength and high modulus S-glass fibers ($d = 10$ µm) is lower than for basalt roving of similar diameter ($d = 9$ µm) (Figure4, curves 1 and 2). The character of surface defects of these fibers close in diameter is different: Glassy S-fiber demonstrate clear extremes on the alkali aging curve, whereas basalt roving's show weakly expressed ones.

At the slow stage of chemical dissolution of basalt or glass (Figure 4) the breaking load decreasing rate will be defined by chemical composition and the fiber diameter. For quantitative estimation of alkali resistance of basalt or glass fibers at the slow stage of chemical dissolution, after passing the maximum of the kinetic curve, an expression resembling Equation (3) [18] is used:

$$\left(\frac{P}{P_0}\right)^{0,5} = \left(\frac{P_{\max}}{P_0}\right)^{0,5} \left(1 - k_V \cdot (t - t_{\max})\right) \tag{5}$$

Where: $t > t_{max}$, t_{max} is time for reaching local maximum on the aging curve; P_{max}/P_0 is the relative breaking load at $t = t_{max}$ (that is in the local maximum).

Table 3 shows values of kinetic parameters of roving aging in NaOH solution, determined from the experiment by Equation (5). As observed, k_V and k_R values indicate higher alkali resistance of basalt roving's compared with S-glass fibers.

3.3.1 ALKALI RESISTANCE OF STRESSED BASALT FIBERS.

In various structures (filters for liquid aggressive media, basalt roving's in concrete, etc.), basalt fibrous materials (woven and nonwoven) are exposed to a simultaneous impact of aggressive medium and mechanical stresses. It was of interest therefore to clear out and assess, even qualitatively, the degree of impact of external mechanical stress applied on alkali resistance of basalt fibers.

The experiments were carried out on a basalt roving with the filament of 9 μm in diameter, modulus of elasticity of 91 GPa and strength of 1.50 GPa. Roving were stressed by winding them on a glass tube of 8 mm external diameter, with perforated side surface (adapter). The roving ends were mechanically fixed to the adapter. Then the samples representing adapter tubes with a single layer of the roving tightly wound on it were placed in NaOH alkali solution (1.25 N) and exposed to temperature of 20 ± 1.5°C for the given time. After that time, the samples were taken out of the solution, detached from the adapter and rinsed with plenty of running water and then with distilled water. Preparation of roving sections for mechanical tests and the tests themselves were performed according to the standard procedure, for details see [14, 15, 18, 22, 28]. Figure 6 shows mean values of breaking load for the test results of 7-10 samples.

The tensile stress affecting the external surface of wound roving's was estimated by the known formula:

$$\sigma = E\, d\, R^{-1} \tag{6}$$

Where, σ is the tensile stress affecting the external surface of wound fibers, *E* is modulus of elasticity, *d* is the filament diameter; *R* is the adapter diameter (perforated glass tube) on which the roving is wound. In this case, the tensile stress was found equal 102.4 MPa. This value is less than 7% of the average roving strength, that is for the selected method of loading the stress level in fibers was found expectedly low.

Figure 6 shows that alkaline aging (breaking load decrease) of stressed basalt fibers differs from unstressed (free) aging.

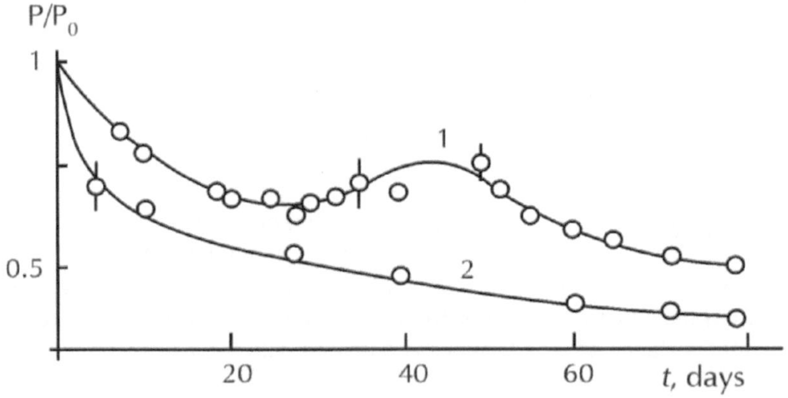

FIGURE 6 Kinetic curves of alkaline aging of 1--unstressed and 2--stressed (102.4 MPa) basalt fibers in 1.25 N NaOH. Vertical dashes show scattering of breaking load values for the roving.

Kinetic curves for aging of stressed basalt fibers have no local extremes, the initial aging rate increases and, principally, the stressed roving aging curve lies beneath that of unstressed roving.

As a hypothesis that explains qualitatively the observed effect, it can be assumed that surface defects present in the fiber (Figure 5) develop additionally under the effect of tensile stress and this facilitates diffusion of the aggressive medium into them. In this connection, faster and greater decrease of strength (breaking load) at the initial adsorptive stage of aging should happen. This very fact was observed in the experiment (Figure 6). On the other hand, however, as mentioned above, the stress impacting the fiber (monofilament) did not exceed 7% of the short-term roving strength. At this stress level, no any dangerous macroscopic deformation of filaments happens and, consequently, high development of defects (that is deformation) may hardly be expected. Observed noticeable differences in alkaline aging of loaded and unloaded fibers (Figure 6) are presumably associated with the occurrence of local overloaded zones near a surface defect. In these zones (in the orifice of sharp-nosed pore or crack and by the edge of its basement), local acting stresses (due to geometry reasons) are often much higher than the average stresses obtained for defect-free body by the Equation (6). As a consequence, local relative deformations in the defect zones will be much higher than the average macroscopic deformation of the fiber.

Apparently, the main reason for observed speeding-up of alkaline aging (its adsorptive stage) of basalt fiber is increased local deformations of defect microzones, that is their opening.

3.3.2 MECHANICAL PROPERTIES OF BASALTOPLASTICS

The mechanical and physicochemical properties of basalt roving's (discrete in the early works [23] and continuous in later works [1, 2, 17, 40-42]) demonstrate prospectivity of their use for the reinforcement of polymers. Table 4 shows mean mechanical characteristics of uniaxial epoxy basaltoplastics and glass fiber plastics produced by wet winding with a roving [17, 40, 42]. Composites were produced [16-18, 22, 40-42] with the use of basalt roving's of various manufacturers, made at different time, having different filament diameter, and treated with various gloss finishes, well-reputed in the previous production of glass fibers and glass fiber plastics.

Table 4 shows that that main mechanical characteristics of epoxy basaltoplastics exceed those of glass fiber plastics analogues, produced from E-glass roving's. Hence, mechanical properties of basaltoplastics are close to properties of S-glass fiber plastic based on the higher strength and higher modulus S-roving's type [9, 11, 21]. As noted above, basalt roving's significantly exceed E-glass ones by the modulus of elasticity and approach them by strength (Table 2). The increase of the modulus of elasticity of basaltoplastic above E-glass fiber plastic therefore, seems quite expectable, but a noticeable increase of basaltoplastic strength observed (Table 4) over E-glass fiber plastic seems interesting and untrivial.

TABLE 4 Comparable physical and mechanical parameters of uniaxial wound epoxy composites [16, 40].

Parameters	Filament diameter in roving, μm	Basaltoplastic	Glass fiber plastic (E-fibers)	Glass fiber plastic (S-fibers)[1]
Tensile strength in reinforcement direction, MPa	9 13	1130-1210 1120-1330	952	962-1450
Tensile modulus of elasticity in reinforcement direction, GPa	9 13	51.16-53.56 53.1–57.89	46.13	50.0-66.30
Compressive strength in reinforcement direction, MPa	9 13	844	736	—
Breaking stress at interlaminar shear, MPa	9 13	51–64.5 54–71	54.70	57.2–76.5
Interlaminar shear modulus, GPa	9 13	5.9	4.3	—
Density, kg/m^3	9 13	1960–2240 2090–2120	2020–2045	1945–2127

1 Filament diameter in S-glass roving is 10 μm.

At the moment, the reasons for this phenomenon are not studied in detail; apparently, they relate to structural features and the surface composition of basalt roving's. Systematic excess of basaltoplasic tensile strength above E-glass fiber plastic is associated [10, 24] with the higher level of adhesive interaction between components in basaltocomposite (even in case of existing industrial gloss finishes, designed for glass fibers processing) and, as a consequence, by somewhat different conditions of stress transfer through the interface. Therefore, interphase interaction control in basaltoplastics (*via* development of gloss finish and elasticizing-modifying layer systems particularly for basalt) seems one of the effective ways for further modification of basaltoplastic properties.

Table 5 shows characteristics of epoxyphenolic basaltoplastics and glass fiber plastics determined from testing results of the samples prepared from pressed plates subproducts 5 mm thick. The plates were prepared from linen-woven basalt and glassy fabrics with mean thickness 0.27 mm, the average surface density 285 g/m^2, the same thread density, and the filament diameter 9 μm [16].

TABLE 5 Comparable characteristics of pressurized fabric composite materials (epoxy novolac binder EP-5122)

Parameters	Basaltoplastic	Glass fiber plastic
Ultimate tensile stress in warp direction, MPa	610	440
Ultimate tensile stress in weft direction, MPa	390	240
Ultimate compressive stress in warp direction, MPa	320	210
Ultimate compressive stress in weft direction, MPa	310	160
Tensile modulus of elasticity in warp direction, GPa	27.5	24
Tensile modulus of elasticity in weft direction, GPa	24	18.5
Density, kg/m^3	1980	1900

For the purpose of studying potential abilities of epoxy basaltoplastic, uniaxial samples 1 mm thick and 15 mm wide have been produced and tested [16, 40]; They were prepared by pultrusion from a roving with linear density of 2,520 tex and the average filament diameter 13 μm. In a series of experiments, tensile strength in the reinforcement direction equaled 1,550–1,850 MPa [40], that is basaltoplastic strength was as good as the strength of S-glass fiber plastic (based on high strength and high modulus roving's).

3.3.3 CHEMICAL RESISTANCE OF BASALTOPLASTICS

When analyzing chemical (corrosion) resistance of composites, it should be taken into account that in the general case, chemical resistance of the composite depends on chemical resistance of polymeric matrix, reinforcing fibers resistance and diffusion-transport properties (penetrability, continuity) of the entire composite [9, 43-45].

Chemical resistance (example waterproof, acid and alkali resistance, etc.) of various polymeric matrices (polyester, vinyl ester and epoxy), applied to manufacturing of articles from the composites, may change in a broad range, even inside each class of the above mentioned binders. Among the above mentioned oligomeric systems, epoxy binders have higher mechanical and thermal properties and, therefore, they are of special interest for production of basaltoplasts. Chemical resistance of epoxy polymers may vary widely and is defined by oligomer and curing agent selection [46].

Moisture and alkali resistance of epoxybasaltoplastics will be considered below in this review and will be compared with these parameters of glass fiber plastics.

From positions of alkali and moisture resistance, it is necessary that the matrix polymer contains no groups and bonds in the structure, capable of alkaline hydrolysis, and polar groups, which increase its hydrophilic behavior and thereby increasing water absorption, and making water and alkali electrolyte transfer to the fiber-matrix interface easier [38, 43]. Epoxypolymers, obtained by curing epoxydiane resins by tertiary amines and other curing agents of the catalytic action, have high resistance to the action of moisture and aqueous medium. In this case, oligomeric molecules incorporate to a three-dimensional network by very stable ether C-O-C bonds [46]. Of this type is the well-known binder EDT-10 [9, 21] derived from ED-20 resin, modified by diethyleneglycol diglycidyl ether (the so-called DEG-1) and the curing agent triethanolaminotitanate (TEAT). This binder has been used for making reinforced plastics [17, 18, 22, 42, 47], see below for the results. Apparently, higher moisture and alkali resistance is demonstrated by product of cycloaliphatic resins (or systems derived from them) curing by catalytic curing agents.

As is already mentioned above, chemical (corrosion) resistance of the composite also depends on its continuity, that is physical penetrability in relation to aggressive medium. In the general case, liquid aggressive medium permeation into the polymeric composite volume happens both at the expense of transportation by pores and capillaries present in the composite and their filling (fast processes) and due to solubility and molecular diffusion of aggressive medium component diffusion [43, 44, 46], water, first of all (slow processes).

In the case of reinforced plastics with high filling degree (required for high mechanical properties), liquid aggressive medium permeation, to a large extend may proceed by defects (pores and cracks) both in the matrix interlayers and on the interface [38, 43, 46]. These defects are formed by curing of the binder resulting from shrinkage at the fiber-matrix interface, as a consequence of non-optimal impregnation of fibers with the binder and similar technological factors [46]. After filling various pores with the aggressive medium, they become something like “depots”, wherefrom this medium continues diffusing into the surrounding volume of the composite. Permeated into interlayer volume of the matrix polymer, water molecules plasticize the polymer that often causes reduction of elastic and strength properties of the composite [43, 44, 46]. Aggressive medium, which permeates to the interface surface, breaks adhesive bonds and reduces the fiber strength as a consequence of adsorptional effects and chemical dissolution (etching) of the fiber. In the aggregate, these processes cause a substantial decrease of the composite strength [22, 43]. At the interface, adhesive strength is the most sensitive parameters of the composite. It is usually estimated by changes in the breaking stress at the interlayer shear.

3.3.4 HEAT HUMIDITY RESISTANCE OF BASALTOPLASTS

The most important characteristic of reinforced polymers derived from cured epoxy, polyester, and other polar polymeric matrices is heat and moisture resistance, that is resistance to simultaneous influence of temperature and humidity.

It has been shown [17, 18, 22, 42, 47] that heat and moisture resistance of epoxy basaltoplasts is usually higher than that of analogous glass fiber plastics based on E-type and S-type glass fibers. For quick estimation and comparison of basaltoplastic and glass fiber plastic resistance to heat and humid aging, a method of accelerated heat and humidity tests, suggested by the authors, has been used [17, 18, 22, 42, 47]. The method consists in impacting the composite material samples with overheated water steam (T >100°C). This method is sufficiently strict comparative test technique, but it allows for quick reaching high water content in "thick" samples of the material and therefore, noticeable changes in strength characteristics of compared composites. For the purpose of these tests the sample were for some time exposed in autoclaves made of stainless steel, equipped with thermostat control system, in which preset temperature and respectively saturated water vapor pressure P was maintained. In the experiments [17, 18, 22, 47], some results of which are shown in Figure 7, these parameters equaled T ≈ 117-120°C, P ≈ 2.18 atm.

As is known [43], the composite porosity and respectively moisture transfer sorptional-diffusional characteristics of the composite increase with the filament diameter, and consequently, reduction of the composite mechanical properties is intensified [18, 22, 42, 47]. Figure 7 shows that at the same filament diameter equal d = 13 μm, and at close values of it (9 and 10 μm), the shear strength was always preserved better for basaltoplastics. Simple estimates made from data in Figure 7 allows for determination of parameter t_{05}, that is time for two fold decrease of the shear strength. Parameter t_{05} equals 0.6 days for E-glass fiber plastic (d = 13 μm), 1.0 and 5.5 days for basaltoplasts with d = 13 μm, 2 days for S-glass fiber plastic (d = 10 μm), and 4 days for basaltoplastics with d = 9 μm. This indicates different organization of the interphase layer (adhesive interaction) in basaltoplastics and glass fiber plastics, and high stability of the interface in basaltoplastics to heat humidity aging. This feature of basaltoplastics has been already mentioned in connection with adhesive properties of fibers and is additionally strengthened by purposeful modification of the gloss finish system that leads to increasing basaltocomposite resistance (Figure 7, curve 1). For instance, Figure 7 shows that basaltoplastic based on 13 μm filaments with the surface treated by modified gloss finish system dupli-color (DC) is beyond both glass fiber plastics and basaltoplastics based on thinner filaments with the surface treated by standard gloss finishes, used for glass fiber treatment, by water proofing (integrity, respectively).

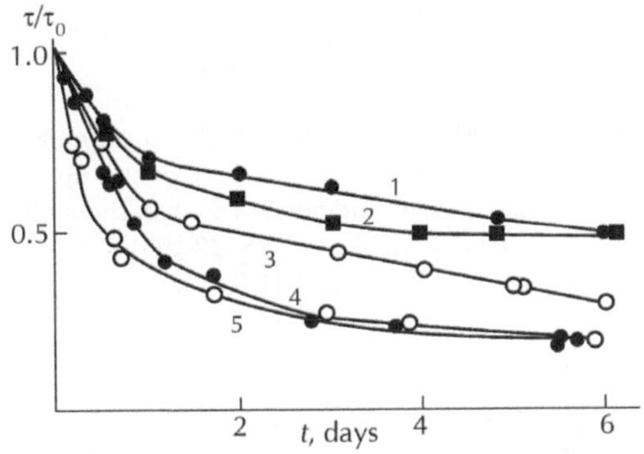

FIGURE 7 Kinetic curves of interlaminar shear strength reduction at highly accelerated heat humidity aging of wound epoxy basaltoplastics (1, 2, 4) and glass fiber plastics (3, 5). 1--basaltoplastic derived from roving Number 7 (d = 13 μm), 2-- basaltoplastic derived from roving Number 1 (d = 9 μm), 3-- glass fiber plastic derived from S-roving Number 4 (d = 10 μm), 4--basaltoplastic derived from roving Number 8 (d = 13 μm), 5--glass fiber plastic derived from E-roving Number 5 (d = 13 m); numbers of roving's represent their positions in Table 2.

Macrokinetic analysis [22, 47] shows that moisture transfer proceeding in these composites and determining regularities of the shear strength decrease (Figure 7), is of combined diffusion-convective mechanism with predominance of non-Fick component in some cases (curves 3, 4 and 5), as well as with the Fick transfer component domination (curves 1 and 2).

3.3.5 ALKALI RESISTANCE OF BASALTOPLASTICS

As for basalt roving's, beside the main indices of chemical resistance of reinforced polymers (water, acid and alkali resistance) in the case of basaltoplastics, the ultimate attention was devoted to their resistance to aqueous alkali media. As in case of basalt roving's, historically this is associated with technical and economic attractiveness of basaltoplastics use (shaped as rods, pipes, profiles, and plates) for concentrated reinforcement in alleviated concrete structures operated under increased aggressiveness conditions (in relation to steel reinforcement): marine or underground structures, bridge conduit components, etc.

The results of comparative studies of alkali aging of basaltoplastics and glass fiber plastics present in the literature [18, 22, 42, 47, 48] indicate higher resistance of basaltoplastics (Figure 8).

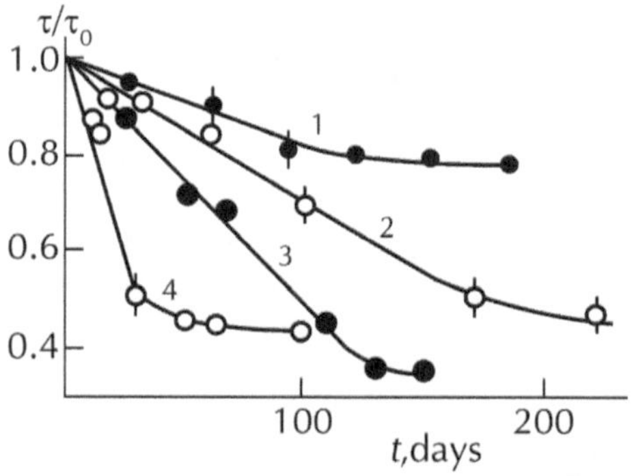

FIGURE 8 Kinetic curves of interlaminar shear strength reduction at alkali aging of wound epoxy basaltoplastics (1, 3) and glass fiber plastics (2, 4). 1 – basaltoplastic derived from roving Number 7 (d = 13 μm), 2 – glass fiber plastic derived from S-roving Number 4 (d = 10 μm), 3 – basaltoplastic derived from roving Number 8 (d = 13 μm), 4 – glass fiber plastic derived from E-roving Number 5 (d = 13 μm); 1.25N NaOH, T = 20°C; numbers of roving's represent their positions in Table 2.

For example, time t_{05}, during which basaltoplast and E-glass fiber plastic shear strength (in both cases fibers had the same diameter d = 13 μm and were treated by the same gloss finish Number 76) two fold decreases, equals 93 days and 38 days, respectively (Figure 8, curves 3 and 4). For higher solid S-glass fiber plastic based on d = 10 μm filaments, $t_{0.5}$ = 190 days was determined. For basaltoplasts with d = 9 μm filament, according to data from reference [18] during ~600 days the value $t_{0.5}$ was not reached. In this case, the criterion can be shear strength decrease at the initial, first part of the kinetic curve of the composite alkali aging. Therefore, according to [18], for this basaltoplast (d= 9 μm) during initial 200 days of exposure in alkali solution (NaOH, 5%) the shear strength equal ~38–40%.

The results of comparative study of basaltoplasts and glass fiber plastics alkali aging, as well as heat-humidity aging results indicate different arrangement of adhesive interaction in these materials and, as a consequence, higher integrity and higher stability of the interface in basaltoplasts to aging factors considered. As for water resistance, alkali resistance of basaltoplasts was significantly increased (Figure 8, curve 1) for the account of application of a new modified gloss agent specially developed with respect to specificity of basal roving surface. In this case, for basaltoplast based on d = 13 μm filament during initial ~200 days of alkali aging the shear strength (with respect to the initial first stage, Figure 8, curve 1) decreased by ~20%. Thus, the comparison of these data with the above considered data [18] shows that application of a gloss finish created "for basalt" allows for two fold increase of alkali resistance of the composite based on filaments with d = 13 μm compared with the composite based on thinner basalt filaments (d = 9 μm), which surface was treated by standard industrial gloss finish used for glass fibers [18]. It ought to be noted that investigations on purposeful creation of hydrophobic adhesive systems (direct-action gloss finishes) for basalt roving's with respect to specificity of their surface are now at the initial stage still. Therefore, the results shown in this part of review are just tentative whilst encouraging. It is obvious that development of work in this direction will lead to broader disclosure of the property potential of basalt roving's and composites on their basis, similar to glass fibers and glass fiber plastics.

3.4 CONCLUSION

The review presents mechanical and physicochemical properties of modern continuous basalt fibers and epoxy polymeric composites on their basis. The data presented characterize modern continuous basalt fibers (roving's) as a novel type of silicate fibers with increased acid and alkali resistance having the strength level close to that of the well-known E-fibers, but with rather high modulus of elasticity typical of high modulus and high strength magnesium aluminosilicate S-fibers.

The mechanical properties of basaltoplastics are closer to properties of analogous high strength and high modulus S-glass fiber plastics and are beyond properties of analogous E-glass fiber plastics. Epoxy basaltoplastic resistance to aging in alkali medium and their heat humidity resistance are far beyond appropriate characteristics of epoxy glass fiber plastics based on E- and S-glass fibers. This is connected with specificity of adhesive properties of basalt fibers. All this in aggregate indicates the possibility and perspective of basaltocomposites application in the composition of various structures, for which high mechanical properties and high resistance to humidity and various aggressive media impacts are required (bridges, internal and external reinforcement of concrete products, marine, river and underground concrete structures, corrosion resistant tanks and pipes, etc.).

KEYWORDS

- **Acid and alkali resistance of basalt fibers**
- **Alkali resistance and heat-humidity resistance of epoxy basaltoplastics**
- **Basalt roving's**
- **Epoxy basaltoplastics**
- **Mechanical properties**
- **Continuous basalt fibers**

REFERENCES

1. Jigiris, D. D. and Makhova, M. F. *Production basis for basalt products and roving's*, (Russian). M.: p. 407 (2002).
2. Kostikov, V. I., Firsova, T. D., Mostovoi, G. E., and Chernenko, N. M. In Coll.: *Theory and practice of production technology for the articles from composite materials and new metal alloys* (PTCMM), (Russian). Proceedings of Moscow International Conference. Moscow, pp. 121–126 (Aug. 27–30, 2003).
3. Igonin, N. G. and Tatarintseva, O. S. In Coll.: *Production technology of thermal insulating materials from mineral raw materials*. Reports of VI All-Russian scientific-practical conference, May 31–June 2, 2006 (Belokurikha, Altaiski Krai) (Russian). M., FGUP "TsNIIKhM", pp. 207–220 (2006).
4. Artemenko, S. E., Kadyikova, Yu. A., Vasilieva, O. G., Ezhov, A. A., et al. *Plast. Massy*, (Russian), (12), pp. 47–51 (2005).
5. Artemenko, S. E., Volkov, Yu. P., Vasilieva, O. G., Baiburin, V. B., Kadyikova, Yu. A., and Leontiev, A. N. Plant laboratory. *Material diagnostics* (Russian), **69**(11), 36–39 (2003).
6. Goncharova, T. P., Artemenko, S. E., and Kadyikova, Yu. A. *Perspective materials*, (Russian), (1), pp. 66–68 (2007).
7. Smirnov, L. N., Karpova, Z. I., Smirnov, A. L., and Kunitsin, Yu. K., *Conversion in machine industry*, (Russian), (5), pp. 24–27 (1999).
8. Gromkov, B. K., Smirnov, L. N., Trofimov, A. N., Zharov, A. I., et al. *Conversion in machine industry*, (Russian), (4), pp. 81–88 (2000).
9. Gurtovnik, I. G., Sokolov, V. I., Trofimov, N. N., and Shalgunov, S. I. *Radiotransparent articles from glass fiber plastics*, (Russian). *M.: Mir*, p. 361 (2003).
10. Trostyanskaya, E. B. In Book: *Reinforced plastics. Reference ai*d. G. S. Golovkin and V. I. Semenov (Eds.), M.: MAI, (Russian), pp. 263–268 (1997).
11. Matthews, F. and Rollings, R. Composite materials. Mechanics and technology, (Russian). (Transl. from Engl.) *M.: Tekhnosfera*, p. 407 (2004).
12. Novitski, A. G. and Mazur, V. L. *Chemical industry of Ukraine*, (Russian), (3), pp. 42–43 (2003).
13. Sokolinskaya, M. A., Zabava, L. K., Tsybulya, T. M., and Medvedev, A. A. *Glass and Ceramics*, (Russian), (10), pp. 8–9 (1991).
14. Dalinkevich, A. A., Gumargalieva, K. Z., Sukhanov, A. V., Aseev, A. V., and Zharov, A. I. *Plast. Massy* (Russian), (3), pp. 7–10 (2002).
15. Dalinkevich, A. A., Gumargalieva, K. Z., Sukhanov, A. V., Aseev, A. V., and Zharov, A. I. *Plast. Massy* (Russian), (12), pp. 23–26 (2002).

16. Dalinkevich, A. A., Sukhanov, A. V., and Aseev, A. V. *Concrete Technology* (Russian), (3), pp. 10–13 (2005).
17. Dalinkevich, A. A., Gumargalieva, K. Z., Sukhanov, A. V., and Aseev, A. V., COBRAE Conference «*Bridge engineering with Polymer Composites*», EMPA, Dubendorf, Switzerland, paper № 7 (30 march–1 april 2005).
18. Dalinkevich, A. A., Gumargalieva, K. Z., Sukhanov, A. V., and Aseev, A. V. *The Third International Conference on Durability & Field Applications of Fiber Reinforced Polymer (FRP) Composites for Construction* (CDCC 2007, Quebec City, Canada) pp.549–556 (May 22–24, 2007).
19. Leskinen, K. *JEC Composites M.*, (28), pp.33–34 (2006).
20. Kalinchev, V. A. and Makarov, M. S. *Wound glass fiber plastics*,(Russian). M.: Khimia, p. 268 (1986).
21. Zelenski, E. S., Kuperman, A. M., Gorbatkina, Yu. A., Ivanova-Mumzhieva, V. L., and Berlin, A. A. Reinforced plastics – modern construction materials, (Russian), Zh. Ros. Khim. Obshch. Im. and D. I. Mendeleeva (Eds.). *Ros. Khim. Zh.*, **45**(2), 56–74 (2001).
22. Dalinkevich, A. A., Gumargalieva, K. Z., Sukhanov, A. V., and Aseev, A. V. In Coll.: *Theory and practice of production technology for the articles from composite materials and new metal alloys (PTCM*M). Corporate nano-CALS-technologies in science-intensive branches of industry, (Russian). Proceedings of 4th International Conference. Moscow, pp. 110–118 (April 26–29, 2005).
23. Trostyanskaya, E. B., et al. *Plast. Massy*, (Russian), (1), pp. 28–31 (1987).
24. Trostyanskaya, E. B. In Book: *Reinforced plastics*. Reference aid. Ed. G. S. Golovkin and V. I. Semenov, M.: MAI, (Russian), pp. 63–91 (1997).
25. Trostyanskaya, E. B. *Structures from composite materials*, (Russian), (1), pp. 11–20 (2000).
26. Trostyanskaya, E. B., Antonov, V. V., and Shibanov, A. N. *Khim. Volokna*, (Russian), (2), p. 4 (1994).
27. Aslanova, M. S., Kolesov, Yu. I., Khazanov, V. E. et al., *Glass fibers*. M. S. Aslanova (Ed.), M.: Khimia, (Russian), pp. 155–180 (1979).
28. Dalinkevich, A. A., Gumargalieva, K. Z., Sukhanov, A. V., and Aseev, A. V. *Oxidation Communication*, **26**(4), pp. 597–604 (2003).
29. Dalinkevich, A. A., Buldakov, V. P., Gumargalieva, K. Z., Marakhovsky, S. S., and Soukhanov, A. V. Kinetics Of Basalt Roving Alkaline Corrosion. Corrosion: Materials, *Protection*, (Russian). (2), pp. 33–42 (2012).
30. Monakov, Yu. B., Zaikov, G. E., and Dalinkevich, A. A. *Trends in Polymer and Composites Research*. New York, Nova Science Publish., p.198 (2004).
31. Tsyibulya, Yu. L., Medvedev, A. A., Smirnov, L. N., Kuziv, P.,et al. *Conversion in machine industry*, (Russian), (4), pp. 75–77 (2000).
32. Pashchenko, A. A., Serbin, V. P., Paslavskaya, A. P., Glukhovski, V. V., Biryukovich, Yu. L., Solodovnik, A. B. *Reinforcement of inorganic binding substances by mineral fibers*, (Russian). M.: Stroiizdat., p. 200 (1988).
33. Gromkov, B. K., Smirnov, L. N., Trofimov, A. N., Zharov, A. I. et al. In Book: *Basalt fibrous materials*, (Russian). M.: Informkonversia, pp. 54–64 (2001).
34. Orlov, D. L. and Gorin, A. S. *Glass and Ceramics* (Russian), (7), pp. 3–7 (1999).
35. Kudryavtsev, M. Yu., Kolesov, Yu. I., Mikhailenko, N. Yu. Khim. *Volokn*a,(Russian), (3), p. 64–66 (2001).
36. Frolov, Yu. G. *Colloid chemistry course*. M.: Khimia (Russian), p. 400 (1981).
37. Moiseev, Yu. V. and Zaikov, G. E. *Chemical resistance of polymers in aggressive med*ia. M.: Khimia (Russian), pp. 162–165 (1977).

38. Frank-Kamenetski, M. D. *Diffusion and heat transfer in chemical kinetics* (Russian). M.: Nauka, p. 486 (1967).
39. Tyuikaev, V. N. *Plastics for constructive purposes*. (Reactoplasts). E. B. M. Trostyanskaya (Ed.), Khimia, (Russian), pp. 120–203 (1974).
40. Sukhanov, A. V., Aseev, A. V., and Sisauri, V. I. *21st century construction materials, equipment, and technology*,(Russian), (12), pp. 20–22 (2003).
41. Bukharov, S. V., Okorokov, V. V., and Stankoi, G. G. In Book: *Theory and practice of production technology for the articles from composite materials and new metal alloys (PTCMM)*, (Russian). Proceedings of Moscow International Conference. Moscow, p. 62 (Aug. 27–30, 2003).
42. Dalinkevich, A. A., Sukhanov, A. V., and Aseev, A. V. *Concrete Technology*, (Russian), (4), pp. 11–16 (2005).
43. Perlin, S. M. and Makarov, M. S. *Chemical resistance of glass fiber plastics*, (Russian). M.: Khimia, p. 184 (1983).
44. Makarov, V. G. and Shevchenko, A. A. *Reliability of articles from glass fiber plastic in chemical industry*, (Russian). M.: Khimia, pp. 68–80 (1993).
45. Trofimov, N. N., Kanovich, M. Z., Kartashev, E. M., et al. *Physics of composite materials* (Russian). M.: Mir, **2**, 315 (2005).
46. Chernin, I. Z., Smekhov, F. M., and Zherdev, Yu. V. E*poxy polymers and compositions*,(Russian). M.: Khimia, pp. 48–52 (1982).
47. Dalinkevich, A. A., Gumargalieva, K. Z., and Sukhanov, A. V. In Book: Theory and practice of production technology for the articles from composite materials and new metal alloys (PTCMM). *Identification and modeling of properties of materials and processes*, (Russian). Proceedings of 5th International Conference. Moscow, pp. 110–118 (April 24–27, 2007).
48. Rozental, N. K., Chekhniy, G. B., Bel'nik, A. R., and Zhilkin, A. P. *Concrete and reinforced concrete*, (Russian), (3), pp. 20–23 (2002).
49. Nikolaev, V. N. and Filippova, E. Yu. *21st century construction materials, equipment, and technology*, (Russian), (11), p. 20 (2004).

CHAPTER 4

CALCULATIONS OF BOND ENERGY IN CLUSTER WATER NANOSTRUCTURES

A. K. HAGHI and G. E. ZAIKOV

CONTENTS

4.1 INTRODUCTION

Water plays the ambiguous role in hydrocarbon fuel of internal combustion engines. On the one hand, simple dilution of petroleum or diesel fuel with water can significantly deteriorate technological characteristics of the fuel. As soon as, water drops get into cylinders, the following happens—in the compression stroke when both valves are closed, the piston bears against the water plug when moving upwards. The pressure inside the cylinder increases multiply. The engine tries to bring the connecting rod to the upper position, continuing the cycle. In fact, the pistons in one or several cylinders stop at once, and the crankshaft which continues rotating takes enormous loads. It bends connecting rods, breaks piston pins and often breaks down itself. On the other hand, optimal water content in hydrocarbon fuel is defined by the standard technological norm of such fuel mixture which is prepared by a special technique. Moreover, water containing fuel can have the potential energy and nevertheless the engines produce the same power as with additional amount of petroleum by the mass equaled to the mass of water added. And you not only have the power gain but you also benefit in fuel technological characteristics, such as fire safety, octane number, application temperature limits, possibility to use cheaper fuels, and so on. Such specificity of technological processes is ultimately defined by the mechanism of physic-chemical transformations occurring on atom-molecular level. In this investigation their possible evaluations are studied based on the concept of spatial-energy parameter (P-parameter).

4.2 FORMATION OF HIGH ENERGY BONDS IN FUEL MIXTURE

The practical use of hydrogen containing fuel is possible only if a number of conditions are fulfilled:

- Introduction of complex additives into the fuel, with alcohols and so-called "hydrogen catalyst" content having the primary meaning.
- Such additives are mixed following the special technique—first, by separate fractions, and in the end all the mixture is intensively stirred by a hydraulic cutting pump (hydraulic shears).

"Hydrogen catalyst" contributes to active dissociation of water molecules with the formation of hydrogen and oxygen which burn in the engine chamber afterwards.

But is not clear how during such a short combustion time of the given amount of mixture introduced into the chamber, water dissociation in this volume and burning of its products can take place. Moreover, as a result of water dissociation by the reaction $H_2O = H^+ + OH^-$ the direct oxygen release is not observed. Obviously, other important mechanisms of physic-chemical transformation of energy are involved. For instance, it is known that as a result of biochemical reactions in the presence of certain ferments the synthesis of adenosine triphosphate (ATP) molecule can take place whose potential energy increases due to the formation of special high energy bonds.

It is possible that similar processes take place during the formation of burning mixture of this type of fuel when nanoclusters in the form of fullerenes can be formed under certain technological conditions. First of all, this is aided by the introduction of alcohols into the fuel mixture that results in the formation of fullerene, for example, $C_{60}(OH)_{10}$. Therefore, the addition of alcohols (up to 20%) just corresponds to the ratio of molar masses of hydroxyl groups OH^- and carbon atoms. At the second stage of fuel preparation, high energy bonds are formed in the systems $C_{60}(OH^-) – n(H_2O)$, first of all, due to the introduction of "hydrogen catalyst" into the mixture, and besides, when filtering water through the coal filter which contributes to the extraction of nanostructured formations of carbon atoms into the mixture.

Similar to ATP hydrolysis, which is accompanied by the release of chemical bond energy, the breakage of high energy bonds and heat energy release occur in hydrogen containing fuel when it is burning in the engine chamber.

The physic-chemical mechanism of the formation of energy saturated bonds in this system is given.

4.3 EXPERIMENTAL

The value of the relative difference of P-parameters of interacting atom components–coefficient of structural interaction α was used as the main numerical characteristic of structural interactions in condensed media [1-2]:

$$\alpha = \frac{P_1 - P_2}{(P_1 + P_2)/2} 100\% \tag{1}$$

Applying the reliable experimental data, we obtain the nomogram of the dependence degree of structural interactions upon coefficient α –unified for the wide range of structures. This approach allows evaluating the degree and directedness of structural interactions of phase formation, isomorphism and solubility processes in multiple systems, including molecular ones. In particular, the features of cluster formation in the system $CaSO_4$–H_2O have been investigated.

To evaluate the directedness and degree of phase formation processes the following equations are used:

Initial values of P-parameters:

$$\frac{1}{q^2/r_i} + \frac{1}{W_i n_i} = \frac{1}{P_E} \tag{2}$$

$$\frac{1}{P_0} = \frac{1}{q^2} + \frac{1}{(Wrn)_i} \tag{3}$$

$$P_E = P_0 / r_i \tag{4}$$

Where, W_i–electron orbital energy, r_i–orbital radius of *i* orbital, $q = Z^*/n^*$–by, n_i–number of electrons of the given orbital, Z^* and n^*–effective nucleus charge and effective main quantum number, P_0 is called as spatial-energy parameter, and P_E–effective P-parameter.

The calculation results by equations for a number of elements are given in Table 4.1, from which it is seen that for hydrogen atom the values of P_E–parameters substantially differ at the distance of orbital (r_i) and covalent (R) radii. The hybridization of valence orbitals of carbon atom is evaluated as the averaged value of P-parameters of $2S^2$ and $2P^2$- orbitals.

Values of P_c-parameter in binary and complex structures:

$$\frac{1}{P_{\tilde{n}}}=\frac{1}{N_1P_1}+\frac{1}{N_2P_2}+\ldots\ldots \tag{5}$$

Where N–number of homogeneous atoms in each subsystem.

The results of such calculations for some systems are given in Table 4.2.

Bond energy (E) in binary and more complex systems:

$$\frac{1}{E}\approx\frac{1}{P_E}=\frac{1}{P_1(N/\kappa)_1}+\frac{1}{P_2(N/\kappa)_2}+\ldots\ldots\ldots \tag{6}$$

Here (as applicable to cluster systems) κ_1 and κ_2 – number of subsystems forming the cluster system, N_1 and N_2 – number of homogeneous clusters.

So for $C_{60}(OH)_{10}$ $\kappa_1 = 60$, $\kappa_2 = 10$.

4.4 CALCULATIONS AND COMPARISONS

It was assumed that structural-stable water cluster (H_2O) can have the same static number of subsystems (κ) as the number of subsystems in the system interacting with it. For example, the water cluster n $(H_2O)_{10}$ is interacting with fullerene $[C_6OH]_{10}$.

Similarly with cluster $(C_6OH)_{10}$ the formation of cluster $[(C_2H_5OH)_6 - H_2O]_{10}$ is apparently possible, which corresponds to the system $(C_2H_5OH)_{60} - (H_2O)_{10}$. The interaction of water clusters was considered as the interaction of subsystems $(H_2O)_{60}$ $- N(H_2O)_{60}$.

Based on such concepts, the bond energies in these systems are calculated by Equation (6), the results are given in Table 4.3.

To compare, the calculation data obtained by with quantum-chemical techniques are given.

Both techniques produce consistent values of bond energy (in eV). Besides, the methodology of P-parameter allows explaining why the energy of cluster bonds of water molecules with fullerene $C_{60}(OH)_{10}$ two times exceed the bond energy between the molecules of cluster water (Table 4.3).

In accordance with the nomogram, the phase formation of structures can take place only if the relative difference of their P-parameters (α) is under 25–30%, and the most stable structures are formed when $\alpha < 6–7\%$.

In Table 4.4 different values of coefficient α in systems H-C, H-OH, and H-H_2O are given, which are within 0.44–7.09%.

But in the system H-C for carbon and hydrogen atoms the interactions at the distances of covalent radii have been taken into account, and for other systems–at the distance of orbital radius.

The interaction in system H-C at the distances of covalent radius plays a role of fermentative action, which results in the transition of dimensional characteristics in water molecules from the orbital radius to the covalent one and formation of system $C_{60}(OH)_{10}$ — $N(H_20)_{10}$ with bond energy between the main components two times greater than between the water molecules (high energy bonds).

TABLE 4.1 P-parameters of atoms calculated *via* the electron bond energy

Atom	Valence Electrons	W (eV)	r_i (Å)	q^2 (eV Å)	P_0(eV Å)	R (Å)	$P_E=P_0/R$ (eV)
H	$1S^1$	13.595	0.5295	14.394	4.7985	0.5295	9.0624
H	$1S^1$					0.28	17.137
C	$2P^1$	11.792	0.596	35.395	5.8680	0.77	7.6208
C	$2P^1$					0.69	8.5043
C	$2P^2$	11.792	0.596	35.395	10.061	0.77	13.066
C	$2S^1$	19.201	0.620	37.240	9.0209	0.77	11.715
C	$2S^2$				14.524	0.77	18.862
C	$2S^2+2P^2$				24.585	0.77	31.929
C	$1/2(2S^2+2P^2)$						15.964
O	$2P^1$	17.195	0.4135	71.383	4.663	0.66	9.7979
O	$2P^2$	17.195	0.4135	71.383	11.858	0.66	17.967
O	$2P^2$					0.59	20.048
O	$2P^4$	17.195	0.4135	71.383	20.338	0.66	30.815

TABLE 4.2 Structural P_c-parameters

Radicals, molecules	P_1 (eV)	P_2 (eV)	P_3 (eV)	P_4 (eV)	P_c (eV)	Orbitals of oxygen atom
OH	17.967	17.137			8.7712	$2P^2$
OH	9.7979	9.0624			4.7080	$2P^1$

TABLE 4.2 *(Continued)*

H_2O	2 × 17.138	17.967			11.788	$2P^2$
H_2O	2 × 9.0624	17.967			9.0226	$2P^2$
C_2H_5OH	2 × 15.964	2 ×9.0624	9.7979	9.0624	3.7622	$2P^1$

TABLE 4.3 Calculation of bond energy – E (eV)

System	C_{60}	$(OH)_{10}$	$(H_2O)_{10}$		P_E (eV)	E (eV)
					(calculation)	
	P_1/κ_1	P_2/κ_2	P_3/κ_3	n_3	By Equation (6)	Quantum-mechanical
$C_{60}(OH)_{10}$–$N(H_2O)_{10}$	15.964/60	8.7712/10	11.788/10	1	0.174	0,176
				2	0.188	0,209
				3	0.193	0,218
				4	0.196	0,212
				5	0.197	0,204
$(H_2O)_{60}$–$N(H_2O)_{60}$	9.0226/60	9.0226/60		n_2		
				1	0.0768	0,0863
				2	0.1020	0,1032
				3	0.1128	0,1101
				4	0.1203	0,1110
				5	0.1274	0,115
$(C_2H_5OH)_{60}$–$(H_2O)_{10}$	3.7622/60	9.0226/10			0.0586	0.0607
$(C_2H_5OH)_{10}$–$(H_2O)_{60}$	3.7622/10	9.0226/60			0.1074	≈0.116

TABLE 4.4 Spatial-energy interactions in the system H-R, where R= C, (OH), H_2O

System	P_1(eV)	P_2(eV)	$\alpha = \frac{\Delta P}{\langle P \rangle} 100\%$	Spatial bond type
H-C	17.137	15.964	7.09	Covalent
H-OH	9.0624	8.7712	3.27	Orbital
H-H_2O	9.0624	9.0226	0.44	Orbital

Thus, broad capabilities of water clusters to change their spatial-energy characteristics apparently explain all the diversity of structural properties of water in its different modifications, including the formation of high energy bonds in water containing fuel for internal combustion engines.

4.5 CONCLUSION

The consistent calculations of bond energy in cluster water nanostructures have been performed following the P-parameter methodology and quantum-mechanical methods.

The formation of high energy bonds in the process of hydrocarbon hydrogen containing fuel preparation has been explained.

The results of bond energy calculations in water cluster nanostructures following the P-parameter methodology agree with quantum-mechanical methods.

The changes which can take place in spatial-energy characteristics of water clusters explain the formation of high energy bonds in the process of hydrocarbon fuel preparation.

Breaking of these bonds with the release of additional amount of heat energy occurs in the combustion chamber.

KEYWORDS

- **Adenosine triphosphate (ATP)**
- **Bond energy**
- **Hydrogen catalyst**
- **Isomorphism**
- **Spatial-energy parameter**

REFERENCES

1. Tewarson, A., Su, P., and Yee, G. G. in the Proceedings of the Interscience Communications Conference—*Hazards of Combustion Products: Toxicity, Opacity, Corrosivity, and Heat Release*. V. Babrauskas, R. Gann, and S. Grayson (Eds.), Greenwich, UK, p.225 (2008).
2. Warneck, P. in *Chemistry of Multiphase Atmospheric Systems*, W. Jaeschle (Ed.), NATO ASI Series, Springer-Verlag, Berlin, Germany, **6G**, 473 (1986).

CHAPTER 5

EXPERIMENTAL RESEARCH ON DESALTING OF INSTABLE GAS

V. P. ZAKHAROV, T. G. UMERGALIN, B. E. MURZABEKOV, F. B. SHEVLYAKOV, AND G. E. ZAIKOV

CONTENTS

5.1 INTRODUCTION

There was revealed a possibility of effective desalting of instable gas distillate before its supplying into stabilization string. The mixing of gas distillate with a little quantity of fresh water (0.5–2% vol.) in tubular turbulent device allows to decrease total of salt from 97 g/m^3 to 15–20 g/m^3. There is optimal load interval for tubular turbulent device on quantity of supplying gas distillate providing effective desalting.

The most popular way of stabilization of gas distillate is its fractioning in the stabilization string. Presence of salts, mainly sodium, potassium, and calcium in gas distillate (to 145–200 g/m^3) results in scaling on heater of stabilization string and surfaces of heating elements (heating intensity decreases, wear (burnout) of fire chamber). Due to the scaling on plates and valves of contact devices stabilization string works in instable mode because of often stopping in case of valve sticking. Through decrease of mass transfer between ascending stream of gas-vapor phase with descending liquid phase take place loses of fraction of hydrocarbon of the gas distillate from the string top together with stabilization gases. As a result, takes place decrease of quality of gas distillate from potentially possible [1, 2].

One of the possible ways of the problem solution is preliminary washing of dehydrated gas distillate from salts of fresh water [2]. In such case it is necessary to provide intensive mixing of gas distillate with little quantity of fresh water.

The purpose of the current work is experimental research of process of desalting of gas distillate while mixing with little quantity of fresh water in tubular turbulent device of divergent-convergent construction [3].

The experimental research of appropriateness of desalting was carried out in a shop of gas and gas distillate preparation at oilfield Borankoli (JSC "Sea Oil Company "KazMunaiTeniz", Kazakhstan). Component content of initial gas distillate and some of its characteristics are shown in Table 1. The total quantity of salts in initial gas distillate was 97 mg/l (ionic composition $[HCO^{3-}]/[SO_4^{2-}]/[Cl^-]/[Ca^{2+}]/[Mg^{2+}]/[Na^+ + K^+]$ = 3/1/210/13,9/1,1/119,8). For washing of the gas distillate from salts there was used fresh water.

TABLE 1 Physical-chemical characteristics of gas distillate

Parameter	Value
Distillate density at 20°C, kg/m^3	759,6
Blend composition, % turn	
Propane	2,973
Isobutane	1,387
n-Butane	3,880
Isopenthane	3,053

TABLE 1 *(Continued)*

n-penthane	3,561
Sum of hexanes + more	80,140
Resins	5,006
Fractional composition of gas distillate	
Boiling start, °C	44
Distilled, %vol.	94,5
Residue in flask, %vol.	3,5
Loses at distillation (volatile), %vol.	2,0
Driven off volume at temperature °C, %vol.	
82	10
95	20
107	30
118	40
130	50
150	60
177	70
209	80
262	90
300	95

The research of appropriateness of gas distillate washing with fresh water from salts was carried out using tubular turbulent device of divergent-convergent construction (Figure 1) with the following parameters—divergent diameter d_d = 0,024 m, convergent diameter d_c = 0,015 m, length of divergent-convergent section L_s = 0,048 m, divergent expansion angle 45°. The experimental assembly and explanation of choosing the construction of tubular turbulent device are shown in chapter [4]. Volume flow of the gas distillate varied within 0.5–1 m^3/hr at fixed percentage composition of fresh water (0,5; 1; 1,5; 2% vol.). The temperature of gas distillate supplied into tubular turbulent device for washing composed 47°C, fresh water temperature–30°C.

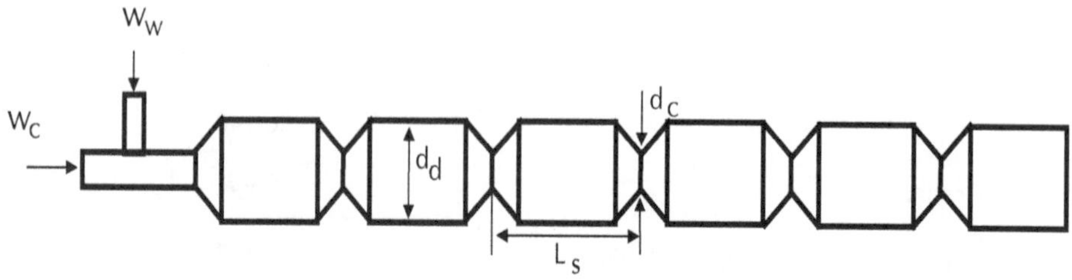

FIGURE 1 General view of tubular turbulent device of divergent-convergent construction for washing from salts of gas distillate with fresh water. w_c, w_w–volume flow of gas distillate and fresh water, correspondingly.

In the whole examined interval of volume flows of gas distillate there has been noticed decrease of salts composition with grow of quantity of supplied for mixing fresh water (Figure 2). At the same time, at fixed correlation of water and gas distillate effectiveness of desalting extremely depends on volume flow of liquid. At low flow of gas distillate (0.5 m^3/hr) in tubular turbulent device there is noticed low effectiveness of mixing with fresh water. Reynolds' criterion was Re = 6000 confirming the fact that hydrodynamic structure of liquid flow motion corresponds to intermediate mode. Further growth of volume flow of gas distillate increases effectiveness of desalting and minimal composition of salts is reached at w_c = 0.76 m^3/hr (Re = 8500). Obviously, that is related with increase of turbulent mixing level in tubular turbulent device of divergent-convergent type.

The exceeding of volume flow of gas distillate on value $w_к$ = 0.76 m^3/hr, independently on quantity of fresh water, is attended by decrease of desalting effectiveness (Figure 2). At that it happens together with reaching of Reynolds criterion of value Re = 12000 that characterizes transfer into sphere of developed turbulence. The possible explanation of noticed dependence can be decrease of time of reagents stay in zone of mixing at increase of volume flow of gas distillate, not providing finishing of process on salt extracting from hydrocarbon phase into aqueous one. The hydrodynamic operational modes of tubular turbulent device let evaluative calculate time necessary for extracting of salts from gas distillate with fresh water what composes about 0.6 s. If decrease time of stay of two-phase flow in device, process of extraction does not manage to finish notwithstanding the growth of turbulence level and, correspondingly, mixing effectiveness.

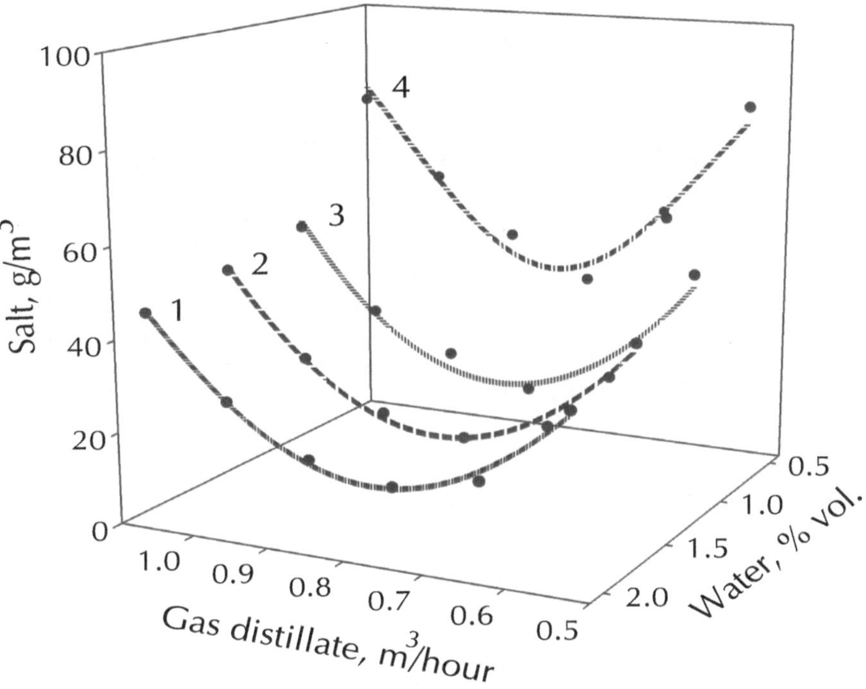

FIGURE 2 Dependence of total composition of salts in gas distillate on its volume flow and quantity of fresh water. Quantity of fresh water at washing state in relation to gas distillate: 2 (1); 1,5 (2); 1 (3); 0,5 (4) % vol.

The significant increase of desalting is reached through heating of fresh water even in spite of the fact that its quantity in relation to gas distillate does not exceed 2% vol. So, at water heating from 30°C to 90°C, supplying of gas distillate in reactor with flow of 0.86 m^3/hr and quantity of supplied water 1.5% volume total composition of salts decreases by 2 times (from 20 g/m^3 to 10 g/m^3). Increase of effectiveness of salts washing with growth of temperature is explained by increase of extraction speed, as well as intensity of mixing through release from gas distillate of light hydrocarbons in form of fine cavities.

In the result of fulfilled research there is proposed a scheme of preliminary desalting of instable gas distillate before its supplying in the stabilization string that includes small tubular turbulent device of divergent-convergent construction for mixing of hydrocarbon fraction with fresh water (Figure 3).

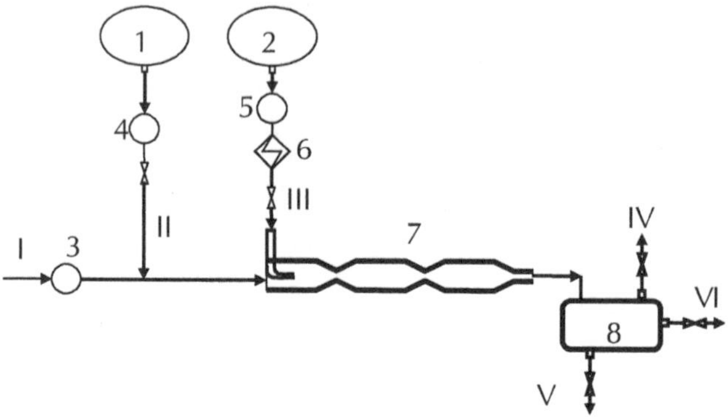

FIGURE 3 Scheme of desalting of gas distillate when mixed with fresh water. (1) vessel with deemulsifying agent, (2) vessel with water, (3), (4), (5) pumps, (6) heat exchanger, (7) tubular turbulent device, and (8) three-phase separator. Streams are—(I) initial gas distillate, (II) deemulsifying agent, (III) fresh water, (IV) gas, (V) draining water, and (VI) gas distillate after desalting.

There is shown possibility of effective desalting of instable gas distillate before its supplying into stabilization string. Mixing of gas distillate with a little quantity of fresh water (0.5–2% vol.) in tubular turbulent device of divergent-convergent construction lets decrease total composition of salts from 97 g/m^3 to 15–20 g/m^3 (at optimal flow of gas distillate 0.76 m^3/hr). The significant increase of effectiveness of gas distillate washing from salts is reached through heating of fresh water. The proposed method will let significantly increase steadiness of operation of stabilization string through decrease of scale technological surfaces.

5.2 CONCLUSION

In such a manner, there should be effective use of turbulent device of divergent-convergent construction at intensification of processes of oil and gas sphere [5].

The work has been carried out with financial support from Grant of the RF President MD -3178.2011.8, RFFI (project № 11-03-97017).

KEYWORDS

- **Desalting**
- **Divergent-convergent construction**
- **Gas distillate**
- **Hydrocarbons**
- **Stabilization**

REFERENCES

1. Rudenko, S. V., Hutoryanskii, F. M., Kapustin, V. M., and Molchanov, Z. V. On technological necessity of washing gas distillate and oil mixture with water at their preparation on BSC-1 of Orenburg PGPD. *Ecologicheskii Vestnik Rossii.*, (6). P. 10–13(2010).
2. Uhalova, N. B., Latyuk, B. I., and Umergalin, T. G. Influence of water on effectiveness of processes of gas and gas distillate fractioning. Material from II International scientific conference. «Theory and practice of mass exchanging processes of chemical technology». Ufa: Edition UGNTU, pp. 151–152 (2001).
3. Zakharov, V. P., Berlin, A. A., Monakov, Yu. B., and Deberdeev, R. Ya. *Physical-chemical basis of fast liquid-phase processes behavior*. M.: Nauka, p. 348 (2008).
4. Zakharov, V. P., Mukhametzyanova, A. G., Takhavutdinov, R. G., Diakonov, G. S., and Minsker, K. S. Creating of homogeneous emulsions in tubular turbulent device of divergent-convergent constructions. *Jurnal Prikladnoi Khimii.*, **75**(9), pp. 1462–1465 (2002).
5. Shevlyakov, F. B., Zakharov, V. P., Kaeem, D. Ch., and Umergalin, T. G. *Improvement of process on pre-extracting of higher boiling hydrocarbons of oil dissolved gas in turbulent device of divergent-convergent construction.* Vestnik Bashkirskogo Universiteta, **13**(4). pp. 916–918 (2008).

CHAPTER 6

A STUDY ON RECTIFICATION OF HYDROCARBONIC MIXES

O. R. ABDURAKHMONOV, Z. S. SALIMOV,
SH. M. SAYDAKHMEDOV, and G. E. ZAIKOV

CONTENTS

6.1 INTRODUCTION

There are resulted the negative consequences of use of water steam (WS) at distillation of hydrocarbon mixes. There are proved the perspective method of process of primary distillation oil and gas condensate raw materials. There are defined mass-change characteristics of process of rectification at use of alternative steaming agents and spent their comparisons with indicators of traditionally applied steaming agent are spent.

The rectification is carried out at interaction of phases with various concentration of a distributed component which is transferred through boundary diffusive layer, under formation between contacting phases. In an interface sharp in the thickness (δ_c) a decrease in a difference of concentration occurs and it might come nearer to linear as in this area of a stream speed of process is defined by molecular diffusion, the role of convective diffusions is small. Such decrease of mass exchange speaks that on border of section of phases the braking action of forces of a friction between phases and forces of a superficial tension on border of a liquid phase amplifies [1-3]. The formation of a hydrodynamic interface close to surface of phases section conducts to occurrence in it diffusive boundary film in the thickness δ, it is usually ready in the smaller thickness than δ_c.

In a kernel of a stream the mass transfer is carried out basically by turbulent pulsations, therefore concentration of distributed substance in a stream kernel is practically constant. Obviously, in an interface between contacting phases the basic resistance to carrying over diffusive mass transfer process is concentrated [2, 3].

6.2 EXPERIMENTAL

One of intensification ways of mass transfer process is reduction of interface of the films. There are number of treatments about occurring processes on border of section of phases and models by means of which they speak. One of extended is Lewis and Whitman's two-film model. According to this model from both parties of phases contact surface motionless or laminar moving films are formed where the substance transfer is carried out only by molecular diffusion. These films separate a surface of contact of phases from a stream kernel in which concentration is practically constant, all changes of concentration of substance occur in a film [2-4].

According to film model equation integration:

$$M = -D \cdot (dc / dn) \cdot F \tag{1}$$

leads to expression:

$$M = F(D/\delta)(c_0 - c_b) \tag{2}$$

where δ is a thickness of a boundary film, c_b and c_b are average concentration in a kernel of a phase and concentration on border of section of phases accordingly, D is factor of molecular diffusion of distributed substance, F is a surface, normal to a diffusion direction, and dc/dn is a gradient of concentration of diffusive substance.

Comparing last equation with the equation:

$$M = \beta\left(y - y_{ep}\right)F \tag{3}$$

we will receive:

$$\beta = D/\delta, \tag{4}$$

where β is a factor of mass transfer of the steaming agent.

Hence, to be deduced from the received expression the equation for definition of a thickness of diffusive film of an interface.

$$\delta = D/\beta \tag{5}$$

From this equation we may see that parameter β is inversely proportional to film's thickness δ.

The thickness of films of an interface created by hydrocarbonic steaming agents (by steams from gas fractionation unit (GFU), of light (LN) and heavy (HN) naphtha and WS (δ_{en}) are defined at steaming of some fractions (with limits of boiling: Fr.1 - 60–80°C, Fr.2 - 120–130°C, and Fr.3 - 170–180°C).

6.3 DISCUSSION AND RESULTS

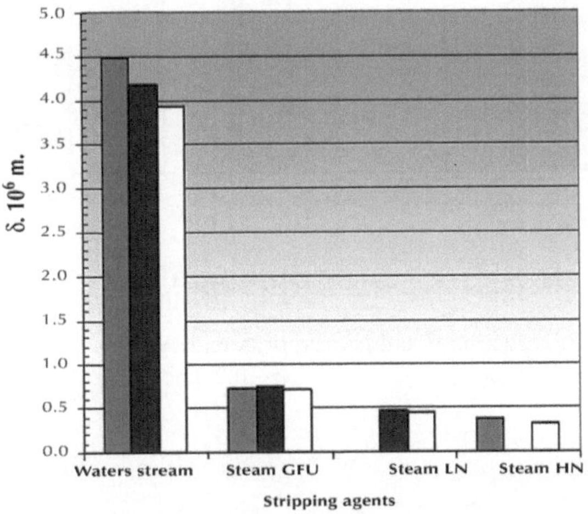

FIGURE 1 Thickness of a film created by steaming agents at division from a mix of fractions— ■– fr.1, ■ – fr.2, and □ – fr. 3.

The results of calculations show (Figure 1), that the thickness of a film created by WS at steaming above the specified fractions fluctuates within 3.95·× 10^{-6}–4.49·× 10^{-6} m, in steams from GFU this indicator has decreased down to 0.71·× 10^{-6}–0.75·× 10^{-6} m, and at steaming in steams of LN to 0.45·× 10^{-6}–0.47·× 10^{-6} m, and in steams of HN to 0.32·× 10^{-6}–0.36·× 10^{-6} m. Parities of a thickness of boundary films created with WS and with steams of the hydrocarbons, showing degree of its reduction $I_\delta = \delta_{\omega s}/\delta_h$, are revealed. The results of calculations show the decreasing of a thickness of boundary films at steaming with application of steams from GFU by 5.8 times, at application of steams of LN and HN it has decreased by 9.3 and 12.6 times accordingly (Figure 2) and (Figure 3).

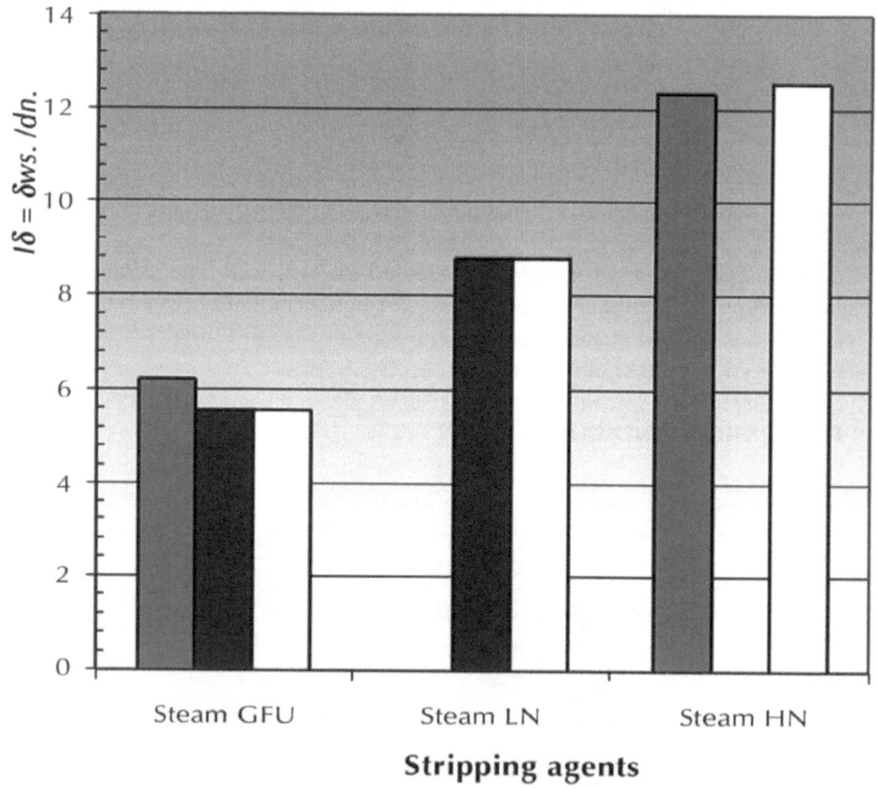

FIGURE 2 Degree of reduction of a thickness of boundary films created by steaming agents in a steam phase at fractions steaming—■ – fr.1, ■ – fr.2, and □ – fr. 3.

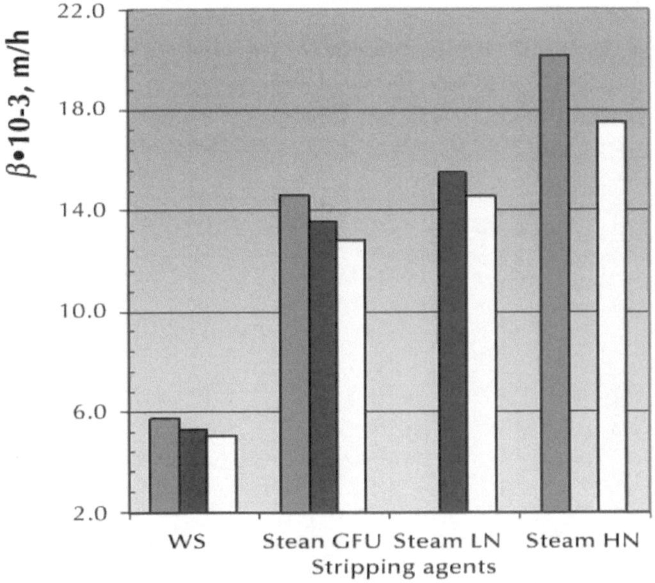

FIGURE 3 Mass exchange factor change in dependence of the steaming agent at division from a mixture of fractions—■ – fr.1, ■ – fr.2, and □ – fr. 3.

6.4 CONCLUSION

The perspectives of application of a method of dry distillation of oil a gas condensate mixes proves to be true by means of the obtained research result which is ascertained by the facts of decreasing of a thickness of a boundary film with application of above resulted hydrocarbonic steams, as the steaming agent, approximately by 9.2 times.

KEYWORDS

- **Boundary diffusive layer**
- **Gas fractionation unit (GFU)**
- **Hydrocarbon**
- **Hydrocarbonic steams**
- **Rectification**

REFERENCES

1. Kasatkin, A. *Basic processes and devices of chemical technology.* 9 pub., the rew. and add., Moscow, Chemistry, pp. 403–453 (1973).
2. Kafarov, V. *Basis mass transfer 3-pub.*, the rew. and add., Moscow, Higher school, pp. 187–194 (1979).

3. Dytnersky, J. *Process and devices of chemical technology.* P. 2., Moscow, Chemistry, pp. 10–42 (1995).
4. Aynshteyn, V. *General course of processes and devices of chemical technology.*, Moscow, «Logos» • "Higher school", pp. 768–805 (2003).

CHAPTER 7

A STUDY ON THE EFFECT OF ANTHROPOGENIC POLLUTION ON THE CHEMICAL COMPOSITION AND ANATOMIC STRUCTURE OF BIOINDICATIVE PLANTS

N. V. ILYASHENKO, A. I. IVANOVA, and YU. G. OLENEVA

CONTENTS

7.1 INTRODUCTION

The effect of anthropogenic pollution on the chemical composition and anatomic structure of bioindicative plants (hydrophytes) was studied with the aid of Fourier transform IR spectroscopy (FTIR), scanning electron microscopy (SEM), and X-ray microanalysis. A correlation is established between the changes existing in the IR spectrum of the plant samples and anthropogenic pollution of the plant inhabitation. The deformation and epidermis cell disruption were revealed in the samples from polluted localities.

The present research is aimed at the experimental study of the chemical composition and anatomic structure of polluted higher aquatic plants making use of combined physical methods of characterization by FTIR spectroscopy, SEM, and X-ray microanalysis.

The recent studies demonstrated that aquatic plants are highly sensitive tools for assessing the status of water bodies [1-4] because they quickly establish themselves with their environment. Further usage of plants as bioindicators requires accumulation of knowledge and development of appropriate methods of their characterization.

The higher hydrophytic plants possess high accumulative properties and find their use in the determination of anthropogenic chemical loads on the water body in the system of environment state biomonitoring [1-4].

The IR spectral analysis data on the chemical composition changes in bioindicative plants may be informative for the estimation of hydrosphere pollution in industrial regions. The exact identification of the types of compounds forming in the plant as a result of accumulation of various pollutants enables the use of FTIR for biomonitoring of acid pollutions (sulfur and nitrogen dioxides) and also petroleum products and organic compounds [5-7].

As far as bioindicative plants are subject to change at chemical and anatomic levels due to the action of anthropogenic environment pollution, these changes may be effectively monitored with the aid of FTIR spectroscopy, SEM, and energy-dispersive X-ray spectroscopy (EDX) [8, 9].

The combination of SEM and EDX makes it possible to determine the element composition in a volume of ~1 cubic micrometer by means of registration the X-ray characteristic irradiation occurring during the primary electron interaction with the sample surface. The EDX is an analytical technique used for the chemical characterization of a sample. Its characterization capabilities are due largely to the fundamental principle that each element has a unique atomic structure allowing X-rays that are characteristic of an element's atomic structure to be identified uniquely from one another. The number and energy of the X-rays emitted from a specimen are measured by an energy-dispersive spectrometer. As the energy of the X-rays are characteristic of the difference in energy between atomic shells, and of the atomic structure of the element from which they were emitted, this allows the elemental composition of the specimen to be measured.

The aim of the present work is to study the effect of pollutants on the chemical composition and anatomic structure of hydrophytic plant *Ceratophyllum demersum* deep green by the methods of FTIR spectroscopy, SEM, and EDX.

7.2 EXPERIMENTAL

The *Ceratophyllum demersum* L. (Hornwort deep green) occurs in Tver region in stagnant reservoirs with slowly flowing water, ponds, peaceful river backwater, cutoff meanders, and is able to grow both in clear and contaminated inhabitants [1, 10]. The collection of the plants and water intake for chemical analysis was performed in water bodies of the city of Tver and Tver region classified by the factor of proximity to the sources of contamination in two groups—background and polluted (Table 1).

In parallel the chemical analysis of the water from the places of *C. demersum* growth was made to ensure the proper interpretation of the element composition and IR spectra of the aquatic plants from industrial regions. The chemical analysis was performed with the aid of a spectrofluorimeter "Fluorat-02-Panorama" and capillar electrophoresis system (Kapel-105) (Lumex). The determination of the contents of inorganic anions, surfactants, petroleum products, phenols in water was made in accordance with standard procedures described in [11-14].

The IR spectra of the samples under study were registered in the range of 400–4000 cm^{-1} by a standard method with potassium bromide [15] making use of the Fourier transform spectrometer (Equinox 55) (Bruker).

TABLE 1 Places of *C. demersum* collection

No.	Water body	Ecological status/Source of pollution [16]		Main pollutants [17]
I	Mezha river, Big Fyodorov village, Nelidov district of Tver region	SPNT* (background area), Central-Forest National Natural Biosphere Park		
II	Coolant lake Udomlya, Udomlya, Tver region	Ecologically stressed nodes	Kalinin nuclear energy station	Petroleum products, small radiation doses
III	Cunette, Redkino township, Konakov district, Tver region		Public corporation (Redkino pilot plant)	Petroleum products, inorganic anions, ammonia, aromatic compounds, anionic surfactants

* SPNT–Special Protected Natural Territory

The SEM studies were made in high vacuum regime with JEOL 6610LV SEM (Japan), EDX was performed with the INCA Energy system (OXFORD INSTRUMENTS, UK). The morphology of the epidermal surfaces was examined at the magnifications of x 500 and x 1000. Samples of the plants for SEM and EDX studies were dried at 30–40°C and fixed by a graphite adhesive tape [8, 9].

7.3 DISCUSSION AND RESULTS

7.3.1 CHEMICAL ANALYSIS OF THE WATER IN RESERVOIRS UNDER STUDY

The results of the water analysis of the places specified in Table 1 confirmed the existence of pollutants listed in Table 2.

Main water indices (Cl^-, NO_2^-, NO_3^-, SO_4^{2-}, phenols, and anionic surfactants) in samples II, III are much greater as compared with the background (I).

In the water sample III the content of phenols and inorganic anions is by a factor of several tens larger than in sample I thus indicating discharges of chemical pollutants from the Redkino pilot plant (Table 2). The existence of various pollutants in the water reservoirs should be taken into account because aquatic plants are able to accumulate the contaminants (Table 2) [18-20].

TABLE 2 Chemical analysis of the water samples from the reservoirs under study

Index, mg/l	Sample		
	I (background)	II	III
Cl^-	1.2	6.2	64.0
NO_2^-	< 0.2	0.88	5.0
SO_4^{2-}	2.8	7.4	42.3
NO_3^-	< 0.2	1.2	< 0.2
PO_4^{3-}	< 0.2	< 0.2	< 0.2
Petroleum products	0.17	–	0.16
Phenols (total)	0.001	–	0.38
Anionic surfactants	0.11	–	0.09

7.3.2 IR SPECTRAL ANALYSIS OF C. DEMERSUM SAMPLES

As shown in Figure 1 are the IR spectra of *C. demersum* samples from different reservoirs. The IR spectral analysis shows that all forms under study have absorption bands corresponding to the main chemical composition of the plant: Carbohydrates ~56% (of the absolute dry weight), proteins ~18%, fat~1% [1]. The existence of carbohydrates in the plant is testified by absorption bands due to valent oscillations of CH_2- groups at a frequency of ~2925 cm^{-1} and OH- groups at ~3400 cm^{-1} [5, 21, 22]. The existence of proteins is evidenced by the absorption bands at ~1640 (Amid I), ~1535 (Amid II), and ~1235 cm^{-1} (Amid III) [7, 22]. The presence of fat may be judged by the existence of absorption bands at ~1735 ($\nu_{C=O}$), ~1446 (δ_{CH2}) (Table 3) [6].

The comparison of the IR spectra of *C. demersum* samples taken from background and contaminated reservoirs demonstrates considerable changes in the chemical composition of the plants. It is important that most significant changes of the absorption bands intensity correspond to the samples collected in the regions of industrial contamination. In samples collected from the reservoirs not subjected to direct contamination the band intensity conforms to the background values.

The spectra of plant samples from contaminated localities (Figure 1) demonstrate essential changes at the following frequencies— ~2514 cm^{-1} (III), due to valent oscillations of S-H groups, ~1794 cm^{-1} because of valent oscillations of C=O groups (III), ~876 cm^{-1} owing to valent symmetrical oscillations of the S-O-C groups (III), and ~712 cm^{-1} by virtue of valent oscillations of C-S-C groups (III) (Table 3) [7, 22].

The absorption band ~1431 cm^{-1}, characterized by $\nu_{as}(SO_2)$, δ(N-H), displays itself in all spectra of samples from contaminated places, but it is mostly intensive in the spectre of sample III on account of high concentration of sulfur-bearing anions absorbed from the water by the plant (Figure 1 and Table 3).

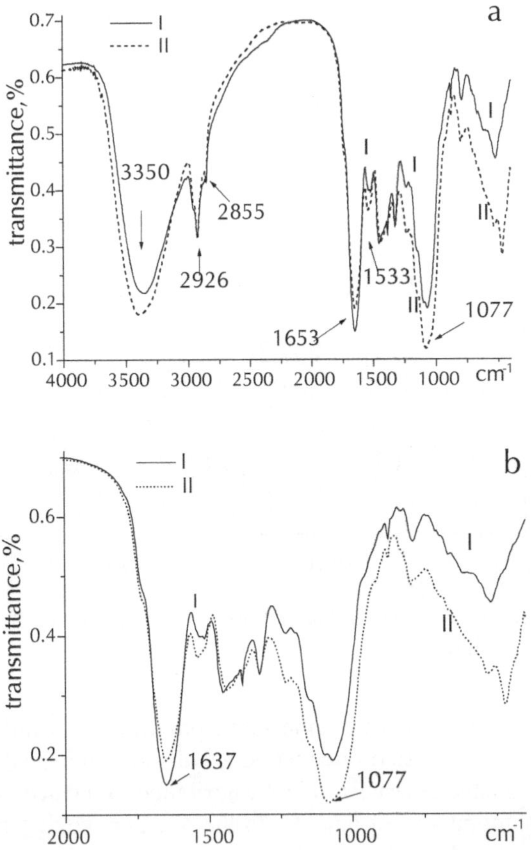

FIGURE 1 *(Continued)*

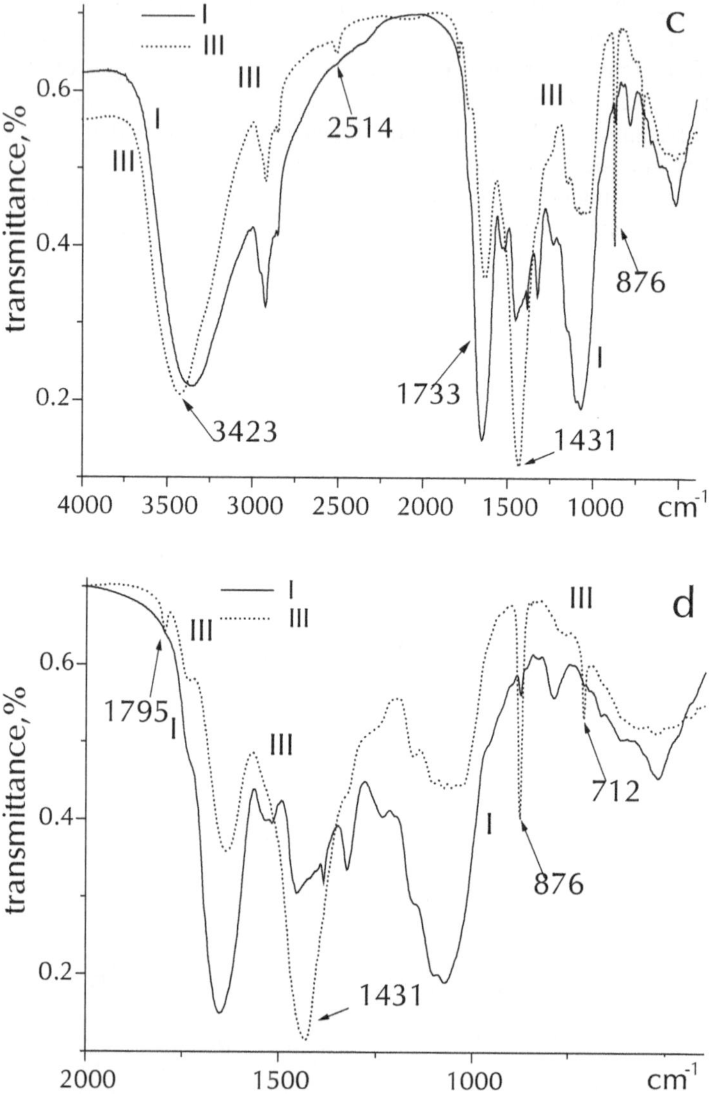

FIGURE 1 IR spectra of *C. demersum* samples—I (backgraund), II (lake Udomlya), III (Cunette, Redkino township).

The comparative analysis of C. demersum from different localities enables to propose that the intensity increase of the mentioned absorption bands of the samples collected from contaminated reservoirs may be attributed to the accumulation and adoption of chemical compounds not inherent to the plant containing thiol-, carbon-, and nitrogen-bearing groups [5].

TABLE 3 Assignment of IR absorption bands of *C. demersum* samples

Band position, cm^{-1}			Assignment	Reference
I	II	III		
3350	3393	3423	ν(OH)	[4, 7, 21]
2926	2925	2925	$\nu_{as}(CH_2)$	[6, 21]
2855	2855	2854	$\nu_s(CH_3)$	
–	–	2514	ν(S-H)	[22]
–	–	1795	ν(C=O)	[22]
1733	1738	1735	ν(C=O)	[4, 6, 21]
1653	1653	1637	Amid I ν(C=O)	[7, 22]
1533	1542	–	Amid II ν(O-C-N)	[21]
–	1438	1431	$\nu_{as}(SO_2)$, $\delta(CH_2)$, δ(N-H)	[22]
1385	1385	–	δ(OH)	[5, 22]
1325	1325	1325	$\delta(CH_2)$	[22]
1233	1230	1265	Amid III δ(N-H)	[7, 22]
1201	1202	1204	ν (C–O), δ(OH)	[15, 21]
1098	1153	1101	ν (C–O), $\nu_{as}(COC)$	[4, 21]
1071	1078	1077	$\nu_{as}(COC)$ chain	[5, 21]
–	–	1052	$\nu_{as}(COC)$ chain	
–	–	1025	ν(OH)	
874	874	876	S-O-C	[6, 7, 22]
–	–	712	ν (C-S-C)	[22]

7.3.3 EXAMINATION OF HYDROPHYTIC PLANT SAMPLES WITH THE AID OF EDX AND SEM

The electron images of *C. demersum* samples studied at different magnifications with the aid of SEM are presented in Figure 2. It is seen that in samples collected in polluted places there are deformation and destruction of epidermis resulting in violation of the epidermal cell layer integrity.

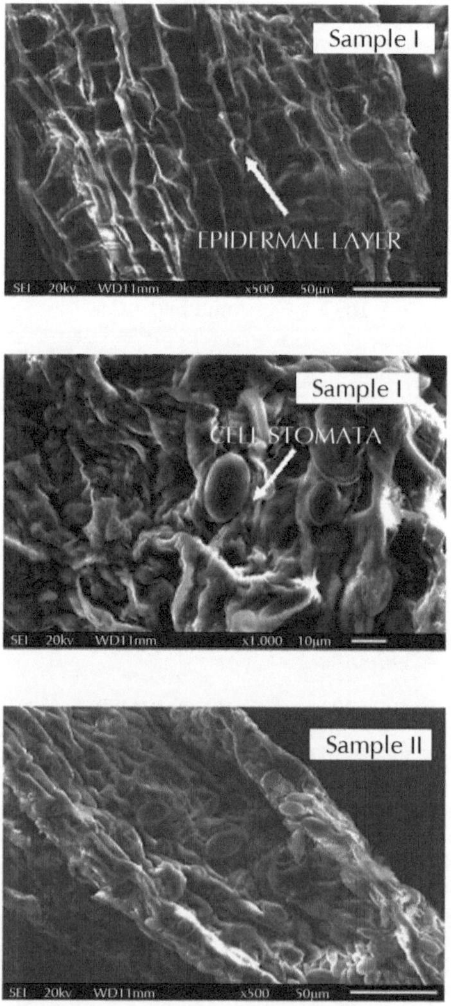

FIGURE 2 *(Continued)*

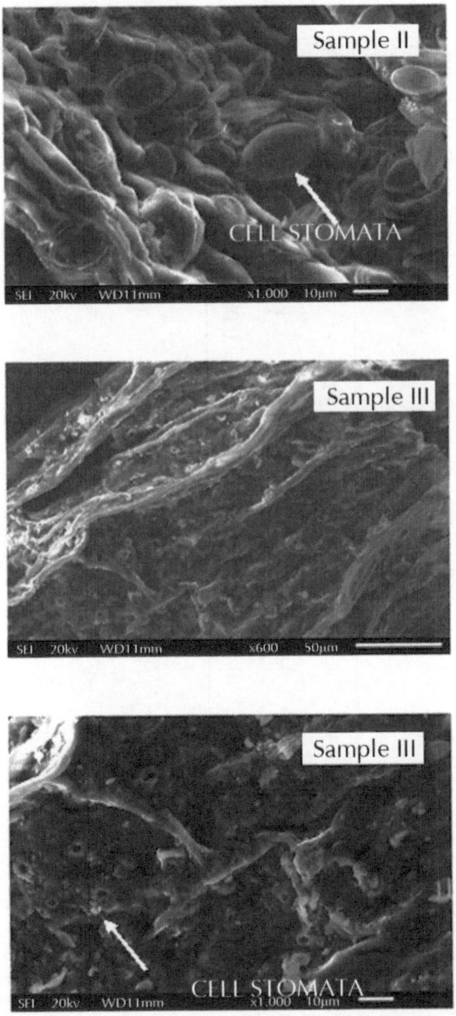

FIGURE 2 The SEM images of *C. demersum* leaf samples I, II, and III at different magnifications (x500 and x1000).

In sample II high density of stomatal cells in the epidermis is notable, presumably related with the adoption properties of the plants living in warm water of the udomlya lake of the kalinin nuclear plant. The increased density of stomata is indicative of heightened heat exchange between the plant and its environment. In addition to the destruction of epidermal formations the arrangement of stomatal cells on the epidermis surface is disturbed. It should be mentioned that the destruction of surface stomatal cells is far less than that of epidermis cells. Thereby, in spite of disintegration and violation of the epidermal surface of the modified leaf the heat transfer function may be partially conserved [8, 23].

TABLE 4 Elemental chemical composition of *C. demersum*

Sample	Chemical Element	C	O	NA	MG	AL	P	S	CL	K	CA	SI	MN	FE	TOTAL
I	wt%	56.35	40.9	0.6	0.27	0.18	0.41	0.1	0.4	0.67	0.12	–	–	–	100
II		48.34	45.31	0.34	0.68	0.11	0.12	0.23	0.5	0.75	–	3.48	0.07	0.06	100
III		53.44	41.92	1.93	0.27	–	0.05	0.14	1.25	0.25	0.26	0.09	0.26	0.11	100

The elemental composition and quantitative chemical analysis of *C. demersum* plants was performed with the aid of EDX. Given (Table 4) are the data for elemental composition and occurrence of various chemical elements in the tissues of *C. demersum* from some water bodies of Tver region.

The EDX data reflect the overall chemical composition of *C. demersum* [1]. However, in samples II and III the manganese, iron, sulfur, and chlorine content is higher than in the background sample, which is in accordance with the corresponding chemical analysis data of the water given in Table 2.

7.4 CONCLUSION

By the use of FTIR it was proved that higher aquatic plants have a capability to respond actively on the water chemical composition changes by the increase of absorption bands intensity related to contaminants. The results of the study show that the method of FTIR spectroscopy may be recommended for the effective application in biomonitoring of contaminated water bodies.

With the aid of SEM by an example of biomonitoring *C. demersum* plants anatomic changes in the leaf platelets were noticed. The deformation and destruction of epidermis cells were observed thus giving an evidence of high accumulation and adoption properties of aquatic plants. The elemental composition of plants under study was determined by means of EDX. The heavy metals were detected in samples collected from the places of anthropogenic pollution.

The described physical methods of analysis may be effectively employed in biomonitoring of the ambient environment.

KEYWORDS

- **Anthropogenic pollution**
- **Fourier IR spectroscopy**
- **Hydrophytic plants**
- **Scanning electron microscopy**
- **X-ray microanalysis**

ACKNOWLEDGMENT

This work was performed within the Federal Target Program "Scientific and Scientific–Pedagogical Personnel of Innovation Russia for 2009–2013", contract № 14.740.11.1281 and supported by the grant of Foundation for Development of Small Enterprises in Sceintific–Technical Sphere under the Program "Participant of the Youth Scientific–Innovation Competition", contract № 8754p /14008. Permission to use the equipment and support of the Sharing Service Center of Tver State University is greatly acknowledged.

REFERENCES

1. Sadchikov, A. P. and Kudryashov, M. A. *Hydrobotany: Coastal Vegetation* (in Russian). Moscow, (2005).
2. Chukina, N. V. *Effect of water contamination on the chemical composition of hydrophytic leafs*(in Russian). Proc. XIII Intern. Youth School, Rybinsk, pp. 290–296 (2007).
3. Goldovskaya, L. F. *Chemistry of Ambient Environment* (in Russian). Moscow, Mir, (2005).
4. Ilyashenko, H. V., Ilyashenko, V. D., Dementieva, S. M., Khizhnyak, S. D., and Pakhomov, P. M. Vestnik TvGU. *Ser. Biologiya i Ekologiya*, (13), pp. 211–221 (2009) (in Russian).
5. Duraees, N., Bobos, I., and Ferreira Da Silva, E.. Chemistry and FT-IR spectroscopic studies of plants from contaminated mining sites in the Iberian Pyrite Belt. *Portugal Mineralogical Magazine*. **72**(1), 405–409 (2008).
6. Ilyashenko, N. V., Khizhnyak, S. D., and Pakhomov, P. M. Effect of anthropogenic factor on the content of biologically active compounds in *Bidens tripartita* L. And *Potentilla erecta* L, Intern. Conf. *Renewable Wood and Plant Resources: Chemistry, Technology, Pharmacology, Medicine*, Book of abstracts, Saint-Petersburg, pp. 265–266 (2011).
7. Meisurova, A. F., Khizhnyak, S. D., and Pakhomov, P. M. IR spectral analysis of the chemical composition of the lichen Hypogymnia physodes to assess atmospheric pollution, *Journ. Applied Spectroscopy, 76, 420–426 (2009).*
8. Pathan, A. K., Bond, J., and Gaskin, R. E. Sample preparation for SEM of plants surfaces, *Materials Today*, **12**, 32–43 (2009).
9. Goldstein, J. I., Newbury, D. E., Echlin, P., Joy, D. C., Fiori C., and Lifshin, E. *Scanning Electron Microscopy and X-ray Microanalsis*, Plenum Press, New York and London, (1981).
10. National Atlas of Russia, Vol. 2: Nature and Ecology. Moscow, 2008 (in Russian).
11. *Determination of inorganic anions in water*, Russian federal standard M 01-30-2009 PND F 14.1:2:4.157-99, (2009).
12. *Depermination of anionic surfactants in water*, Russian federal stabdard M 01-06-2009 PND F 14.1:2:4.158-2000 GOST P 51211-98, (2009).
13. *Determination of etroleum in water*, Russian federal standard M 01-05-2007 PND F 14.1:2:4.128-98 MYK 4.1.1262-03, (2007).
14. *Depermination of phenols in water*, Russian federal standard MVI M 01-07-2006 PND F 14.1:2:4.182-02 MYK 4.1.1263-03, (2006).
15. Smith, A. *Applied IR Spectroscopy–Bases, Techniques, and Analytical Applications*. (1982).
16. A.S. Kurbatov (Ed.), (in Russian), City Ecology, Moscow, (2004).
17. Federal Report "*On the Status and Ambient Environment Protection in Russian Federation in 2009*" (in Russian), Moscow, (2009).
18. Ilkun, G. M. *Air Filtering of Pollutants with the Aid of Woody Plants* (in Russian), Tallin, (1982).
19. Ilkun, G. M. *Atmosphere Contamination and Plants* (in Russian), Naukova dumka, Kiev, (1978).
20. Smolov, A. P. and Opanasenko, V. K. Ammonium factor in the vital activity of plant cell, *Cytology*, **51**, 358–366 (2009).
21. Bazarnova, N. G., Karpova, E. V., and Katrakov, I. B. *Methods of Study of Wood and its Derivatives* (in Russian), Altai State University, Barnaul, (2002).
22. G. Socrates (Ed.), *Infrared Characteristic Group Frequencies: Tables and Charts*. John Wiley & Sons, London, (1994).
23. Garty, J., Kunin, P., Delarea, J., and Weiner, S. Calcium oxalate and sulphate-containing structures on the thallial surface of the lichen *Ramalina lacera*: response to polluted air and simulated acid rain, Plant, Cell and Environment, **25,** 1591–1604 (2002).

CHAPTER 8

PHASE TRANSITIONS IN WATER-IN-WATER BSA/DEXTRAN EMULSION IN THE PRESENCE OF STRONG POLYELECTROLYTE

Y. A. ANTONOV and P. MOLDENAERS

CONTENTS

8.1 INTRODUCTION

We examine whether a small amount of strong polyelectrolyte (dextran sulfate sodium (DSS) salt) can induce mixing of bovine serum albumin (BSA) and dextran in aqueous water-in-water BSA/dextran emulsion and how intermacromolecular interactions affect its the rheological properties. Addition of DSS to water-in-water emulsion at pH 5.4 leads to its mixing, a noticeable increase in viscosity and module (G′). Mixing is observed at the DSS/BSA weight ratio, $q_{BSA} \geq 0.07$. Increasing the ionic strength in the resulting single-phase system induces phase separation. Our results show that the increase in viscoelasticity results from the interaction of DSS with both macromolecular components. The interaction of DSS with BSA leads to the screening of BSA tryptophanyls from the aqueous environment. Such interaction is not accompanied by the polarization of the protein, whereas the affinity of DSS to dextran results in an increase of viscoelasticity and in an appreciable change in the microstructure of the DSS/dextran mixture. It was assumed that similar to compatibilization of polymer blends by diblock copolymers, the driving force for inducing mixing of water/BSA/dextran emulsions by DSS results from the affinity of strong polyelectrolytes to both macromolecular components of the mixture.

Biopolymer incompatibility and complex formation are fundamental physicochemical phenomena that determine the structure and physical properties of biopolymers mixtures [1-3] playing an important role in protein processing [6-8]. Polymer incompatibility is a consequence of the normally unfavorable interactions between polymer species. Even small positive values of the Flory–Huggins interaction parameter between different polymer species can result in phase separation, due to the small entropy gain upon mixing these macromolecules [9-11]. Therefore, even minute differences in the structure of macromolecules may result in phase separation. The introduction of charges on a water-insoluble polymer generally increases its solubility in water. Similarly, the introduction of charges on one of the two polymers in an aqueous polymer mixture increases the miscibility of the two polymers [12, 13]. Added salt reduces the effects of the charges. For ternary systems solvent/polymer A/polymer B the introduction of charges of the same sign on both polymers reduces the favorable miscibility effect seen when charging only one of the polymers [14-16]. These observations indicate that the mixing entropy of the small counter ions of the polyelectrolytes plays an important role for the solubility and for the miscibility in polyelectrolyte systems, and they have inspired theoretical investigations on the Flory-Huggins level of the solubility and miscibility of polyelectrolyte solutions. The behavior of blends of neutral and polyelectrolyte chains in a common solvent has been investigated theoretically by Khokhlov and co-workers [12, 17]. They studied the dependence of free energy on concentration fluctuations, systematically accounting for the translational entropy of polymers and counter ions, the interfacial tension, and the electrostatic interactions between the charged species.

Their study shows that increasing the charge of the polyelectrolyte has the following effects:

- It stabilizes the mixed phase, moving the spinodal point to lower temperatures, and increases the extension of the homogeneous region

- It changes the character of the transition from macro to micro phase separation, the characteristic length of the micro phase separation being dependent on the polyion charge and on the amount of added salt.

The former effect is due to counter ion entropy, whereas the latter results from the new length scale introduced by the electrostatic interactions, the Debye screening length.

Independently, Nilsson [14, 15] and later Johansson et al.[18] numerically solved models based on an extended version of the Flory–Huggins theory, where the counter ions were explicitly included as a separate component on the same level as the solvent and the chain molecules. In the spirit of the random-mixing approximation, the charges of each component in a phase are uniformly distributed; leading to zero electrostatic energy and the effect of the electrostatic interactions among the charges enters the model only by imposing that each phase should be electroneutral. Both approaches predict results in qualitative agreement with experimental data [14-17].

The thermodynamic behavior of aqueous systems containing one or two types of charged biopolymers is less clearly analyzable, because the structure of theses macromolecules is more complicated than that of synthetic polymers. In contrast to ternary solvent/polymer/polymer systems, an introduction of charges of the same sign on both biopolymers frequently does not reduce the miscibility. For example, similarly charged ternary water/linear anionic polysaccharide-1/linear anionic polysaccharide-2 systems are single phase one in all the concentration range, pH and ionic strength [19]. Moreover, data obtained for water-protein-uncharged polysaccharide systems show a variety of different phase behaviors. Some show single phase behavior at a low ionic strength. However, unsimilar to ternary polymer systems the dominant mechanism responsible for single phase state of such systems at low ionic strength (below 0.01) involves the formation of water-soluble weak interpolymer complexes, which may be destroyed by increasing ionic strength [20]. The other water/protein/uncharged polysaccharide systems are two phase ones at a low ionic strength at relatively high concentrations of biopolymers. Typical examples of such systems are BSA/dextran system. This system is two phase one at pH 5.4 and at relatively high enough concentrations [5]. At low concentrations dextran molecules are able to form interpolymeric complexes with BSA in water if the polysaccharide is in excess and if the protein exists in its associated state [21]. The other example is water/gelatin/dextran system. This system is two phases in water and single phase in the acidic or alkaline range. The possibility of complex formation in this system in the acidic range has been discussed by Woodside and colleagues [22, 23], and by Grinberg and Tolstoguzov [24]. These authors analyzed the thermodynamic behavior of gelatin/D-glucan mixtures. The existence of gelatin- D-glucan complexes was inferred from the considerable solubility of D-glucans in acidic ethanol in the presence of gelatin [22, 23] and from nephelometric and viscosimetric data [24]. Thus, phase behavior in the system water-charged biopolymer 1/uncharged biopolymer-2 depends strongly on weak intermacromolecular interactions, the origin of which is not quite clear till now.

Although, phase separation in biopolymer systems frequently allows control of the morphology and, hence, the rheology/texture of biopolymer systems [1, 25-27], in many cases it leads also to a spontaneous separation into two layers, which is not desired in many food and biotechnology processes, for example, blood substitutes production.

In polymer processing compatibilizers are frequently used [28] to stabilize a fine microstructure of synthetic polymer blends and to increase the interfacial adhesion between their phases. Normally these compatibilizers are copolymers, containing two blocks each compatible with one of the polymers, or so called ionomers containing both nonionic repeat units, and a small amount of ion-containing repeat units. It is less clear if it is possible to compatibilize two phase biopolymer mixtures, because experimental observations in this field are lacking.

The recent studies show [29, 30] that intermacromolecular interactions caused by the presence of a complexing agent (DSS salt) in semi dilute single-phase protein-anionic polysaccharide mixtures can induce phase separation and structure formation at pH values above the isoelectric point of the protein, due to water soluble protein/DSS associations resulting in network formation.

The aim of this work is to examine whether a strong polyelectrolyte can lead to the opposite phenomenon, that is mixing in concentrated globular protein-polysaccharide mixtures, if the capacity of the protein to form large interpolymer associates is weak and if the polysaccharide does not contain charged functional groups. Assuming that the question of weak interactions between different macromolecular species is important for understanding the phenomena of the incompatibility of biopolymers, and taking into account the lack of experimental data in this area, the present study deals with the relationship between the phase state of protein–neutral polysaccharide mixtures in the presence of a strong polyelectrolyte, and possible interactions between these biomacromolecules.

It was focused on aqueous two-phase systems formed by BSA and high molecular weight dextran (a highly branched nonionic polysaccharide). The water/BSA/dextran system is chosen because of the low compatibility of these macromolecules under the selected conditions. Moreover, these biopolymers are widely used for biomedical and bioseparation purposes [33], and the thermodynamic behavior and phase diagram of the ternary water/BSA/dextran system has been described in past [5].

The effect of a sulfate polysaccharide DSS on phase equilibrium, macromolecular interactions, and structure of the two-phase water-gelatin-dextran system was studied using a rheo-small angle light scattering (SALS) device, optical microscopy (OM), phase analysis, static light scattering (SLS), rheology, and isoelectric focusing.

8.2 MATERIALS AND METHODS

8.2.1 MATERIALS

The BSA Fraction V, pH 5 (Lot A018080301), was obtained from Across Organics Chemical Co. (protein content = 98–99%; trace analysis, Na< 5000 ppm, Cl< 3000 ppm, no fat acids were detected). The isoelectric point of the protein is about 4.8–5.0 [34], and the radius of gyration at pH 5.4 is equal to 30.6 Å [34]. The water used for solution preparation was distilled three times. Measurements were performed at pH 5.4, because serum albumin undergoes conformational isomerization and changes in the conformation state and secondary structure with changes in pH from pH 5–5.5 to acid and alkaline region [35]. The extinction of 1% BSA solution at 279 nm was A^{1cm}_{279} = 6.70, and that value is very close to the tabulated value of 6.67 [35]. The high molecular weight dextran T-2000 sample was purchased from Amersham Pharmacia Biotech AB. Its radius of gyration, Stokes radius, intrinsic viscosity in water at 20°C, and weight average molecular mass, reported by the manufacturer, are 380 Å, 270 Å, 0.9 dL/g, and 2×10^6 Da respectively. Dextran 2000 behaves in solution as a highly branched expanded coil. Dextran sulfate DSS (M_W = 500 kDa, M_n = 166 kDa, intrinsic viscosity η (in 0.01 M NaCl) = 0.5dL/g, 17% sulfate content, and free SO_4 less than 0.5%) was produced by Fluka, Sweden (Reg. No. 61708061 A, Lot No. 438892/1).

PREPARATION OF THE PROTEIN/POLYSACCHARIDE MIXTURES

To prepare BSA and dextran stock solutions, the biopolymer was gradually added to the three times distilled water and stirred at 298 K for 2 hr. The solutions were centrifuged at 13,000 g for 1 hr at 298 K to remove insoluble particles. Subsequently, the concentration of the biopolymer was determined by measuring the dry weight residue. In the case of BSA the content of the protein nitrogen in dry BSA sample was always taken into account in order to calculate the concentration of protein in solution. In some cases, the final protein concentration was determined also by spectrophotometric measurements.

All the measurements were performed after equilibrating the biopolymer solutions and their mixtures for12–15 hr. Both, the BSA and the dextran solutions show Newtonian behavior (at temperatures of 298 K and at shear rates up to 30 s^{-1}) in the concentration range from 25–30 wt% within which the hydrodynamic volumes of the different macromolecules already overlap [36].

8.2.2 METHODS

TURBIDIMETRY AT REST

Cloud points of aqueous solutions of BSA plus dextran (DEX), and BSA plus DEX plus DSS were determined by measuring their light transmittance I either as a function of temperature or of composition. A laser beam (5 mW, He-Ne laser, 632.8 nm) of intensity I_0 passes the solution and the intensity, I of the transmitted light is determined.

The apparatuses and the procedures for the variation of temperature have already been described earlier in detail [37]. The demixing temperatures, T_1, are defined as the intercept of the tangent to the transmittance curves (I/I_0 as a function of T) in the point of inflection, with $I/I_0 = 1$. Isothermal measurements were performed by titrating solutions of BSA with solutions of DEX, and (BSA and DSS) + (DEX + DSS). In this case one determines the phase separation conditions from plots of the transmittance as a function of the amount the DEX solution added. Characteristic demixing compositions were determined as the intercept of the tangent to the transmittance curves (I/I_0 as a function of the concentration of the DEX solution added) in the point of inflection, with $I/I_0 = 1$.

Rheo-optical study A rheo-optical methodology based on SALS during flow, is applied to study *in-situ* and on a time-resolved basis the structure evolution. Light scattering experiments were conducted using a Linkam CSS450 flow cell with parallel plate geometry. The CSS450 uses two highly polished quartz plates that are parallel to within 2 um. Each plate is in thermal contact with an independently controlled pure silver heater utilizing platinum resistors sensitive to 0.1°C. A 5 mW He-Ne laser (wavelength of 633 nm) was used as light source. The 2D scattering patterns were collected on a screen by semitransparent paper with a beam stop and recorded with a 10-bit progressive scan digital camera (Pulnix TM-1300). Images were stored on a computer with the help of a digital frame grabber (Coreco Tci-Digital SE). The optical acquisition setup has been validated for scattering angles up to 18°. The gap between the plates has been set at 1 mm and the temperature was kept 20°C by means of a temperature-controlled water bath. In house developed software was used to obtain intensity profiles and contour plots of the images (New SALS SOFT-WARE-K.U.L.). Turbidity measurements have been performed by means of a photo diode.

Bright light microscopy is used to visualize particles distributed in the coexisting phases, using an Olympus BX51W1 fixed stage microscope equipped with a high resolution charge-coupled devices (CCD)-camera, (1000x1000 pixels, C-8800-21, Hamamatsu).

DLS. Determination of Intensity-size distribution, and volume-size distribution functions, as well as zeta potentials of BSA, DSS, and BSA + DSS particles were performed, by the Malvern Zetasizer Nano instrument (England), using a rectangular quartz capillary cell. The concentration of BSA in the water BSA/DSS mixtures was kept 0.20% (w/w). For each sample the measurement was repeated three times. The samples were filtered before measurement through DISMIC-25cs (cellulose acetate) filters (sizes hole of 0.22 μm for the binary water-protein solutions. Subsequently the samples were centrifuged for 30 s at 4000 g to remove air bubbles, and placed in the cuvette housing. The detected scattering light intensity was processed by Malvern Zetasizer Nano software.

Fluorescence emission spectra between 280 and 420 nm were recorded on a RF 5301 PC Spectro-fluorimeter (Shimadzu, Japan) at 25°C with the excitation wavelength set to 270 nm, slit widths of 3 nm for both excitation and emission, and an integration time of 0.5 s. The experimental error was approximately 2%.

Rheological measurements were performed using a Physica rheometer, type MCR 501, with a cone/plate geometry CP50-1/Ti (cone diameter 5 cm, angle 0.993°), Anton Paar. The temperature was controlled at 18°C by using a Peltier element. For each sample, flow curves were recorded in the shear rate range 0.1 to 150 s^{-1}. The ramp mode was logarithmic. The mechanical spectra were recorded over the 0.1–200 rad/s frequency range; the dynamic moduli G' (storage) and G'' (loss) were measured at constant strain amplitude of 3%, which was checked as being within the linear regime.

During the rheological measurements, all samples were covered with paraffin oil to avoid drying.

ENVIRONMENT SCANNING ELECTRON MICROSCOPY (ESEM)

Micro structural investigation was performed with the environment scanning electron microscope (ESEM) Philips XL30 ESEM FEG. The samples were freeze-fractured in Freon and immediately placed in the ESEM. Relative humidity in the ESEM chamber (100%) was maintained using a Peltier stage. Such conditions were applied to minimize solvent loss and condensation, and control etching of the sample. Images were obtained within less than 5 mins of the sample reaching the chamber. The ESEM images were recorded multiple times and on multiple samples in order to test reproducibility.

Iso-electric focusing (IEF) was performed with the IPGphor isoelectric focusing system (Amersham Biosciences) at 45°C using 2 μl IEF Standard (Bio-rad #161–0310) range pI 4.45–9.6. Run conditions: 51 V for 1 hr, 200 V for 1.5 hr, 200 V for 1 hr, and 500 V for 30 mins. Protein were fixed in a bath containing 40% methanol, 10% trichloro acetic acid. Gels were stained overnight in a Sypro Ruby fluorescent protein stain (Invitrogen) bath and then scanned with Typhoon Imager Analyzer (GE Healthcare).

8.3 RESULTS AND DISCUSSION

8.3.1 DSS-INDUCED MIXING

The experimental results shown in this section have been obtained on water (76.90 wt%) /BSA(7.70 wt%)/dextran (15.40 wt% 2000 systems located in the two-phase region far from the binodal line [5]. The volume fraction of the BSA-enriched disperse phase was 0.11 [38]. To study the effect DSS on the phase behavior, the mixtures water (76.90–X wt%)/BSA (7.7 wt%)/dextran (15.4 wt%) /DSS (X wt%) were prepared. Here X corresponds to a variable amount of DSS. The system was hand mixed prior to loading into the rheo-optical device. First, a preshear of 0.5 s^{-1} was applied for 1000 s (500 strain units). Subsequently, this preshear was stopped, and the sample was allowed to relax for 30 s, leaving enough time for full relaxation of deformed droplets. Then SALS patterns are monitored.

The SALS patterns and the scattering intensity upon adding different amount of DSS are shown in Figure 1, starting from of DSS concentrations as low as 0.01 wt%. In each experiment, a freshly loaded sample has been used. The presence of 0.03 wt% of DSS in the two phase system led to appreciable decrease in the SALS signal (Figure 1, curves a,b). The higher the concentration of DSS, the weaker the scattering intensity becomes (Figure 1, curves b,c,d). This indicates that the sample is less and less heterogeneous on the length scales probed by light scattering (in the order of 0.5 micron).

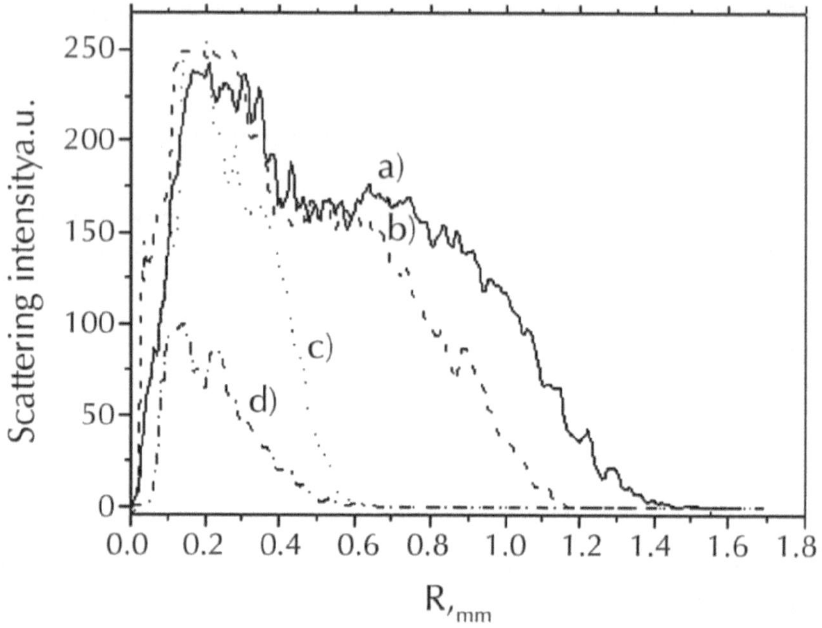

FIGURE 1 The SALS patterns and scattering intensity of water (76.9 %)/BSA (7.7 wt%)/ dextran (15,4 wt%) emulsion. upon adding different amount of DSS. 20°C Concentration of DSS (wt%) : a) 0.0, b) 0.03, c) 0.05, d) 0.08. The SALS patterns were obtained after shearing at 0.5 s^{-1} for 1000 s and subsequently left to rest for 30 s before measurements.

It can be argued that the scattering power of small structures decreases significantly and therefore the sensitivity of the camera becomes insufficient to pick up the scattering patterns. Therefore, in addition to SALS experiments, the effect of the presence of DSS on the absorption values (A_{500}) of water/BSA (variable)/dextran (15.4 wt%), containing constant concentration of dextran and different concentrations of BSA, was examined (Figure 2). The concentration of DSS at which the absorption value reaches the same value as the one measured for the external/continuous/phase is in agreement with the concentration of DSS at which the SALS pattern disappears. The presence of DSS in the water-in-water BSA/dextran emulsions results in dramatic changes in its morphology. The corresponding microscopy images for the same concentrations DSS and the same flow conditions are shown in Figure 3.

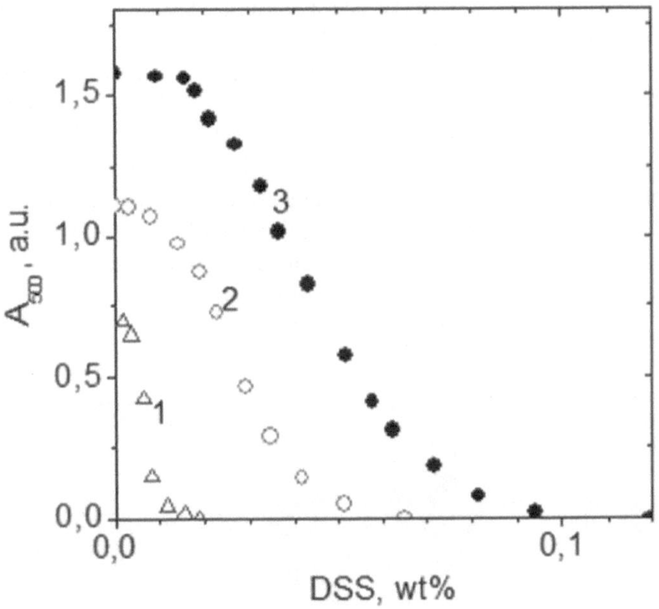

FIGURE 2 Dependences of light absorbance at 500 nm on concentration of DSS in the emulsion for BSA(variable)/dextran(15.4 wt%) emulsions. Concentration of BSA in the emulsion: 1)–10 wt%; 2)–5.0 wt%; 3)–2.5 wt%. The absorbance of pure BSA solution at the same concentrations was subtracted. 20°C.

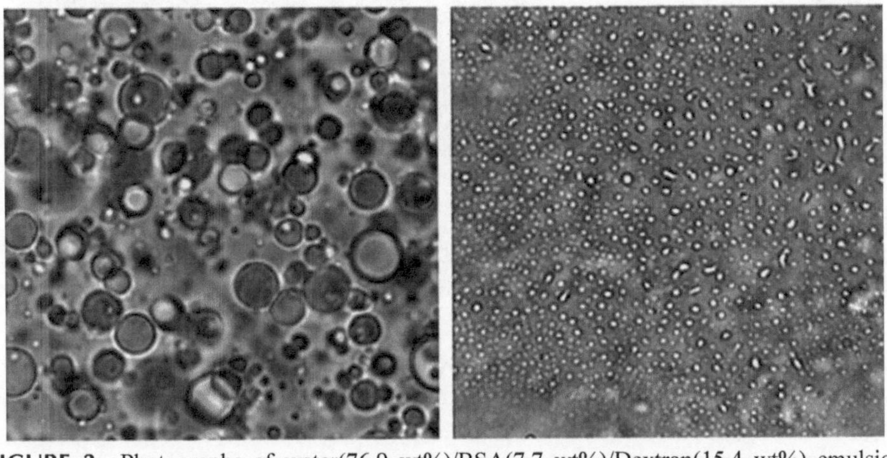

FIGURE 3 Photographs of water(76.9 wt%)/BSA(7.7 wt%)/Dextran(15.4 wt%) emulsion without DSS (Left); with 0.07 wt% of DSS (Right). The full length of each image corresponds to 0.2 mm. Images obtained after stop of preshear at 0.5 s^{-1} for 1000 s and subsequent 30 s rest period. The full length of each image corresponds to 0.2 mm.

After the addition of 0.07 wt% DSS to the emulsion, the droplets became much smaller and their volume fraction decreased, meaning a strong reduction of the interfacial tension and an increasing the thermodynamic compatibility of the biopolymer pair. The ESEM images of water/dextran (15.4 wt%) system, water/DSS (2.0 wt%) system, water/BSA (7.7 wt%)/dextran (15.4 wt%)–DSS (0.03 wt%) system, and water/BSA (7.7 wt%)/dextran (15.4 wt%)/DSS (0.14 wt%) system are shown in Figure 4. Dextran develops a skeleton-like structure (Figure 4a) whereas DSS shows the skeleton-like structure with the cobweb of the sulfate functional groups (Figure 4b). In the joint solution of BSA and dextran containing small amount of DSS (0.03 wt%) skeleton-like structure of polysaccharides and close to spherical structure of BSA were registered, whereas at a higher concentrations of DSS the system develops (Figure 4d) an amorphous structure similar to that observed for compatibilized polymer blends [38].

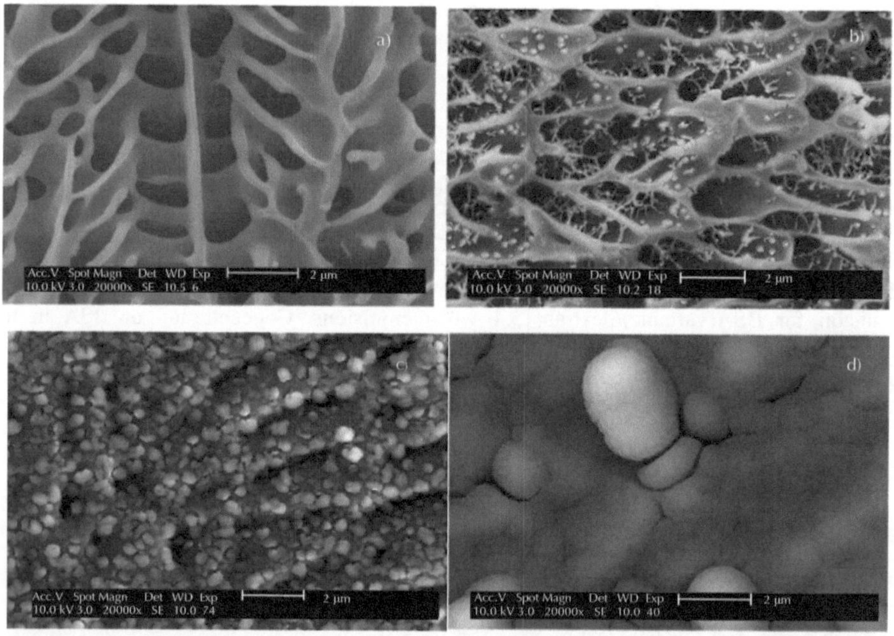

FIGURE 4 ESEM images of the water (76.9 wt%)/dextran (15.4 wt%) system, (a); water (98 wt%) /DSS (2 wt%) system, (b); water (92.3 wt%)/BSA (7.7 wt%)/DSS (0.03 wt%) system, (c); water/BSA (7.7 wt%)/dextran (15.4 wt%)/DSS (0.14 wt%) system, (d). pH 5.4.

Centrifugation of the water–BSA (7.7 wt%)/dextran (15.4 wt%)/DSS (0.14 wt%) system (120 min. 60,000 g, 45°C) prepared in the same way did not result in macroscopic phase separation. This indicates that the presence of DSS moved the system on the phase diagram outside the two-phase range.

In order to quantify the effect of DSS on phase equilibrium in the system BSA/dextran, the cloud point curves of the water/BSA/dextran system were determined at different concentrations of DSS at pH 5.4 and 25°C. (Figure 5). It is important to note that our preliminary experiments showed that DSS is contained almost quantitatively in the BSA-rich phase. The total concentration of the biopolymers at the threshold point increases from 8.76 wt% in the absence of DSS to 24.4 wt% as the DSS concentration reaches 0.045 wt%. At higher DSS concentrations, phase separation was not observed, whatever the total biopolymers concentration in the range studied (6.0 to 30.0 wt%).

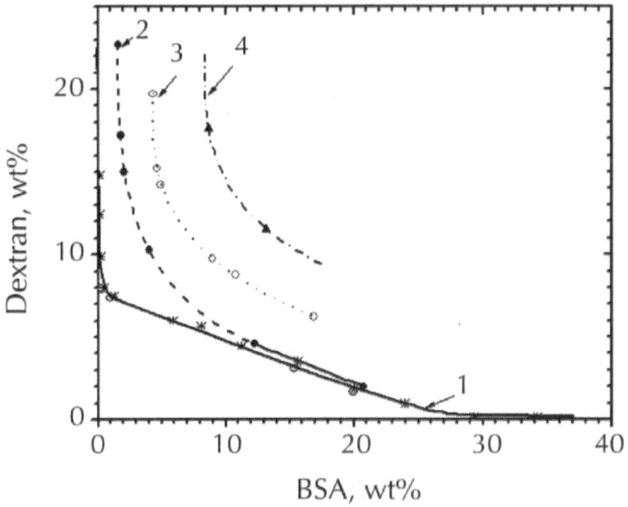

FIGURE 5 Cloud point curves of the water/BSA/dextran/DSS systen at 25°C. pH = 5.4. Concentration of DSS in the water/BSA/DSS and water/dextran/DSS solutions (wt%) before mixing are 0.0, (1); 0.015,(2); 0.03, (3); and 0.08 (4) respectively.

8.3.2 EFFECT OF DSS ON RHEOLOGICAL PROPERTIES OF BSA/ DEXTRAN WATER IN WATER EMULSIONS

We also characterized the phase state of the systems by means of their viscoelastic behavior. It has been shown [25] that at moderately low shear rates, the biopolymer emulsions can be regarded as conventional emulsions and various structural models are available in the literature to prediction the morphology. The evolution of the mechanical spectrum was investigated as a function of DSS concentration. The rheology of the water/BSA (8.46 wt%)/Dextran (11.8 wt%) system containing 15 vol% BSA enriched phase was examined before and after addition of DSS at concentrations from 0.2 to 2.13 wt% (q_{BSA} values from 0.02 to 0.25). The experimental flow protocol was the same as the one used for the rheo-SALS measurements. The viscosity of the system as a function of the concentration of DSS at a shear rate 0.1 s^{-1} and 1.04 s^{-1}, and the values of G' of the system as a function of the concentration of DSS at a frequency 0.2 rad/s and 1.26 rad/s are shown in Figure 6a and Figure 6b. In the presence of DSS, the system undergoes mixing, and this transition leads to an appreciable increase of the moduli. The viscoelastcity of the water/BSA/dextran system, which is very low in absence of DSS, increases markedly by the presence of DSS.

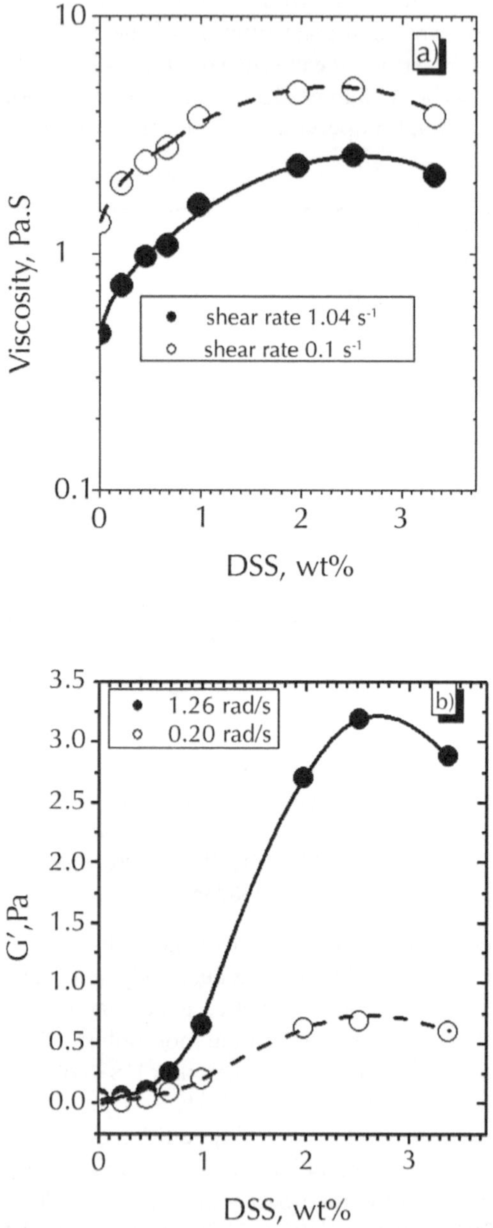

FIGURE 6 Dynamic viscosity of the water (79.74 wt%)/BSA (8.46 wt%)/dextran (11.8 wt%)/DSS (variable) system as a function of the concentration of DSS at a shear rate 0.1 s^{-1} and 1.04 s^{-1}, (a); the values of G' of the same system as a function of the concentration of DSS at a frequency 0.2 rad/s and 1.26 rad/s. (b). 18°C, pH 5.4.

The growth of viscoelasticity peaks for 2.5 wt% DSS (q_{BSA} = 0.295), concentration which four times higher then, according to rheo-SALS results (Figure 1), to the transition of the system from two-phase state to single-phase state. Thus, at 1 rad/s, G′ value is more than 300 times higher than that of the emulsion without DSS (Figure 6a). In the presence of high ionic strength (0.25/NaCl), that is when electrostatic interactions are suppressed, the mechanical spectrum of the system becomes insensitive to the presence of DSS (data are not presented). Similar changes were observed for the viscosity. At q_{BSA} = 0.04 and a shear rate of 1.0 s^{-1} the viscosity is 5.8 times higher compared with that of the single-phase system with almost the same composition (Figure 6b). It is important to note that in the shear rate range from 0.1 to 150 s^{-1} flow curves obtained with ascending and descending ramps superimposed (data not presented).

8.3.3 INTERMACROMOLECULAR INTERACTIONS AS A DRIVING FORCE OF MIXING

To understand the reasons for such dramatic effects, the dynamic modules and the viscosity of the water–BSA–DSS and water/dextran/DSS systems were measured as a function of the DSS/BSA, and DSS/dextran weight ratio (q_{BSA} and q_{DEX} respectively)-(Figures 7, and Figure 8).

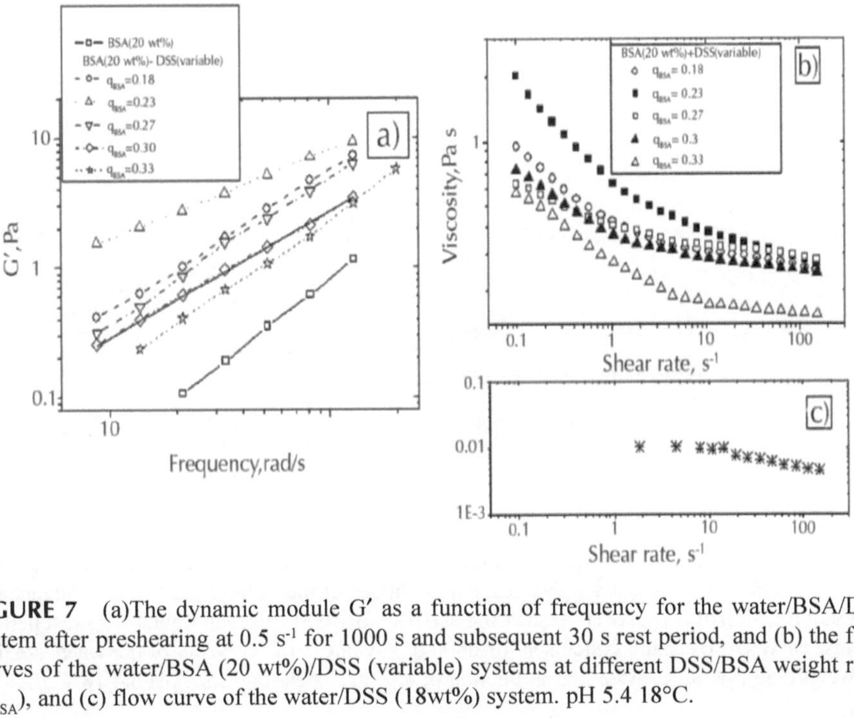

FIGURE 7 (a)The dynamic module G′ as a function of frequency for the water/BSA/DSS system after preshearing at 0.5 s^{-1} for 1000 s and subsequent 30 s rest period, and (b) the flow curves of the water/BSA (20 wt%)/DSS (variable) systems at different DSS/BSA weight ratio (q_{BSA}), and (c) flow curve of the water/DSS (18wt%) system. pH 5.4 18°C.

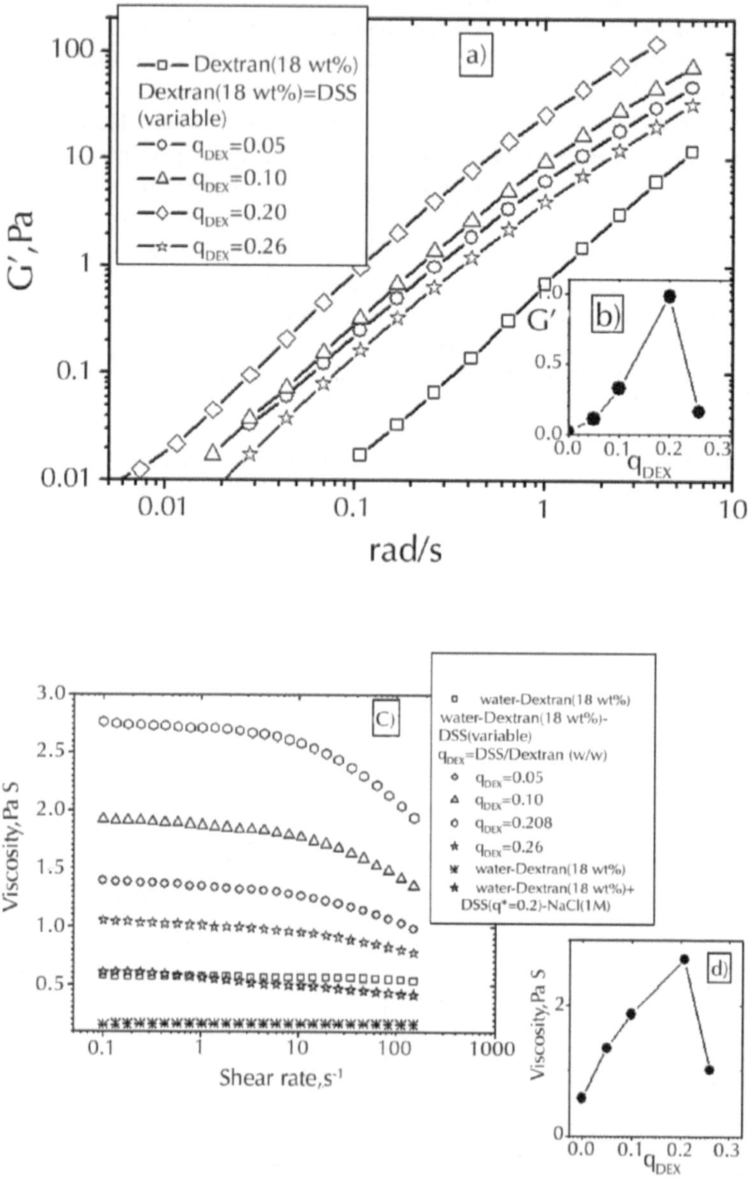

FIGURE 8 (a) The G′ values as a function of frequency for the water/dextran (18 wt%)/DSS (variable) system at different q_{DEX} values. (b) G′ values of the same system at a frequence 0.1 rad/s as a function of q_{DEX} . After preshearing at 0.5 s^{-1} for 1000 s and subsequent 30 s rest period. (c) flow curves of the same systems at different q_{DEX} values , (d) viscosity of the water/dextran (18 wt%)/DSS (variable) system as a function of q_{DEX} at a shear rate 1 s^{-1}. pH 5.4 18°C.

The dynamic module G′, and the viscosity of the water–BSA–DSS systems were measured as a function of the DSS/BSA weight ratio, q_{BSA} (Figures 7). As can be seen, (Figure 7 a,b,c) the G′, and the viscosity of BSA in presence of DSS depends pronouncedly on q_{BSA}, and shows a maximum at $q_{BSA} \cong 0.25$. At $q_{BSA} = 0.25$. The viscosity at shear rate 1 s^{-1}, and G′ values at 0.1 rad/s are more than 60 and 27 times higher, respectively, than those of the pure BSA solution. Unexpectedly, similar dependences of the G, and viscosity of the water/dextran/DSS system as a function of q_{DEX} were detected (Figure 8). These dependences show a maximum at $q_{DEX} = 0.2$. At $q_{DEX} = 0.2$, the viscosity at shear rate 1 s^{-1}, and G′ values at 0.1 rad/s are more than 17 and 57 times, respectively, those of the pure dextran solution.

From theory [39], we know that the dependences of G′ and of η for aqueous polymer system as a function of the concentration of the other polymer in the same solution is typical for the formation of inter-polymer complexes Therefore, it can be assumed that the dramatic changes in rheological behaviour of the water–BSA and water–dextran systems (Figure 7, and Figure 8) in the presence of DSS are due to interactions of DSS with BSA and with dextran.

Since intermacromolecular interactions takes place in the water–BSA–dextran-DSS system it will be useful to understand how such interactions affect the size particles, structure development and possible polarization of BSA in the presence of DSS.

The intensity size distributions functions for BSA (0.20 wt%), DSS (0.20 wt%), and their mixtures at different q_{BSA} values. pH 5.4. 25°C. are presented in Figure 9. It can be seen that at $q_{BSA} = 0.07$ and 0.14, when according to SALS data the water-BSA-dextran system is homogenized (Figure 1), the average size of the complex particles is only slightly higher than that of the DSS particles. It means that the size of the complex particles is determined mainly by the size of the DSS molecules.

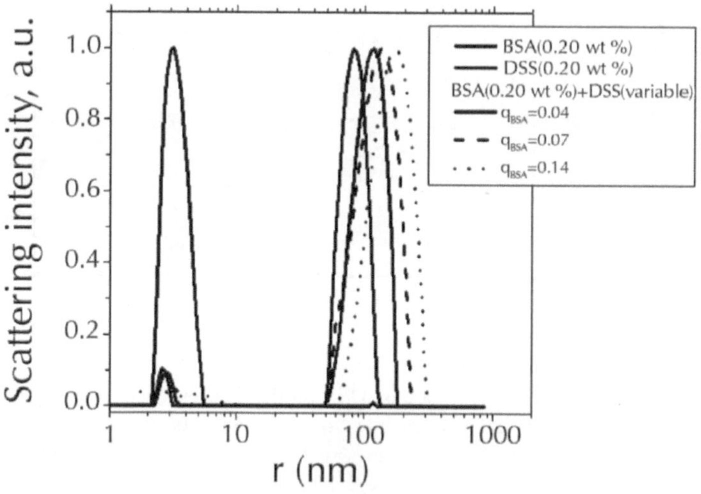

FIGURE 9 The intensity size distributions function for BSA (0.20 wt %), DSS (0.20 wt %), and their mixtures at different q_{BSA} values. pH 5.4. 20°C.

Such intermacromolecular interactions, can affect the structure of the BSA macroions and their isoelectric point due to polarization, and as consequence, change its compatibility with dextran. Let us to analyze this assumption in detail.

The measurements of the fluorescence intensity are frequently used to study possible structural changes of proteins in processes of their complex formation with other polymers. The changes of protein fluorescence may be characterized by the wavelength at the maximum emission (λ_{max}) and the maximum fluorescence intensity (I_{max}). The fluorescence intensity of proteins can be decreased by a variety of molecular interactions, including excited-state reactions, molecular rearrangements, energy transfer, ground state complex formation, and collision quenching [40].

The fluorescence spectra of pure BSA, and BSA in the presence of DSS are given in Figure 10. It is well-known that tryptophan (Trp) fluorescence of proteins varies with conformational changes of these biopolymers resulting in changing of fluorescence parameters, such as emission maximum (λ_{max}), quantum yield, lifetime, and others [40, 41]. The wavelength of maximum emission (λ_{max}) of pure BSA was found to be 339–340 nm. The emission maximum is usually shifted from shorter wavelengths to about 350 nm upon protein unfolding, which corresponds to the fluorescence maximum of pure tryptophan in aqueous solution.

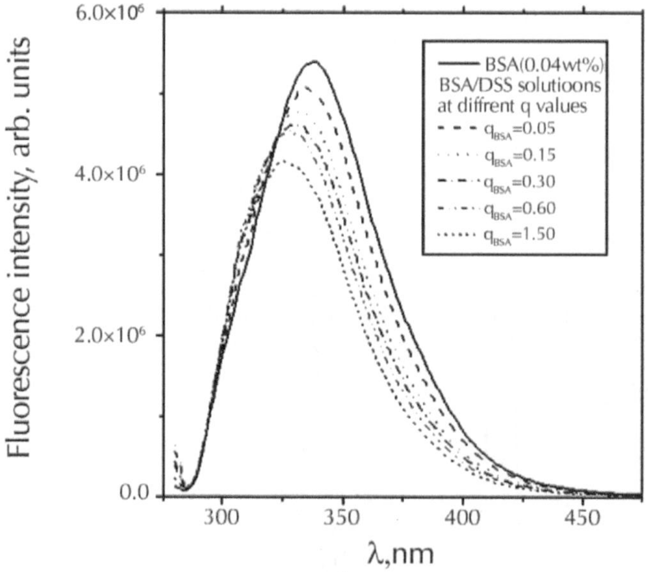

FIGURE 10 The effect of the presence of DSS on the fluorescence intensity of BSA at the excitation wavelength 270 nm. 20°C.Concentration of BSA = 0.04 wt%.

As shown in Figure 10 the fluorescence intensity (I_{max}) of BSA decreases (quenching) in the presence of DSS and λ_{max} shows a clear shift as the weight ratio of DSS/BSA increases in mixture. At the maximal q_{BSA} values, blue spectral shift of BSA fluorescence reaches λ_{max} = 324 nm, which suggests the screening of BSA tryptophanyls from water environment [41]. The BSA tryptophanyls become less accessible in water solution, which may be the result of increasingly tight binding of DSS with protein. It was interpreted these blue shifts as the result of shielding of tryptophan residues from aqueous media by the complexation of protein globules with DSS chains.

The most intriguing type of affinity is that of dextran and DSS shown in Figure 8. We did not find any literature reports on the existence of such type of complexes. However, there are indications for some interaction of dextran with polyampholytes. The possibilities of complex formation in the systems containing the charge and neutral polysaccharide have been discussed by Woodside and colleagues [22, 23], Grinberg and Tolstoguzov [24], and Antonov [42] on the basis of the analysis of the phase behavior of dextran-gelatin, and dextran-caseinate mixtures. Unfortunately most data in this field stem from studies performed a long time ago, when structural methods where less available [22-24, 44, 6]. Note, that it has been shown long ago [6] that at a low ionic strength most proteins accumulate in the dextran rich phase of water-dextran-poly(ethy1ene glycol) (PEG) system, whereas at a high ionic strength situation is reverse; proteins concentrate in the PEG enriched phase. In order to understand the origin of the interaction in the solutions containing dextran and DSS, the viscosity of the water/dextran/DSS systems was measured as a function of shear rate at different concentrations of NaCl (Figures 11). At ionic strengths equal and higher 0.14, the water/dextran/DSS system becomes insensitive to the presence of DSS. It means that affinity of DSS to dextran has electrostatic origin. In order to see how microstracture of dextran changes in the presence of DSS the ESEM images of water/dextran (15.4 wt%)/DSS (3.08 wt%) system (q_{DEX} = 0.2) were obtained (Figure 12).

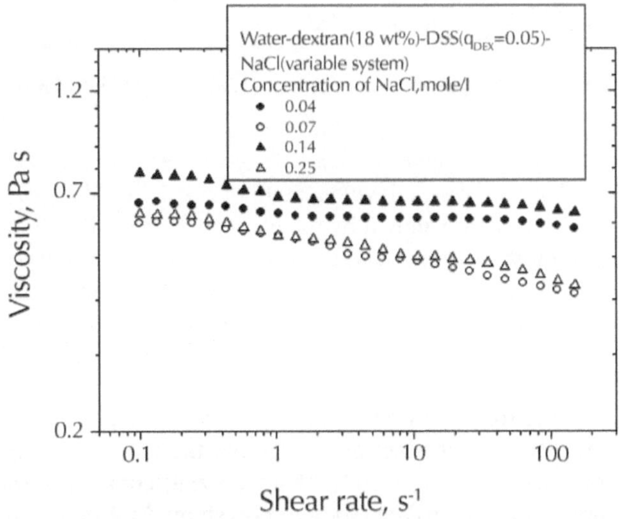

FIGURE 11 Flow curves of the water/dextran (18wt%)/DSS (q_{DEX} = 0.05)/NaCl (variable) systems. pH 5.4 18°C.

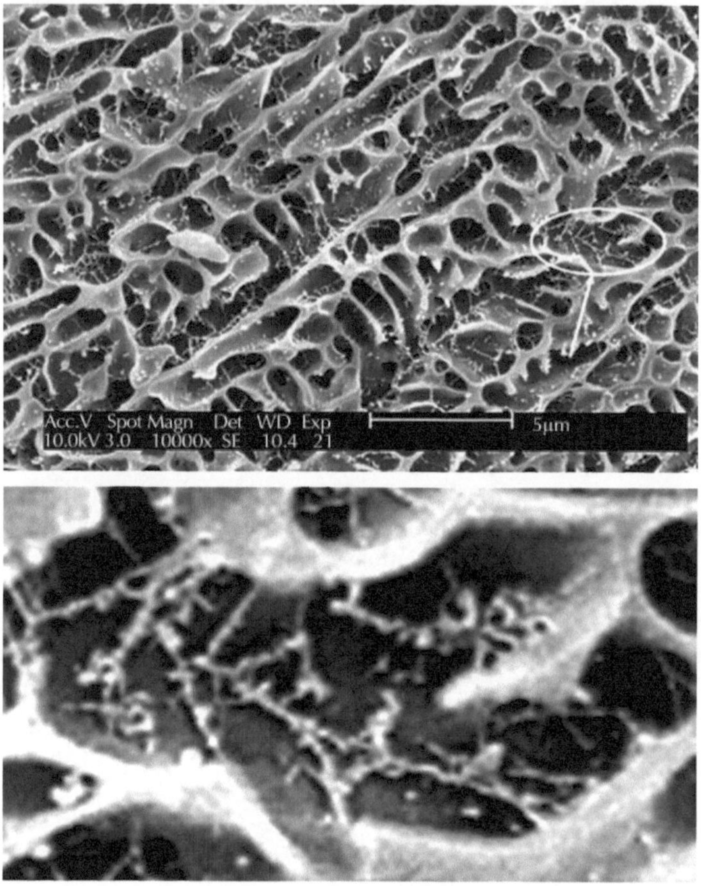

FIGURE 12 The ESEM images of water/dextran (15.4 wt%)/DSS (3.08 wt%) system (q_{DEX} = 0.2).

It can be seen that the system develops both the skeleton-like structure of dextran and the skeleton-like structure shown by the cobweb of the sulfate functional groups. On the enlarge photography we observed formation of some kind of weak network on the basis of negatively charged DSS and dextran. At present time it is difficult to say something definite about the origin of such interaction. Nevertheless, one can imagine that adding small amounts of DSS could modify the state of dextran in highly concentrated solutions (Figure 8,and Figure 12).

What is the driving force for DSS-induced mixing in water–BSA–dextran systems how protein-polyelectrolyte interaction affects the thermodynamic compatibility of BSA with dextran? Bowman et al. [43], characterizing complex formation between a negatively charged polyelectrolyte (sodium polystyrene sulfonate) and a negatively charged gelatin, suggested that the protein is polarized in the presence of strong polyelectrolyte. Therefore, if polarization of BSA in the presence DSS takes place, that is if surface and total charges of the BSA increase in the presence of DSS, then increased compatibility in BSA–dextran systems could be explained by the theory developed by Khoklov and coworkers [12,17]. Let us to consider this possibility.

Figure 13 shows the isoelectric focusing patterns BSA before and after adding different amounts of DSS. The BSA bands remains stable and do not move appreciably as the DSS/BSA ratio increases from 0.07 to 0.14. The results demonstrate that the presence of DSS does not leads to significant polarization of BSA in solution probably due to compact structure of BSA and screening of the BSA charged functional groups in the presence of DSS. Moreover, as it can be seen from Figure 10, interaction of DSS with BSA leads to screening of BSA tryptophanyls from water environment. Therefore polarization of BSA in the presence of DSS is unprobable.

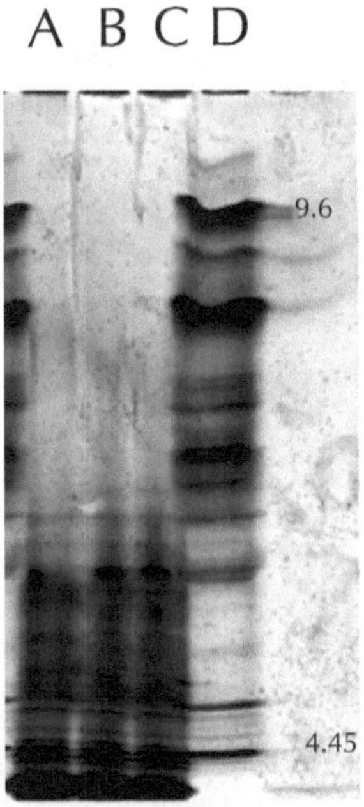

FIGURE 13 Iso-electric focusing patterns of BSA solutions before and after addition of different amounts of DSS. Concentration of BSA 0.5 wt%. (A)- pure BSA, (B,C) –BSA/DSS systems with q_{BSA} values 0.07, and 0.14, respectively, D- Standard protein kit.

Based on results, it is assume that the phenomenon of the DSS induced homogenization in water BSA-dextran emulsion is the result of the affinity of DSS molecules to both, BSA and dextran molecules. It seems, there is a clear analogy in the mechanism of compatibilization for polymer blends by diblock copolymers and homogenization of aqueous BSA/dextran emulsion in the presence of sulfate polysaccharide.

8.4 CONCLUSION

We established experimental evidence for mixing of aqueous concentrated BSA/dextran system at pH 5.4 (slightly above the isoelectric point of BSA) in the presence of a strong polyelectrolyte, DSS, for the DSS/BSA weight ratio $q_{BSA} \geq 0.07$. Homogenization leads to a noticeable increase in viscosity and module (G). The effect of mixing is reversible: increasing the ionic strength leads to phase separation of the water/BSA/dextran/DSS system. Increase in viscoelasticity is the result of interaction of DSS with the both macromolecular components of emulsion. Interaction of DSS with BSA leads also to the screening of BSA tryptophanyls from water environment, and is not accompanied by the polarization of the protein, whereas the affinity of DSS to dextran results in increase of viscoelasticity of dextran + DSS mixtures and an appreciable change in microstructure of DSS/dextran mixture. The driving force for mixing in water-BSA-dextran system in the presence of DSS is the affinity of the strong polyelectrolyte to both macromolecular components of the emulsion. Thus, we have obtained a new compatible biopolymer mixture that exhibits favorable rheological performance.

KEYWORDS

- **Bovine serum albumin (BSA)**
- **Dextran (DEX)**
- **Dextran sulfate sodium (DSS)**
- **Poly(ethy1ene glycol) (PEG)**
- **Small angle light scattering (SALS)**

REFERENCE

1. Tolstoguzov, V. B. Functional properties of protein-polysaccharide mixtures. In: *Functional properties of food macromolecules*, S. E. Hill, D. A. Ledward, and J. R. Mitchel (Eds), Aspen Publishers Inc., Gaitherburg, Maryland (1998).
2. Xia, J. and Dubin, P. L. Protein-polyelectrolyte complexes. In: *Macromolecular Complexes in Chemistry and Biology*, P. L. Dubin, R. M. Davis, D. Schultz, and C. Thies (Eds), Springer-Verlag, Berlin, Chapter 15 (1994).
3. Antonov, Yu. A. *Applied Biochemistry and Microbiology* (Russia) Engl.Transl., **36**, 382–396 (2000).
4. Antonov, Y. A., Puyvelde, P. Van, and Moldenaers, P. *Biomacromolecules*, **5**, 276–283 (2004).
5. Antonov, Y. A. and Wolf, B. A. *Biomacromolecules*, **7**(5), 1582–1567 (2006).
6. Albertsson, P. A. *Partition of Cell, Particles and Macromolecules*, 3rd edn., John Wiley-Interscience, New York, (1986).
7. Antonov, Y. A., Grinberg, V. Y., Zhuravskaya, N. A., Tolstoguzov, V. B., *Carbohydr. Polym.*, **2**, 81–90 (1982).
8. Antonov, Y. A., Grinberg, V. Y., and Tolstoguzov, V. B. *J. of Texture studies*, **11**, 199–215 (1980).

9. Flory, P. J. *Principles of Polymer Chemistry*, Cornell University, Press, Ithaca, New York, (1953).
10. Scott, R. L. *J. Chem. Phys.*, **17**, 279–284 (1949).
11. Tompa, H. *Trans. Faraday Soc.*, **45**, 1142–1152 (1949).
12. Vasilevskaya, V. V., Starodoubtsev, S. G., and Khokhlov, A. R. *Vysokomolec. Soed.*, **29B**, 390–397 (1987).
13. Iliopoulos, I., Frugier, D., and Audebert, R. *Polym. Prepr. (Am. Chem. Soc, Div. Polym. Chem.)*, **30**, 371–273 (1989).
14. Piculell, L., Nilsson, S., Falck, L., and Tjemeld, F. *Polym. Commun.* **32**, 158–160 (1991).
15. Piculell, L., Ilioupolos, I., Linse, P., Nilsson, S., Turquois, T., Viebke, C., and Zhang, W. *Gums and Stabililisers for the Food Industry*, Oxford University Press, Oxford, **7**, 309 (1994).
16. Bergfeldt, K. and Piculell, L. *J. Phys. Chem.*, **100**, 5935–5940 (1996).
17. Khokhlov, A. R. and Nyrkova, I. A. *Macromolecules*, **25**, 1493–1502 (1992).
18. Johansson, H. O., Karlstrom, G., Mattiasson, B., and Tjerneld, F. *Bioseparation*, **5**, 269–279 (1995).
19. Antonov, Y. A., Pletenko, M. G., and Tolstoguzov, V. B. Vysokomolec. Soed., **29 A**(12), 2477–2481 (1987).
20. Antonov, Y. A., Lefebvre, J., and Doublier, J. L. *J. Appl. Polym. Sci.*, **71**, 471–482 (1999)
21. Antonov, Y. A. and Wolf, B. A. *Biomacromolecules*, **6**, 2980–2989 (2005).
22. Woodside, E. E., Trott, G. F., Doyle, R. J., and Fishel, C. W. *Arch. Biochem. Biophys.*, **117**, 125–133 (1966).
23. Woodside, E. E., Trott, G. F., Doyle, and Fishel, C. W. *Carbohydr.Res.*, **6**, 449–457 (1968).
24. Grinberg, V. Y. and Tolstoguzov, V. B. *Carbohydr. Res.*, **25**, 313–321 (1972).
25. Puyvelde, P. Van, Antonov, Y. A., and Moldenaers, P. *Food Hydrocolloids*, **17** 327–332 (2003).
26. Stokes, J. R., Wolf, B., and Frith, W. J. *J. Rheol.*, **45**, 1173–1191 (2001).
27. Wolf, B. and Frith, W.J. *J. Rheol.*, **47**, 1151–1170 (2003).
28. Utracki, L. A. *Polymer Blends Handbook.* 2nd Volume Set (2003).
29. Antonov, Y. A. and Moldenaers, P. *Biomacromolecules*, **10**(12), 3235–3245 (2009).
30. Antonov, Y. A. and Moldenaers, P. *Food Hydrocolloids*, **25**(3), 350–360 (2011).
31. Abbott, N. L., Blankschtein, D., and Hatton, T. A. *Macromolecules*, **24**, 4334–4348 (1991).
32. Abbott, N. L., Blankschtein, D., and Hatton, T. A. *Macromolecules*, **25**, 3917–3931 (1992).
33. Abbott, N. L., Blankschtein, D., and Hatton, T. A. *Macromolecules*, **26**, 825–828 (1993).
34. Foster, J. F. In Albumin Structure, Function and Uses; V. M. Roesnoer, M. Oratz, M. A. Rothschild (Eds.) *Pergamon*, Oxford, 53–84 (1977).
35. Kirschenbaum, D. M. *Anal. Biochem.* **81**, 220–246 (1977).
36. Lefebvre, J. *Rheol. Acta*, **21**, 620-625 (1982).
37. Djokpe, E. and Vogt, W. *Macromol. Chem. Phys.*, **202**, 750–757 (2001).
38. Yeung, C. and Herrmann, K. *Macromolecules*, **36**, 229 (2003).
39. Overbeek, J. T. G. and Voorn, M. J. *J. Cell. Comp. Physiol.*, **49**, 7–39 (1957).
40. Lakowicz, J. R. *Principles of Fluorescence Spectroscopy*, Springer Science/Business Media, New York, (2006).
41. Lakowicz, J. R. *Principles of fluorescence specytros*copy, Plenum, New York and London, p. 496 (1986).
42. Antonov, Y. A. *Thermodynamic compatibility of casein and soya bean globulins with acidic and neutral polysaccharides in aqueous medi*um. PhD thesis, Institute of Organoelement compounds, USSR Academy of sciences. Moscow, USSR (1977).
43. Bowman, W. A., Rubinstein, M., and Tan, J. S. *Macromolecules*, **30**, 3262–3270 (1997).

CHAPTER 9

DEVELOPMENT OF STABLE POLYMER-BITUMEN BINDERS FOR ASPHALTS PAVING MATERIALS

M. A. POLDUSHOV and Y. P. MIROSHNIKOV

CONTENTS

9.1 INTRODUCTION

Bitumen binders which have been used commercially for the road asphalts, roofing, and other pavements for many decades manifest several serious disadvantages. Low elasticity of bitumen results in brittleness and cracking at low temperatures while low softening point (SP) gives rise to low modulus and rutting in the summer season. The modification of bitumen with small portions of linear elastomeric polymers and block copolymers which demonstrate, as a rule, swelling or even solubility in bitumen helps to improve the low temperature properties but unable to resolve the most part of high temperature problems. Compared to these polymers, semi-crystalline polyethylenes (PE) having very low glass transition temperatures and high enough melting points could be reasonable alternative provided some obstacles are overcame. It is known that the PE/Bitumen pair is immiscible and has high interfacial tension in melt. Therefore, it is hard to prepare fine emulsion by mixing and to prevent coalescence and creaming of the PE phase in a quiescent state at high temperatures. It is clear that some surfactants must be added to this blend to solve the outlined problems.

This study represents one of such attempts focused on the increase of stability of the low density polyethylene (LDPE)/bitumen emulsions with the help of a graft copolymer compatibilizer synthesized *in situ* upon high temperature mixing of the main components in presence of reactive additives. Compared to plane bitumen, the samples of the binders obtained have superior phase stability and low and high temperature properties.

The aim of this chapter is to develop stable polymer-bitumen binders for asphalts paving materials and other applications. The compositions comprising conventional road bitumen, LDPE and low-molecular weight reactive additives were prepared in a one-step mixing procedure at 180°C. The dependences of basic low and high temperature properties and phase morphologies on binder composition have been investigated. The data obtained show that LDPE/bitumen graft-copolymers synthesized *in situ* upon blending which play the role of a compatibilizer produce strong positive effect on time stability of emulsions. Compared to virgin bitumen the modified compositions studied demonstrate higher elasticity at low temperatures and much higher SP.

Bitumen has been used for road construction for many decades. However, it has serious inherent disadvantages like high brittleness at low temperature and low SP [1]. The steady increase in high traffic intensity in terms of heavy trucks, and the significant variation in daily and seasonal temperature demand improved road characteristics. It has been a common current practice to modify bitumen by addition of moderate amounts of polymers [1-10]. The purpose of such modification is to achieve desired engineering properties such as increased shear modulus and reduced plastic flow at high temperatures and/or increased resistance to thermal fracture at low temperatures. Homopolymers, like high and low density PE [1-7] and polypropylene [8], as well as random and block copolymers, like ethylene-vinyl acetate, ethylene-propylene, styrene-butadiene copolymer [9], industrial rubbers butadiene-styrene rubber, ethylene propylene diene monomer (EPDM), and tyre reclaim [9, 10] have been used as bitumen modifiers. Unfortunately, the major obstacle to usage of polymer-modified bitumen in paving practice has been their tendency towards fast macro phase separation by coalescence and subsequent creaming under quiescent conditions at elevated temperatures.

The semi-crystalline PE are accounted to be good modifiers for bitumen [1-7]. The LDPE has glass transition temperature of about 100°C. Hence, being added to bitumen in the form of fine dispersion, it can increase elasticity of pavements and prevent cracking at low temperatures. From the other side, LDPE has high elastic modulus up to its melting point temperature which is about 122–125°C. Unlike non-cured elastomeric modifiers including styrene-butadiene-styrene (SBS) block copolymers, LDPE is not miscible with bitumen. This allows retaining high modulus of binders up to 80–90°C and preventing rutting. The economical and ecological factors are also of primary importance, because LDPE can be collected from waste.

However, these properties can be attained provided a proper effective compatibilizer is present during mixing [1-3]. Otherwise, it is hard to prepare fine emulsion and to avoid coalescence and creaming of the LDPE phase in a quiescent state at high temperatures. Hesp and Woodhams in their representative review [1] have analyzed a vast amount of literature concerning various stabilization mechanisms incidental to polyolefin-bitumen emulsions. Not attempting to observe innumerable techniques proposed to solve the problem of instability we agree with the point of view [1-3] that the systems in which graft copolymers LDPE/bitumen are formed *in situ* upon mixing at high temperatures are of primary interest.

In this chapter we are going to test two low-molecular weight additives which being added to the LDPE/bitumen melt blend can presumably react with both components at elevated temperatures with the formation of graft copolymers which can play the role of compatibilizers.

9.2 EXPERIMENTAL

9.2.1 MATERIALS AND PROCEDURES

The commercial materials used in this study were road bitumen BND 60/90 with characteristics listed in Table 1 and LDPE in the form of granules with density of 0.9185 g/cm^3 and melt flow index (MFI) = 2.0 g/10 min at 190°C. Two types of reactive additives enable to chemically react with both LDPE and bitumen at elevated temperatures with the formation of corresponding graft copolymers are also used in this work under the names Additive 1 and Additive 2.

TABLE 1 Properties of bitumen BND 60/90

Parameter	Value	
	Plain	**Annealed at 180°C for 3 hrs**
Softening point (°C)	57	63
Breaking point (°C)	– 11	– 8

TABLE 1 *(Continued)*

Needle penetration depth (0.1 mm)	78	–
Ductility (cm) at 25°C	55	–
at 0 °C	3,5	
Penetration index	(– 1) ÷ (+ 1)	–

The binders including virgin bitumen, bitumen/LDPE, and bitumen/LDPE/additive compositions were blended at 180°C for 3 hr in a propeller mixer operated at150 rpm. Thus prepared compositions were immediately poured into small cans made from aluminum foil and rapidly cooled with iced water to prevent coalescence and creaming of the LDPE droplets. These samples named "as prepared" (AP) were used for further treatment and investigations.

The hardness of the samples was measured on a Heppler consistometer at 21°C with the use of the standard conic indenter loaded with the weight of 0.25 kg for 1 min. The values of hardness were calculated in accord with the following formula:

$$D = \frac{4W}{2\pi h}$$

where W is a load in N and h is the depth of penetration of indenter in m. At least 10 parallel independent measurements were made for each composition.

The following procedure was adopted for measuring phase stability of binders. The AP samples in the cans were annealed at 180°C for 3 hrs and then quenched rapidly to room temperature for conservation of phase morphology. The cylindrical samples were released from the cans followed by measuring hardness of the top and bottom surfaces at 21°C. In an unstable emulsion, the LDPE droplets float upward and enrich the top layers with high modulus polymer. Therefore, at room temperature, the hardness of the sample at its top becomes much higher compared to that at its bottom. As far as well stabilized emulsion is concerned the LDPE droplets are dispersed uniformly through the volume of the sample, and the hardness of the top and bottom layers equalizes.

The SP [11] and breaking point [12] temperatures were measured in accordance with the corresponding Standards of the Russian Federation.

The elastic properties of binders in terms of permanent set were studied on a Heppler consistometer at 21°C in the following way. The cylindrical sample of 8 mm in diameter and initial height h_0 = 8 mm was loaded with the stress of 0.53 MPa for 30 sec in a compression mode with subsequent registration of the height of the deformed sample h_d. After that the load was removed and the sample was allowed to recover for another 30 sec, and the height of the sample after the rest h_r was measured. The permanent set value was calculated as follows:

$$\varepsilon_r = \frac{h_0 - h_d}{h_r - h_d} 100\%$$

Phase morphologies of the samples were studied in a transmitted light on a biological optical microscope MBI-6 equipped with a digital photo camera. Small portions of bitumen compositions were immersed in large excess of toluene for dissolution of the bitumen phase. Floated particles of LDPE were collected, washed again with toluene, dried and analyzed under the microscope. The digital photos were processed on a PC with a homemade program "Image" enabled to calculate particle size distribution curves together with nine other structural parameters of the systems.

9.3 DISCUSSION AND RESULTS

As it was mentioned earlier the method of measuring hardness of the top and bottom surfaces of binder samples was adopted as a representative procedure of estimation of the stability of emulsions. The same method was used to get information about efficiency of the reactive additives used in this study. First, optimal concentrations of the reactive additives needed compatibilization of emulsions were estimated. Figure 1(a) and Figure 1(b) show the dependences of the hardness of the top and bottom layers of the binder samples on concentration of the reactive additives for both AP (dotted lines) and annealed (solid lines) compositions. The AP samples do not display any creaming because they were quenched immediately after melt mixing. Therefore, irrespective of the content of the additives, the hardness of the top and bottom portions are the same within an experimental error.

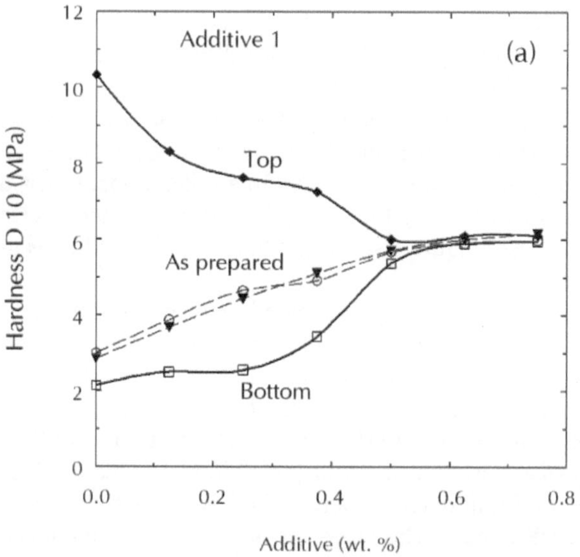

FIGURE 1 *(Continued)*

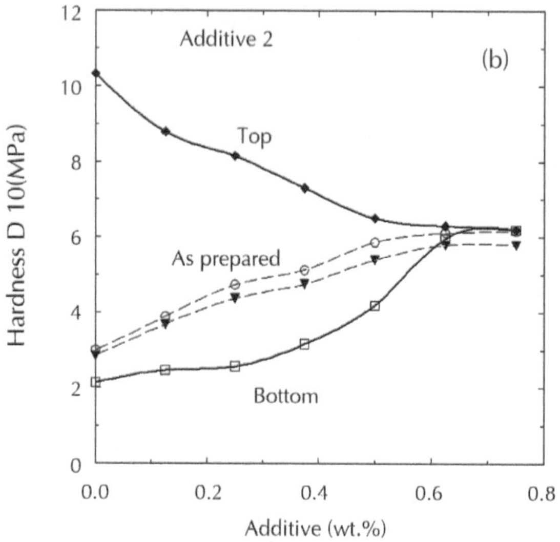

FIGURE 1 Hardness of the top and bottom layers versus concentration of reactive Additive 1 (a) and Additive 2 (b) in the samples of bitumen binders contained 5 wt% of LDPE measured at 21°C. Two similar curves in the middle of the graphs belong to AP blends. Two other curves depict the hardness of the top and bottom layers in the AP specimens annealed for 3 hr at 180°C.

A distinct stabilizing affect produced by the reactive additives is developed when these AP specimens were subjected to subsequent annealing for 3 hr at 180°C. At no or low content of the additives, annealing is followed by severe coalescence and floatation of the LDPE drops. As a result (Figure 1, solid curves), the hardness, *D*, of the upper and bottom layers of the samples differs markedly. Alternatively, when concentration of the additives is close to optimal 0.6–0.7 wt% the hardness of the top and bottom layers is practically equalized. These data indicate that bitumen compositions comprising 5 wt% of LDPE and 0.6–0.7 wt% of the additives retain stability for at least 3 hr of a quiescent storage at 180°C. In all likelihood, this is due to formation of graft copolymers which being located at the LDPE/bitumen interface play the role of compatibilizers. Also, Figure 1 show that both reactive additives used in this study manifest similar efficiencies.

Having measured optimal concentration of the reactive additives providing formation of a sufficient quantity of the compatibilizer, further studies on annealing kinetics of the AP emulsions of different compositions were made. After each 30 min of quiescent heating at 180°C, the emulsions in the cylindrical cans were quenched rapidly followed by measuring hardness of the top and bottom surfaces of the samples at 21°C. Figure 2(a) shows that the hardness of the plain bitumen is barely altered by annealing. At the same time, LDPE/bitumen binders without reactive additives undergo intensive creaming of the LDPE phase. Therefore, the hardness of the top layers of the samples increases rapidly. Appearingly, a complete phase separation after 80 min of heating takes place.

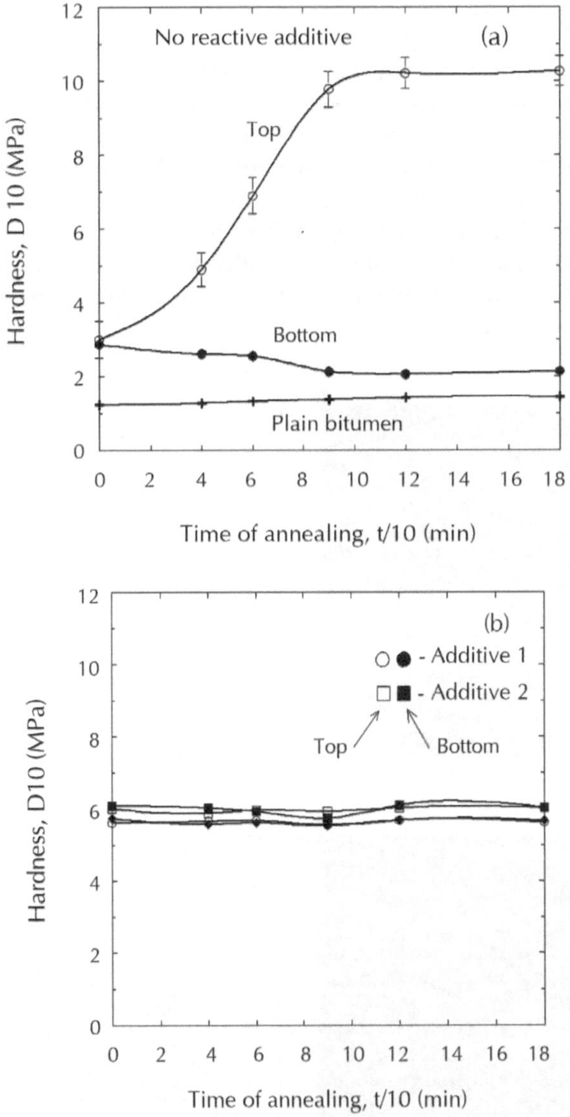

FIGURE 2 Effect of annealing time on the stability of LDPE/bitumen compositions at 180°C in the absence (a) and presence of 0.7 wt% (b) of the reactive additives.

In the presence of 0.7 wt% of the reactive additives (Figure 2(b)), the hardness of both top and bottom layers of the samples was comparable and did not change noticeably up to 180 min of annealing indicating formation of graft copolymers which increase stability of the emulsions considerably. Again, these data evidence an equal efficiency of both reactive additives.

9.3.1 MORPHOLOGY

Optical micrographs together with distribution curves in Figure 3 represent morphologies of the LDPE domains leached from the bitumen compositions. In the additive-free LDPE/bitumen binders the increase of the LDPE phase from 5 to 10 wt% is followed by the increase of the average particle sizes from 9 to 24 μm. The addition of 0.7 wt% of Additive 1 to a binder loaded with 5 wt% of LDPE (Figure 3(b)) resulted in substantial reduction of both domain sizes (to about 5 μm) and polydispersity. It has to be taken in mind that some portions of much smaller domains of LDPE can be formed in the compatibilized emulsions but they cannot be visualized because of low resolution power of optical microscopy.

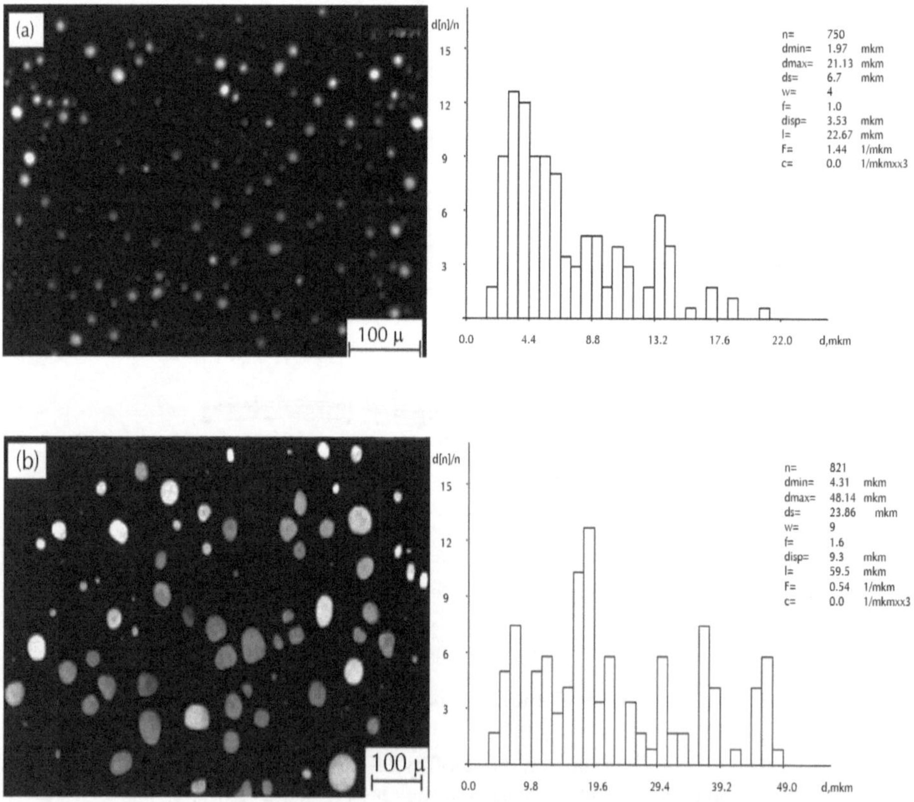

FIGURE 3 *(Continued)*

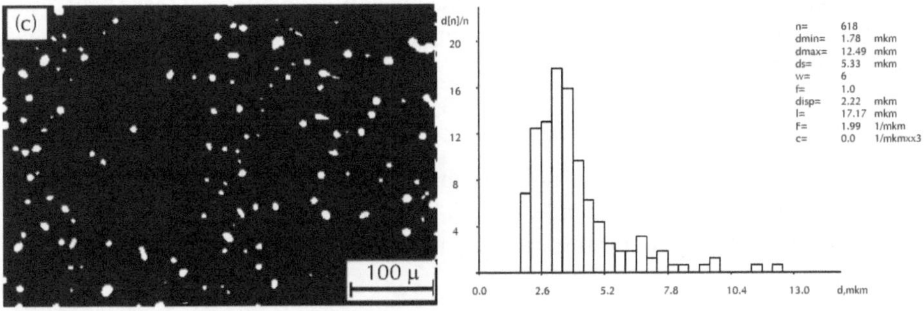

FIGURE 3 Optical photographs and corresponding particle size distribution curves for the LDPE domains leached from bitumen binders contained—(a) 5 wt% LDPE, (b)10 wt% LDPE, and (c) 5 wt% LDPE + 0.7 wt% Additive 1.

It was important to analyze the kinetics of mixing of bitumen compositions. Figure 4 shows that without reactive additive the equilibrium sizes of the LDPE droplets are not reached even after 120 min of mixing (curve 2). Resulting domain sizes remain pretty large. In the presence of the reactive Additive 1 (curve 1) blending cycle goes shorter, that is the domain sizes do not change noticeably after 80 min of mixing. Besides, average particle sizes go down to about 5 μm.

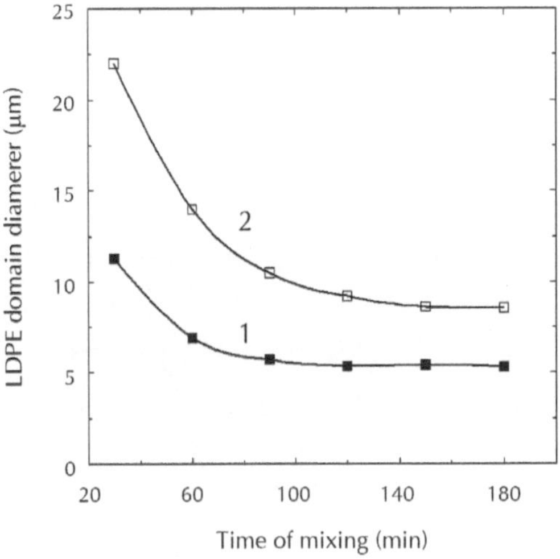

FIGURE 4 Kinetics of mixing at 180°C for bitumen binders containing 5 wt% of LDPE—(1) 0.7 wt% of Additive 1 is added and (2) no Additive.

These data indicate that mixing cycle of 180 min which was used through this study could probably be reduced to 80–100 min. In any case, in actual asphalt compositions containing inorganic fillers like sand, crush stone, and so on shear stresses acting upon mixing will be certainly higher than in unfilled binder. All these may result in reduction of the LDPE particle sizes, mixing time, rate of creaming, and can change final properties of asphalt pavements.

9.3.2 THERMAL PROPERTIES

The main motive of the usage of polymer modified bitumens is to enhance low and high temperature properties of binders. High brittleness and cracking at low temperatures can be reduced by enhancing elasticity of binders.

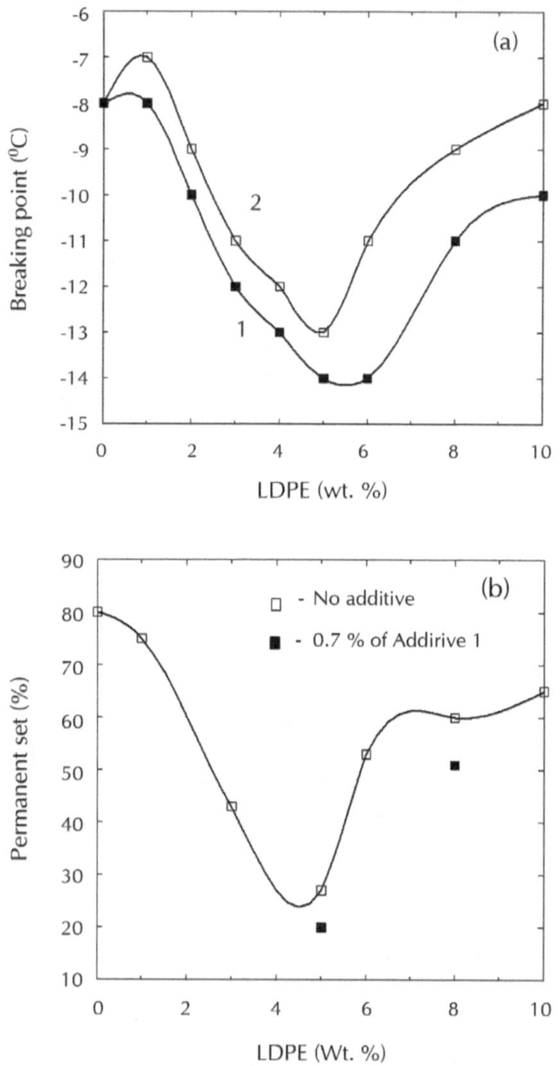

FIGURE 5 Breaking point (a) and permanent set at 21°C (b) versus LDPE content in bitumen—(1) 0.7 wt% of Additive 1 and (2) no Additive.

Figure 5(a) shows that the brittleness of the polymer modified bitumen is a strong function of the LDPE content. An optimal concentration of LDPE of approximately 5 wt% exists for both stabilized and non-stabilized binders which provides decrease of the breaking point from – 8 to – 14°C. The stabilized compositions demonstrate some superiority with respect to unstable blends over full range of LDPE loadings used.

As mentioned, reduction of brittleness is a consequence of enhancement of elasticity (ductility) of materials. Comparing data depicted in Figure 5(a) and Figure 5(b) one can see that minimum of brittleness corresponds to minimum of the permanent set (or residual strain). It should be pointed out that in contrast with moderate reduction of the breaking point a considerable four time increase of elasticity takes place at the optimal content of LDPE. The possible explanation of this effect can be as follows—At 5 wt%, PDPE forms finest and uniform dispersion which contribute to the elasticity itself and induces increase of the ductility of the bitumen matrix as well.

The high temperature properties of the samples were studied with the use of a standard SP procedure. As Figure 6 shows, almost linear increase of the SP with LDPE concentration takes place. Compared to plain bitumen with SP of 63°C a modified sample comprising 5 wt% of LDPE and 0.7 wt% of Additive 1 has SP of around 80°C.

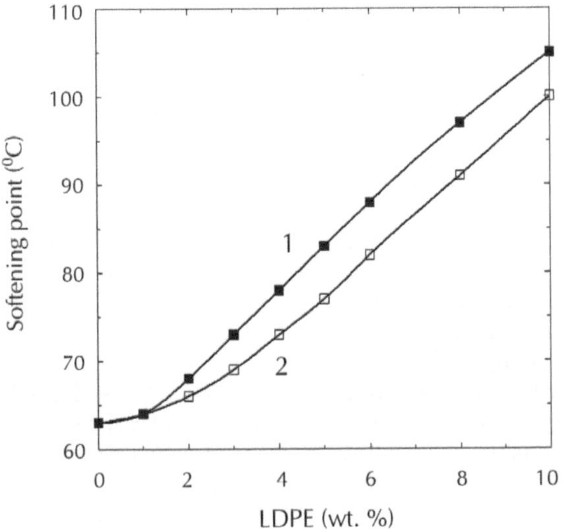

FIGURE 6 SPs of binders versus PDPE content—(1) bitumen + 5 wt% LDPE + 0.7 wt% Additive 1 and (2) bitumen + 5 wt% LDPE.

9.4 CONCLUSION

Since, the 1950's many authors have attempted to solve the problem of instability of polyolefin/bitumen emulsions by a variety of procedures. Accumulated experimental data and experience show that the LDPE/bitumen binders are among the most promising materials for the asphalt pavements. However, the major obstacle to usage of these compositions in paving practice has been their tendency towards gross phase separation under quiescent conditions at elevated temperatures. It has been known that effective compatibilizer of graft copolymer type which enables to interact with both phases through the interface can improve stability of such emulsions. This study represents an attempt focused on stabilization of the LDPE/bitumen emulsions with graft copolymer compatibilizer synthesized *in situ* upon high temperature mixing of the main components in presence of reactive additives. The formation of graft copolymers was proved by an indirect method based on measuring stability of emulsions at 180°C. It was found that 5 wt% of LDPE and 0.7 wt% of one of the additives mixed with bitumen provide superb stability in a quiescent state for more than 3 hr at 180°C. Compared to plane bitumen, the samples of LDPE-modified binders obtained display superior low and high temperature properties.

KEYWORDS

- **Homopolymers**
- **Low density polyethylene (LDPE)**
- **Polyethylenes (PE)**
- **Softening point (SP)**
- **Styrene-butadiene-styrene (SBS)**

REFERENCES

1. Hesp, S. A. M. and Woodhams, R. T. Stabilization mechanisms in polyolefin-asphalt emulsions, *polymer modified asphalt binders.* K. R. Wardlaw and S. Shuler (Eds.), American Society for Testing and Materials, Philadelphia, PA, pp. 1–19 (1992).
2. Liang, Z-Z., Woodhams, R. T., Wang, Z. N., and Harbinson, B. F. Utilization of recycled polyethylene in the preparation of stabilized, high performance modified asphalt binders, *use of waste materials in hot-mix asphalt.* H. Fred Waller (Ed.), American Society for Testing and Materials, pp. 197–210 (1993).
3. Woodhams, R. T. *Bitumen-polyolefin compositions*, U.S. Pat. US 4978698, (1990).
4. Hesp, S., Liang, Z., and Woodhams, R. T. *Bitumen-polymer stabilizer, stabilized bitumen-polymer compositions and methods for the preparation thereof.* U.S. Pat. US 5280064, (1994).
5. Firoozifar, S. H., Alamdary, Y. A., and Farzaneh, O. *Investigation of novel methods to improve the storage stability and low temperature susceptibility of polyethylene modified bitumens.* Petroleum & Coal, **52**(2), 123–128 (2010).
6. Little, D. N. Analysis of the influence of low density polyethylene modification (novophalt) of asphalt concrete on mixture shear strength and creep deformation potential, *Polymer Modified Asphalt Binders.* K. R. Wardlaw and S. Shuler (Eds.), American Society for Testing and Materials, Philadelphia, PA, pp. 186–203 (1992).
7. Moran, L. E. *Polyethylene modified asphalts.* U.S. Pat. US 4868233.
8. Nekhoroshev, V. P., Nekhorosheva, A. V., Popov, E. A., and Gossen, L. P. Influence of the products of chemical modification of atactic polypropylene on properties of bitumen binders. *Rus. J. Applied Chem.*,**74**(8), 1368–1373 (2001).

9. Isacsson, U. and Lu, X. Characterization of bitumens modified with SEBS, EVA and EBA polymers. *J. Mater. Sci.*, **34**, 3737–3745 (1999).
10. Arzamastsev, S. V., Chechulin, D. V., Artemenko, C. E., Ionov, I. A., and Patronov, V. P. Polymer modified industrial bitumens, *Plast. Massy*,(11), pp. 40–44 (2004).

CHAPTER 10

STRONG POLYELECTROLYTER-EFFECT ON STRUCTURE FORMATION AND PHASE-SEPARATION BEHAVIOR OF AQUEOUS BIOPOLYMER EMULSION

Y. A. ANTONOV and PAULA MOLDENAERS

CONTENTS

10.1 INTRODUCTION

Rheo-small angle light scattering (SALS), optical microscopy (OM), rheology, phase composition analysis, and environment scanning electron microscopy (ESEM) were applied to characterize the effect of strong polyelectrolyte on the structure formation and phase-separation behavior in aqueous sodium caseinate sodium alginate (W-SC-SA) water-in-water emulsion with the droplet morphology in the concentration ranges both near and far from the critical point (CP). The addition of 0.04–1.5 wt% of dextran sulfate sodium (DSS) salt to the emulsion placed on the phase diagram close to the CP leads at pH 7.0 and ionic strength (I) = 0.002 to a considerable decrease in the compatibility of sodium caseinate (SC) with sodium alginate (SA), disappearance of the droplet morphology, change in solvent distribution between coexisting phases, and a dramatic increase in moduli (G' and G'') and viscosity, due to formation of water soluble intermacromolecular SC-DSS associates with the pronounced network structure. The maximal relative changes in rheological properties takes place at the 10/1 "molar" ratio SC/DSS, and at 6–10 wt% content of SC in the water-in-water emulsion. Beta casein interacts with DSS in less degree than other casein fractions. Near the CP of phase separation, when the concentration of SC is not so high (6 wt%), the complexes are formed and stabilized *via* electrostatic interactions, but at higher SC concentrations the contribution of secondary hydrogen bonds and hydrophobic interactions becomes obvious at a higher ionic strength (I = 1.13). An addition of DSS to the water-in-water emulsion placed on the phase diagram far from the CP leads to reinforcement of the SC-DSS network, partial inclusion of SA in the SC-DSS network, and as a consequence, increases in cosolubility of SC with SA. The molecular origin of the unusually strong interaction of SC with DSS at a high concentration range on the "wrong side" of pH values, far from isoelectric point of caseins is discussed. The experimental observations suggest that use of small amount of sulfate polysaccharide in aqueous water-in-water biopolymer emulsions is a promising tool for regulation their structure, phase behavior, and rheological properties.

In biological and food systems, polysaccharides and proteins are often jointly present. The mixtures of these biopolymers are very common in daily life, since they are used in the food, pharmaceutical, and cosmetics industries for the manufacture of a variety of products [1-4]. They are very often present in dairy products in which casein constitutes indeed the major protein component [5]. The mixtures of these natural polymers phase separate at definite conditions [6-8] and this is used to create different morphologies and structures with specific properties. Although the phase separation and morphologies of many water-in-water biopolymer emulsions have been studied extensively, and there is a number of reviews concerning this subject [7-10], a little it is known how to adjust thermodynamic compatibility of biopolymers. This especially concern protein-acid polysaccharide systems the phase state of which is not sensitive to changes in pH, ionic strength and temperature under quiescent conditions [7, 8]. It has been demonstrated [11] that single-phase water - gelatin–pectin system containing interacting macromolecules undergo phase separation in the presence of dextran. The phase separation of such system was explained by the blockage of the functional groups of gelatin due to their competitive interactions with dextran. In general, the types of biopolymer interactions can vary widely due to substantial variations in biopolymer structure and solvent conditions (pH, ionic strength, and temperature) [12-15]. The recent study shows [16] that intermacromolecular interactions caused by the presence of complexing agent in two-phase biopolymer mixture can affect the phase equilibrium in semidilute biopolymer mixtures.

The aim of the present study to characterize the effect of strong polyelectrolyte (dextran sulphate) on the structure formation and phase-separation behavior in water-in-water casein-alginate emulsion by rheo SALS, OM, rheology, phase analysis, and ESEM, and to isolate physico-chemical parameters which control thermodynamic compatibility and morphology development in such system. An aqueous two-phase sodium caseinate/sodium alginate system (W-SC-SA) located close and far from the CP of the phase diagram were chosen to achieve phase equilibrium changes with DSS salt at pH 7.0, above the isoelectrical point of caseins. These materials have been chosen because of their relatively large difference in refractive index of the coexisting phases which provides a large contrast for the SALS experiment. It is important to note that complex formation of SC with DSS in dilute and semidilute solutions has been studied for the first time long ago [17] and recently in details [16, 18, 19].

Since, the W-SC-SA mixtures are two phase ones at a high concentrations of biopolymers independently on in pH, ionic strength, and temperature under quiescent conditions [7, 8] as well as under shear flow [20] the effect of demixing that can be reached for this system can easily be reproduced for other water-in-water emulsions in which the phase state is more sensitive to physicochemical parameters. The casein-alginate system is typical with regard to thermodynamic behavior and it is relevant to applications in the food industry for its textural and structuring properties [4, 21, 22]. Alginate is an anionic polysaccharide consisting of linear chains of (1–4)-linked ß-D-mannuronic and-α-L-guluronic acid residues. These residues are arranged in blocks of mannuronic or guluronic acid residues linked by blocks in which the sequence of the two acid residues is predominantly alternating [23]. Casein is a protein composed of a heterogeneous group of phosphoproteins organized in micelles. The SC is a salt of casein, obtained by disrupting the supra-molecular organization. This means that the polymer is a random coil and that the internal structure of the micelles is lost [21, 24, 25]. The isoelectric point is around pH = 4.7–5.2 [26]. The caseinate at neutral pH is thus negatively charged, like alginate, and DSS. The thermodynamic [7, 8, 21] and rheological [21] behavior of the ternary water–caseinate–alginate systems is known from literature.

Most experiments were performed in the very dilute phosphate buffer (ionic strength, $I = 0.002$), in the absence of other low molecular salts at 23°C.

10.2 MATERIALS AND METHODS

The caseinate at neutral pH is negatively charged, like alginate and DSS. The SC sample (90% protein, 5.5% water, 3.8% ash, and 0.02% calcium) was purchased from Sigma Chemical Co. The isoelectric point is around pH = 4.7–5.2 [26]. The weight average molecular mass (M_w) of the SC in 0.15 M NaCl solutions at pH 7.0 is 320 kDa [16]. The α and β-casein fractions from bovine milk, chromatographically purified were produced by Sigma. The medium viscosity sodium alginate, extracted from brown seaweed (*Macrocystis pirifera*), was purchased from Sigma. The M_w of the sample in 0.15 M NaCI, is 390 kDa. The DSS salt, DSS (M_W = 500 kDa, M_n = 166 kDa, intrinsic viscosity η (in 0.01 M NaCl) = 50 mL/g, 17% sulfate content, free SO_4 less than 0.5%) was produced by Fluka, Sweden (Reg. No. 61708061 A, Lot No. 438892/1).

10.2.1 PREPARATION OF THE PROTEIN/POLYSACCHARIDE MIXTURES

To prepare molecularly dispersed solutions of SC, SA, or DSS with the required concentrations, phosphate buffer (Na_2HPO_4/NaH_2PO_4, pH 7.0, and I = 0.002) was gradually added to the weighed amount of biopolymer sample at 25°C, and stirred, first for 1 hr at this temperature and then for 1 hr at 45°C. The solutions of SC, SA, and DSS were then cooled to 23°C and stirred again for 1 hr. The required pH value (7.0) was adjusted by addition of 0.1–0.5M NaOH or HCl. The resulting solutions were centrifuged at 60,000 *g* for 1 hr at 23°C, to remove insoluble particles. The concentrations of the solutions are determined by drying at 100°C up to constant weight. The ternary water–SC–SA systems with required compositions were prepared by mixing solutions of each biopolymer at 23°C. After mixing for 1 hr, the systems were centrifuged at 60,000 *g* for 1 hr at 23°C to separate the phases using a temperature-controlled rotor.

10.2.2 DETERMINATION OF THE PHASE DIAGRAM

The effect of the presence of DSS on the isothermal phase diagrams of the SC-SA system was investigated using a methodology described elsewhere [27]. The procedure is adapted from Koningsveld and Staverman, [28] and Polyakov, Grinberg, Tolstoguzov [29].The weight DSS/SC ratio in the system, (q) was kept at 0.14. The threshold point was determined as the point where the line with the slope − 1 is tangent to the binodal. The CP of the system was defined as the point where the binodal intersects the rectilinear diameter, which is the line joining the centre of the tie lines.

RHEO-OPTICAL STUDY

A rheo-optical methodology based on SALS during flow, is applied to study *in situ* and on a time-resolved basis the structure evolution. The light scattering experiments were conducted using a Linkam CSS450 flow cell with parallel-plate geometry. The CSS450 uses two highly polished quartz plates that are parallel to within 2um. Each plate is in thermal contact with an independently controlled pure silver heater utilizing platinum resistors sensitive to 0.1°C.

A 5 mW He-Ne laser (wavelength of 633 nm) was used as light source. The 2D scattering patterns were collected on a screen by semi transparent paper with a beam stop and recorded with a 10-bit progressive scan digital camera (Pulnix TM-1300). Images were stored on a computer with the help of a digital frame grabber (Coreco

Tci-Digital SE). The optical acquisition setup has been validated for scattering angles up to 18°. The gap between the plates has been set at 1 mm and the temperature was kept 23°C by means of a temperature controlled water bath. In house developed software was used to obtain intensity profiles and contour plots of the images (New SALS SOFT-WARE-K.U.L.). Turbidity measurements have been performed by means of a photo diode.

Microscopy observations during flow have been performed on a Linkham shearing cell mounted on a Leitz Laborlux 12 PolS optical microscope using different magnifications.

Rheological measurements were performed using a Physica Rheometer, type CSL2 500 A/G H/R, with a cone-plate geometry CP50-1/Ti ~diameter 5 cm, angle 0,993°, Anton Paar. The temperature was controlled at 23°C using a Peltier element. For each sample, flow curves were measured at increasing shear rate from 0.1 to 150 s^{-1}. Frequency sweeps ~0.1–200 rad/s was carried out as well for a strain of 3.0%, which was in the linear response regime. During the rheological measurements, the edges of the samples were covered with paraffin oil to avoid drying.

ESEM

The micro structural investigation was performed with the ESEM Philips XL30 ESEM FEG. The samples were freeze-fractured in freon and immediately placed in the ESEM. The relative humidity in the ESEM chamber (100%) was maintained using a Peltier stage. Such conditions were applied to minimize solvent loss and condensation, and control etching of the sample. Images were obtained within less than 5 min of the sample reaching the chamber. The ESEM images were recorded multiple times and on multiple samples in order to test reproducibility.

10.3 DISCUSSION AND RESULTS

10.3.1 WATER-IN-WATER EMULSIONS WITH COMPOSITION CLOSE TO THE CP—DSS-INDUCED DECOMPATIBILIZATION

The experimental results shown have been obtained on water (91.35 wt%)-SC (8.54 wt%)-SA (0.11 wt%) and water (91.9 wt%)-SC (7.9 wt%)-SA (0.19 wt%) two phase systems containing 3% vol. and 15% vol. alginate-enriched dispersed phase correspondingly (points A′ and A′′ on Figure 1).

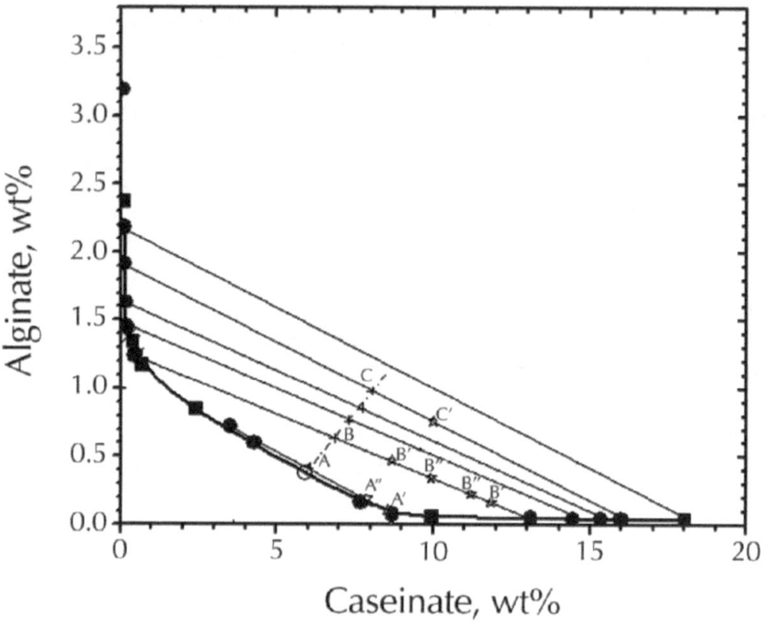

FIGURE 1 Isothermal phase diagram of water- SC-SA system at 23°C, pH 7.0, I = 0.002. • composition of coexisting phases; ▪- other points on the phase diagram; ○ critical point; A, B, C – points placed on the middle of the tie lines. A′, A′′, B′, B′′, B′′′, B′′′′, C′- compositions of the emulsions studied. A′, A′′- emulsions containing 3 wt% and 15 wt% sodium alginate enriched phase correspondingly. B′,B′′, B′′′,B′′′- emulsions containing 10%vol., 15%vol., 25%vol., and 35%vol. of the SA enriched phase C′- emulsion containing 35 % vol. alginate enriched phase.

To study the effect DSS on the phase equilibrium, a flow history consisting of three shear zones is used. First, a preshear of 0.5 s^{-1} is applied for 900 s (450 strain units) to ensure a reproducible initial morphology. Subsequently, this preshear is stopped, and the sample is allowed to relax for 30 s leaving enough time for full relaxation of deformed droplets. Then SALS patterns are monitored.

The evolution of the SALS patterns and the scattering intensity upon applying different amount of DSS are shown in Figure 2 (a) for the emulsion containing 3 wt% SA enriched phase, and in Figure 2 (b) for the emulsion containing 15 wt% SA enriched phase. In each experiment, a freshly loaded sample has been used. The presence of DSS in the emulsions led to appreciable increase of the SALS pattern (the pattern occupy more area around the bean stop) and accordingly, the light scattering intensity. This effect especially appreciable in the case of dilute (3%) emulsion (Figure 2 (a)). It is important to note that W-SC-DSS and W-SA-DSS systems remain homogeneous in the DSS concentration range studied (from 0.25 to 2.0 wt%). When the DSS concentration increases the SALS pattern and the scattering intensity of the system grows also. This indicates that the position of the binodal line on the phase diagram change in the direction to lower concentrations of biopolymers. If phase separation takes place, the driving force for decomposition must overcome the accompanying increase in interfacial free energy, which equals the product of the interfacial tension and the total interfacial area associated with the phase separation. The order of magnitude of the interfacial tension can be estimated from the scaling relation $\gamma = O\ (kT/\xi^2)$ [30, 31] where γ is the interfacial tension and ξ is the width of the region in which the concentration of the components differ from their bulk values in the coexisting phases.

The interfacial tension of the W-SC-SA emulsion, with composition close to the CP determined by a rheo-optical method, amounted to $\sim 10^{-8}$ N/m, and increases to a values of 5.2×10^{-6} N/m–8.8×10^{-6} N/m farther from the CP [21, 32, 33]. The presence of DSS in such emulsion changes the position of the system on the phase diagram deeply into the two phase range. Although direct measurement of the interfacial tension of W-SC-SA-DSS system by rheo-optical methods is difficult due to the great difference in viscosity of coexisting phases, it is logical to assume that decompatibilization of W-SC-SA system lead to similar increase the interfacial tension, because the magnitude of the interfacial tension depends on how far the system deviates from the CP [34, 35]. The presence of DSS in the water-in-water emulsions results in dramatic changes in its morphology. The corresponding microscopy images for the same concentrations DSS, and the same flow conditions are shown in Figure 3. One can see that the decrease in compatibility in the presence of DSS leads also to disappearance the droplet morphology.

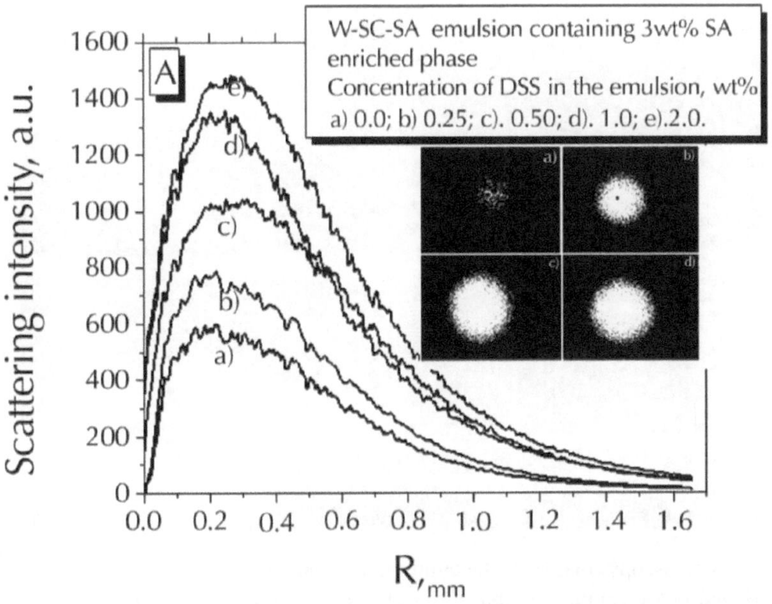

FIGURE 2 *(Continued)*

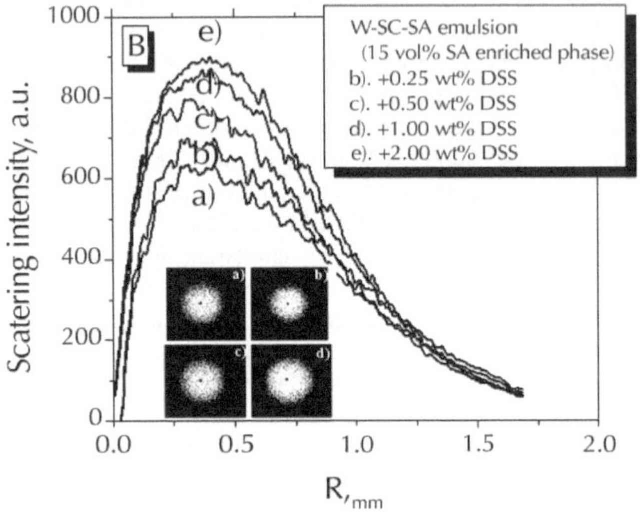

FIGURE 2 Effect of the presence of DSS on the SALS patterns and the scattering intensity of water-SC-SA emulsion containing 3% vol. sodium alginate enriched phase (point A' on Figure 1)-(a), and 15 wt% sodium alginate enriched phase (point A" on Figure 1)-(b). 23°C, pH 7.0, I = 0.002.Concentrations of DSS, wt%—(a) 0.0, (b) 0.25, (c) 0.5, (d) 1.0, (e) 2.0. Images obtained after stop of preshear at 0.5 s^{-1} for 900 s.

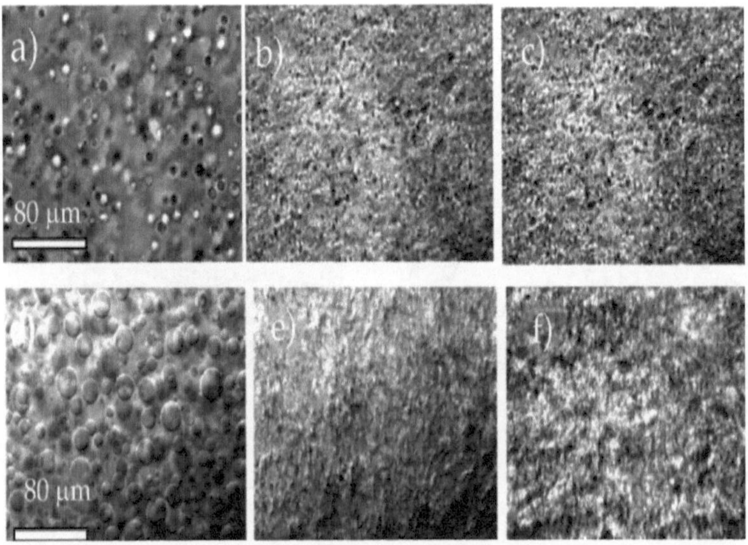

FIGURE 3 Microscopy images of the emulsions containing 3% (a), (b),(c) and 15% (d), (e), (f) SA enriched phase (point A' , and A" on the Figure 1) before and after addition of DSS. Concentration of DSS in the emulsion, wt%—(a), (d) – 0.0, (b), (e) -0.5, (c) ,(f)-2.0. Images obtained after stop of preshear at 0.5 s^{-1} for 900 s. 23°C, pH 7.0, I = 0.002.

10.3.2 WATER-IN-WATER EMULSIONS WITH COMPOSITIONS FAR FROM THE CP—DSS-INDUCED COMPATIBILIZATION

Figure 4 shows the effect of the presence of dextran sulfate on the phase diagram of SC-SA system at q = 0.14. One can see that DSS greatly effect phase separation. At low concentration of the biopolymers compatibility decreases significantly. Inside two-phase range, close to the CP compatibility decrease a little, whereas, at higher biopolymers concentrations co solubility of alginate in SC enriched phase increase considerably. Point B on the phase diagram, Figure 4 present the two-phase W (92.53%)-SC (6.84%)-SA (0.63%) system. In the absence of DSS, the system separated into two coexisting phases, with preferential concentration of casein in the bottom phase (13.1 wt% SC and 0.05 wt% SA) and alginate in the upper phase(0.45 wt% SC and 1.24 wt% SA). The volumes of these coexisting phases were equal to each other. The presence of 1.0% DSS in the system led to a considerable decrease in volume of the SA-enriched phase (from 50 to 15% vol.). At the same time, the concentration of casein in the SC-enriched phase decreased significantly (10.5 wt% SC and 0.29 wt% SA). Both these facts give an indication of the increase in compatibility of SC and SA.

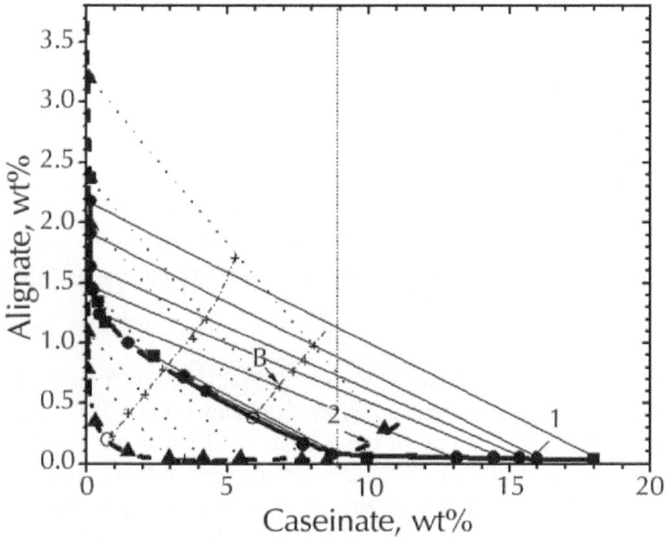

FIGURE 4 Effect of dextran sulfate on the phase diagram of W-SC-SA system.●, ▲-composition of coexisting phases, ■- other points on the phase diagram, ○ CP; + middle of the tie lines. ▬ 1 -Binodal line of the system in the absence of DSS; --- 2 –binodal line of the system in the presence of DSS. – Tie line of the system in the absence of DSS; tie line of the system in the presence of DSS.

The two-phase W (92.53%)-SC (6.84%)-SA (0.63%) system (point B on the phase diagram, Figure 1 was separated by centrifugation, and emulsions containing 15% vol. and 35% vol. of SA enriched phase were prepared and the morphological changes of these systems after an addition of 1.0 wt% DSS were studied (points B′- B′′′′ on Figure 1).

The scattering intensities of the system B" (15% vol. SA enriched phase), before and after addition of DSS, are shown in Figures 5. One can seen, that in the presence of DSS, the scattering intensity is observed mainly in the vicinity of the bean stop. The microscopy image obtained for this system shows absence of the dispersed phase.

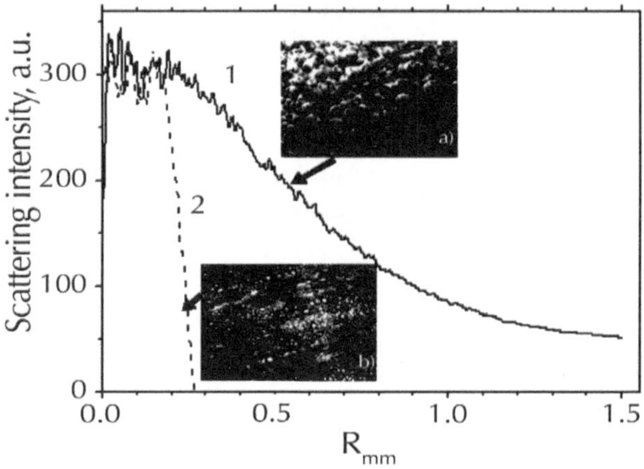

FIGURE 5 The scattering intensities of W-SC(11.2 wt%)-SA(0.23 wt%) emulsion containing 15 wt% SA enriched phase (point B' on Figure 1) as a function of the distance from the bean stop before (curve 1) and after addition of DSS at q = 0.14 (curves 2), and microscopy images the same systems (images a and b correspondingly). Images obtained after stop of preshear at 0.5 s^{-1} for 900 s. 23°C, pH 7.0, $I = 0.002$.

The morphology of such system is almost the same as the morphology of the SC enriched phase without DSS (images Figure 5(a) and Figure 5(b)). In other words, in the presence of DSS, ~15% vol. of the SA enriched phase is solubilized in the SC enriched phase. The presence of DSS also significantly affects the angle of tie line slope on the phase diagram characterizing change in concentrations of biopolymers in coexisting phases, and, consequently, water distribution between coexisting phases. Since the interplay of segregative and associative processes determines the shape of the phase diagram of the system, we can assume that at higher concentrations the contribution of a weak complexation becomes noticeable, and (or) the difference in the intensities of the interaction between SC and solvent and SA and solvent becomes smaller.

The morphology of the water-in-water W-SC (8.67 wt%)–SA(0.47 wt%) emulsions containing 15% vol. and 35-% vol. SA enriched phase F (points B" and B"" on Figure 1) after an addition of 1.0 wt% DSS and preshearing at 0.5 s^{-1} for 900 s is shown in Figure 6. It has an irregular lamellae type of morphology. After keeping the emulsion at rest for 40 min, the droplet morphology is partially recovered. The SC(10 wt%)-SA (0.76 wt%) emulsion containing the same 35% vol. SA enriched phase but with higher concentrations of biopolymers (point C′ on Figure 1), in the presence of 1.0 wt% DSS do not display the droplet morphology at least during 24 hr (Figure 6 (c)).

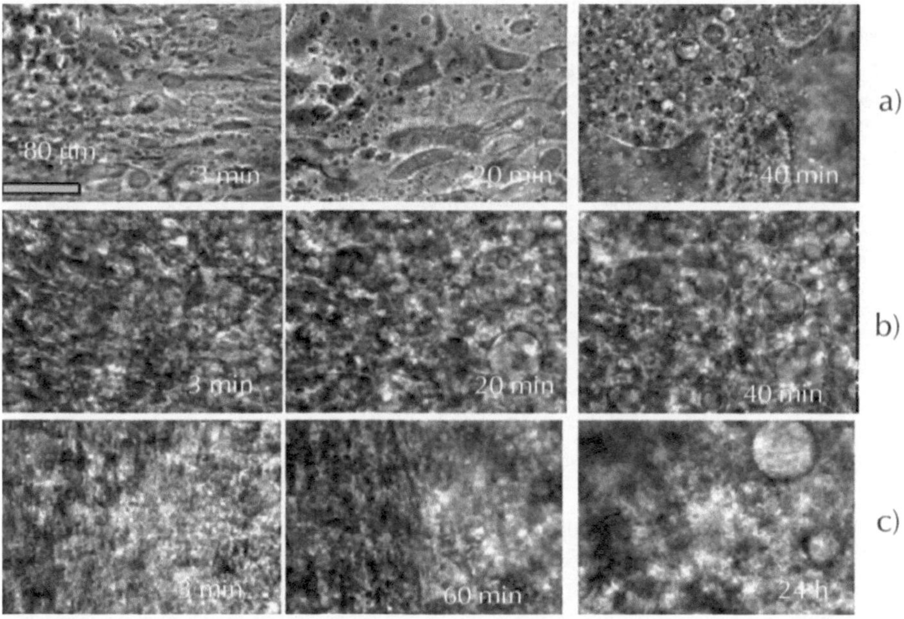

FIGURE 6 Microscopy images of the emulsions containing 15% vol. (point B" on Figure 1), and 35% vol. SA enriched phase (B"" and C' on Figure 1) after addition of 1.0 wt% DSS ((a), (b), and (c) correspondingly). Images obtained after stopping of preshear at 0.5 s^{-1} for 900 s. 23°C, pH 7.0, $I = 0.002$.

10.3.3 RHEOLOGY OF W-SC-SA WATER-IN-WATER EMULSION

For the rheological investigations of the two-phase W-SC-SA system, we separated the two-phase system (point B on the phase diagram, Figure 1) by centrifugation. The coexisting phases and the emulsion containing 10% vol. alginate-enriched dispersed phase were then characterized through their viscoelastic behaviors (point B′ on Figure 1). This emulsion is located in the two-phase region, far from the CP. The emulsion was hand-mixed prior to loading into the rheometer device. The temperature used in this study was 23°C, that is both phases were in a liquid state. The interfacial tension of the W-SC-SA system changes from ~10^{-5} to 10^{-6} N/m when the concentrations of the biopolymers decreases [27, 33]. The interfacial tension observed in our experiments was the same order of magnitude ($5 \cdot 10^{-5}$ N/m), similar to the results of [21, 33]. It can also be noted that the viscosities of the alginate-enriched and the casein-enriched phases at 23°C and a shear rate of 10 s^{-1} were 1.36 and 0.03 Pa s, respectively.

It has been shown [27, 33, 36] that at low shear rates, the biopolymer emulsion can be regarded as a conventional emulsion, and various structural models that are available in the literature for prediction of the morphology of emulsions can be used for aqueous biopolymer emulsions. First, the mixture was presheared at 500 s^{-1} for 900 s, in order to wipe out any previous mixing history. The sample was allowed to relax for 30 s, leaving enough time for full relaxation of the deformed droplets. Finally, the dynamic spectrum was measured in order to characterize the state of the emulsion. The results are presented in Figure 7. In the case of the investigated emulsion, the appearance of a "shoulder" in the low frequency region of G′ was noticed. A shoulder in the mechanical spectrum of a casein alginate system has been shown and studied in detail [21].

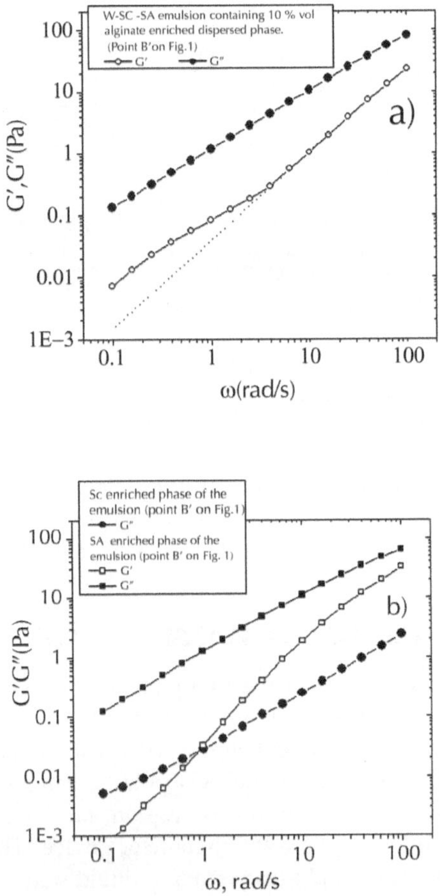

FIGURE 7 Dynamic moduli as a function of frequency of the W-SC (11.84%)-SA (0.17 wt%) system (point B′ on Figure 1) containing 10% vol. alginate- enriched phase and 90% vol. casein-enriched phase -(a) and dynamic moduli of the coexisting phases of this system, after shearing at 500 s^{-1} for 900 s and subsequent stop in shearing for 30 s. (b). 23°C, pH 7.0, $I = 0.002$.

The "shoulder" indicates the terminal relaxation time of the blend of liquids and is attributed to the shape relaxation of the droplets. Changes in size of the droplets induce a shift of the characteristic frequencies or relaxation times. It is important to note that our previous experiments shown [20] that the casein-alginate emulsion remains as a two-phase system over a wide concentration range of biopolymers, at all shear rate values measured (from 0.3 s^{-1} to 500 s^{-1}). This point is mentioned in order to clarify the comparison with other phase-separated biopolymer emulsions that undergo shear-induced mixing at shear rates above 60 s^{-1}.

10.3.4 EFFECT OF DSS ON RHEOLOGY OF SC-SA WATER-IN-WATER EMULSIONS

Figure 8 (a) and Figure 8 (b) presents the dynamic moduli as a function of frequency of the W-SC -SA -DSS emulsions (at a DSS/SC weight ratio, q = 0.14), containing 15%, 25% and 35 % vol. of the SA enriched phase (points B′′,B′′′, and B′′′′ on Figure 1), prepared after separation of the coexisting phases of W(92.53 wt%) -SC (6.84wt%)-SA (0.63 wt%) system (point B on Figure 1) (a) and the coexisting phases of this system (b), after shearing at 1.0 s^{-1} for 900 s and subsequent stop in shearing for 30 s. As reported in Figure 8 (a), in the presence DSS, a dramatic increase in the viscoelasticity of the emulsion takes place. The growth of viscoelasticity is especially outstanding at low frequency. Thus, at frequency values equal to 0.1 rad/s, G′ and G′′ values are more than 160 and 38 times, respectively, higher compared with those of the emulsion without DSS.

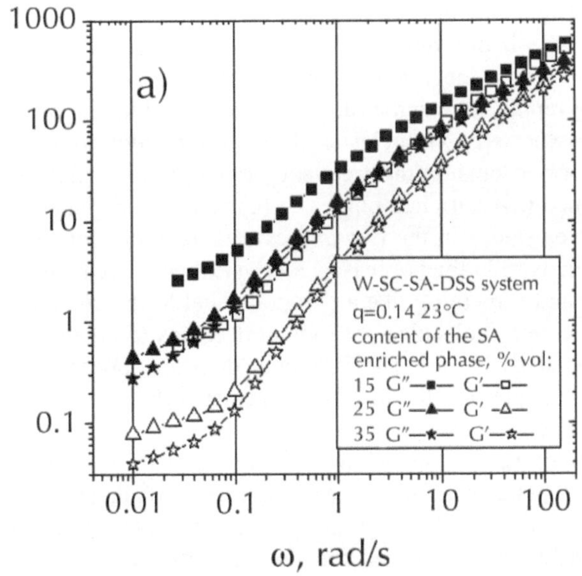

FIGURE 8 *(Continued)*

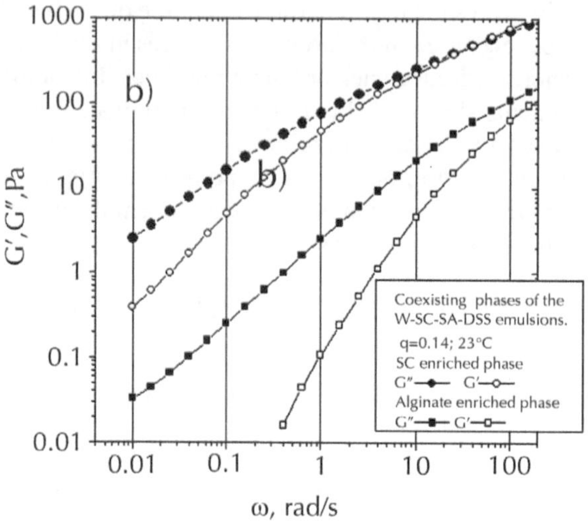

FIGURE 8 Dynamic moduli as a function of frequency of the water-SC -SA -DSS emulsions, containing 15, 25 and 35 % vol. of the SA enriched phase (point B", B'", and B"" on Figure 1)- (a) and dynamic moduli as a function of frequency of the coexisting phases of these emulsions, after shearing at 1.0 s^{-1} for 900 s and subsequent stop in shearing for 30 s.- (b). 23°C, pH 7.0, I = 0.002.

The morphology of such a system is characterized by the presence of lamellae (Figure 6 (a)). The behavior of the coexisting phases of the W-SC-SA emulsion in the presence of DSS is different; the SC-enriched phase is extremely sensitive to the presence of DSS, while the viscoelastic properties of the SA-enriched phase in the presence of DSS remain almost the same as before an addition of DSS. Both separated phases show as expected a viscoelastic plateau on the high frequency side. Actually, the plateau is located outside the frequency range, but the first G', G" crossover is clearly approached (SA enriched phase) or reached (SC enriched phase). These plateaus reflect the existence in the two phases of a network structure, but the network is not of the same type in the two cases; a classical entanglement network for the SA enriched phase, and something like a particulate gel for the SC enriched phase since, with one, hand casein molecules are too short chains to entangle and, on the other hand, they aggregated [37, 38]. The difference does not appear when just considering the spectra. For both phases, the time life of the network is short, since the viscoelastic plateaus begin above 10 rad/s.

The spectrum of the emulsion shows without doubt the existence of a low frequency plateau. Such a plateau would be the signature of a weak (small G' values) network structure at the emulsion scale (lower frequencies = larger structure scale).

Similar changes were observed for viscosity (data are not presented). The viscosities of the emulsions and the SC enriched phase in the presence of DSS are much higher than those without DSS.

10.3.5 INTERACTION OF DSS WITH THE PROTEIN AS A DRIVING FORCE OF DECOMPATIBILIZATION

To understand the reasons for such dramatic effects, the dynamic modules and the viscosity of the W-SC-DSS systems were measured as a function of the DSS/casein weight ratio, q (Figure 9 and Figure 10). As can be seen, the dependences of the modules and viscosity on q have an extreme character, with a maximum at q 0.14. At such conditions, the viscosity and G′ values are more than 2000 and 10^7 times, respectively, higher compared with those of the emulsion without DSS (Figure 10). The maximal relative changes of the viscosity and G′ in the presence of DSS takes place at SC concentration range from 5 wt% to 10 wt%. From theory, we know that such extremal dependences of physico-chemical properties are typical for the formation of inter-polymer complexes [39]. Therefore, it can be assumed that the dramatic changes in rheological and thermodynamical behavior of the casein-alginate system in the presence of DSS are due to interactions of casein molecules with DSS molecule.

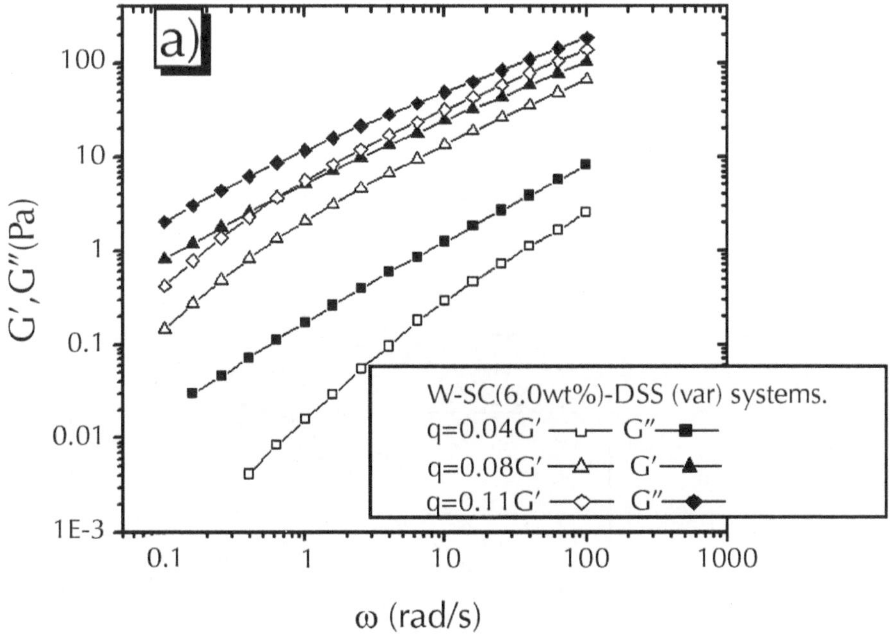

FIGURE 9 Dynamic moduli as a function of frequency of the water-SC(6.0 wt%) –DSS (var) systems at different q values, after shearing at 1.0 s^{-1} for 900 s and subsequent stop in shearing for 30 s. 23°C. (a) q = 0.04÷0.11, (b) q = 0.144÷0.3. 23°C, pH 7.0, I = 0.002.

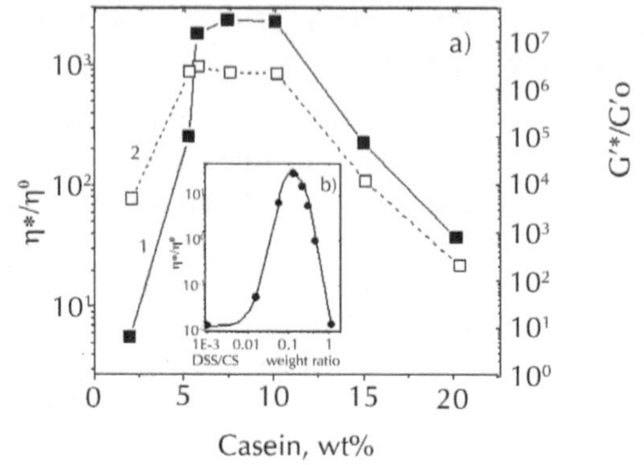

FIGURE 10 Ratio of viscosities of the W-SC-DSS/water-SC systems (η^*/η^0) as a function of SC concentration, determined at a shear rate of 1.0 s^{-1} and q = 0.14 (curve 1), ratio of the storage module of water-SC-DSS-DSS/water-SC systems (G'^*/G'^o) as a function of SC concentration, determined at 1.0 rad/s and q = 0.14 (curve 2) (a), and ratio of viscosities of the W-SC-DSS/W-SC systems as a function q, determined at a shear rate of 1.0 s^{-1} (b). 23°C, pH 7.0, $I = 0.002$.

The dynamic moduli as a function of frequency of aqueous SC (4.5 wt%)-DSS; casein (4.5 wt%) -DSS, and β casein (4.5 wt%)-DSS systems, and viscosity of the same systems as a function of the shear rate are shown in Figure 11. The G′, and G′′ values, and viscosities are maximal for the systems containing alpha casein and SC, whereas for the mixture containing beta casein these values are much smaller.

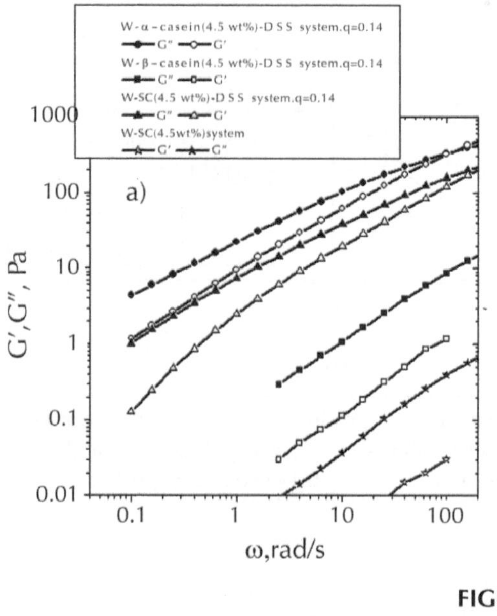

FIGURE 11 *(Continued)*

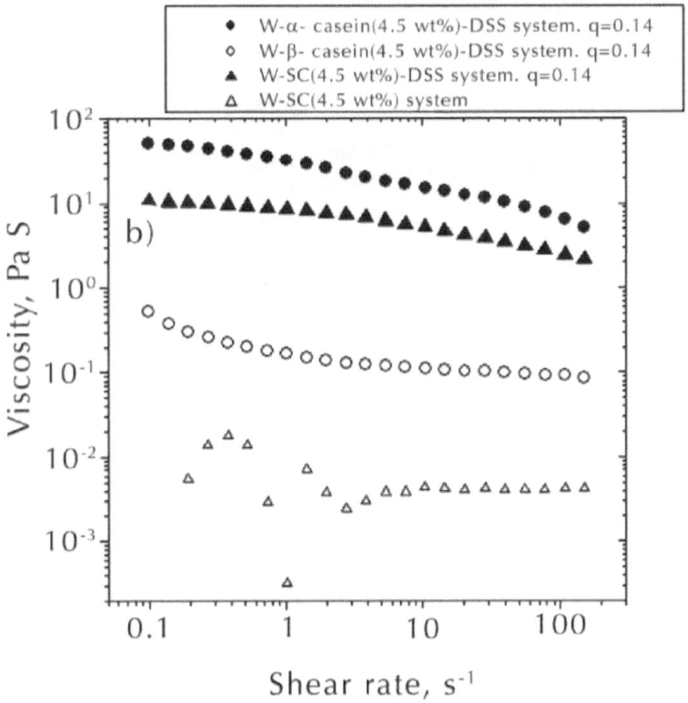

FIGURE 11 Dynamic moduli as a function of frequency of aqueous SC (4.5 wt%)-DSS, α casein (4.5 wt%) -DSS, and β casein (4.5 wt%)-DSS systems (a), and viscosity of the same systems as a function of the shear rate (b). q = 0.14, 23°C, pH 7.0, *I* = 0.002.

Since, the relative content of alpha and beta caseins in SC are 50% vol. and 35 % vol. correspondingly, and the most part of beta casein interact with DSS, we can reliably suppose that at concentrations studied the only smaller part of all casein fractions do not interact with DSS.

One can formulate the most important questions—what is the nature of intermacromolecular interaction of casein and DSS in concentrated solutions at pH 7.0, on the "wrong side" of the isoelectrical point, that is, for pH values above the iep of the protein under which the polyelectrolyte and the protein are like-charged?

The viscosity measurements of the ternary W-SC (6 wt%)- DSS systems as a function of the shear rate performed at different temperatures and in the presence of 1 M NaCl or 6 M urea are an attempt to answer to this question (Figure 12 (a)). An addition of NaCl in the initial buffer results in full insensitivity of the casein solutions to the presence of DSS in all the shear rate range studied (data are not shown). On the other hand, an addition of 1 M NaCl in the W-SC-DSS system results in a sharp decrease of the system viscosity to that of SC solution alone. This shows that the complexes are formed and stabilized *via* electrostatic interaction, rather than through hydrogen bonds formation or hydrophobic interaction.

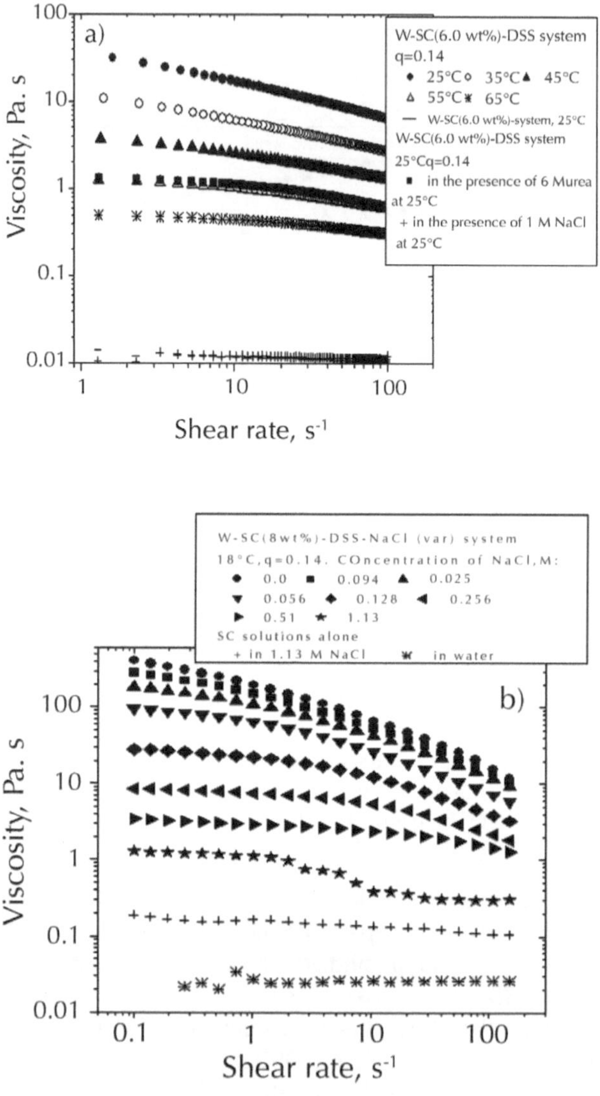

FIGURE 12 The flow curves of the single-phase W-SC (6.0 wt%)-DSS and W-SC (8.0 wt%)-DSS system and obtained at q = 0.14 at different temperatures and solvents. 23°C, pH 7.0, I = 0.002.

The viscosity of the W-SC-DSS system in the presence of 6 M urea is much smaller than that obtained in the absence of urea (in buffer), and it much higher than viscosity of the system in the presence of 1 M NaCI or viscosity of SC solution alone. These results indicate that urea prevent complex formation only partially. The viscosity is high enough and only reduces by half compared with that for the system in absence of 6 M urea. This gives proof that the presence of 6 M urea is not a factor, which completely weakens the SC-DSS interaction, and consequently the interaction is mainly electrostatic. On the other hand, the presence of an appreciable effect of the change of temperature from 25°C to 65°C on the viscosity values of the system (Figure 12 (a)) can be an indication of some role of hydrogen forces in the process of complex formation.

At higher concentration of SC in the system role of the secondary forces in DSS-SC interaction becomes stronger. Figure 12 (b) show the effect of ionic strength on the flow viscosity of W-SC-DSS system (q = 0.14) containing higher concentration of SC (8 wt%). It can be seen that at higher SC concentration in the ternary W-SC (8 wt%)-DSS system , viscosity values remains higher than those, determined for the W-SC(8 wt%) system even in the presence of 1,13 M NaCI, although ionic strength decrease considerably interaction between these macromolecules. The difference in viscosity values for the W-SC (8 wt%) and W-SC (8 wt%) reflect contribution of the secondary forces to complex formation. Similar changes were observed for mechanical spectra of W-SC-DSS system (q = 0.14) containing higher concentration of SC (8 wt%) (Data are not presented).

The total charge and the size of DSS molecule are greater than the total positive charge and size of the relatively small casein molecule 16 .This gives the possibility to regard complex formation between these biopolymers (similarly to other weak-polyelectrolyte–strong-polyelectrolyte interactions [40, 41] as a mononuclear association in which the DSS molecule is the nucleus and the casein molecule is a ligand. Therefore, the formation of SC–DSS complex associates can be regarded as the reaction of few casein molecules successively joining one molecule of DSS matrix. Taking into account the molecular weights of the biopolymers studied (300 kDa and 500 kDa respectively), and the fact that the maximal changes in G′,G′′, and viscosity of the W-SC-DSS system observe at a DSS/casein ratio equal to ~0.14 we can approximately evaluate the molar DSS/SC ratio in complex. Simple calculations give values about 1:10 (mole/mole).

It has been recognized that proteins may interact with polyelectrolytes even on the wrong side of iep, that is under pH were polyelectrolyte and the proteins are like-charged [42, 43, 44]. However, the data obtained in the present study cannot be explained only on the basis of known theories of polyelectrolyte reactions.

First of all, according to Dubin and co-workers, complex formation on the "wrong side" of the isoelectric point can be realized only in the vicinity of iep, whereas, caseins interact with DSS at pH values 7.0, that is far away from iep of caseins (4.4–4.6). Moreover, all the known theories assert that such complex formation can be realized only at a low ionic strength, less than 0.1–0.5, whereas, our data gives evidence of the strong interaction between SC and DSS even at a higher ionic strength (Figure 12). Finally, the main disagreement our data with the known approaches concern a possible contribution of the "concentration factor to the Coulomb interactions of two polyelectrolyte's at pH values above IEP. It is well known, from the theory of polyelectrolyte reactions that the increase of concentrations of the polymer components leads to their suppression of dissociation due to the rise in the electrostatic repulsion within inter- and intra-macromolecules. Moreover, increasing the total biopolymers concentrations favors the release of more counter ions in solution (Na^+ ions in the case of DSS-SC system), which screen the charges of the biopolymers, suppressing complex formation [45]. For these reasons, the binodal lines of the phase diagrams describing complex coacervation phenomenon in protein-polyelectrolyte systems are always vicious oval lines; at concentrations of biopolymers (~5 wt%) complex coacervation is suppressed even on the "right side" of pH values (see, for example, 45-7). However, in the case of W-DSS-SC systems, intermacromolecular interaction according to rheological data, (Figure 10) is strong in despite of electrostatic and steric repulsions, even at concentration of SC equal to 20 wt%. The main question is arises, what is the origin of such strong interaction on the "wrong side" of iep at neutral pH values? All data published so far demonstrate that protein-polyelectrolyte adsorption at low ionic strength becomes much stronger compared with the interaction of protein with linear anionic polyelectrolytes if the polyelectrolyte chains are grafted onto solid surfaces to form polyelectrolyte brushes. This phenomenon was named polyelectrolyte-mediated protein adsorption (PMPA) [48]. However, even in the case of formation of spherical polyelectrolyte brushes no adsorption with protein takes place at high (>0.35) ionic strength [48], whereas caseins interact with DSS even at higher ionic strength (>1.0). In the absence of any others reasonable explanations we suggest, that the driving forces of such interactions can be the presence of two factors which together create cumulative effect for interaction in W-SC-DSS and W-SC-SA-DSS systems. First one is a strong association of casein molecules in soluble DSS/SC complex at high concentrations.

At a high protein concentration, the contribution of the secondary hydrophobic and hydrogen forces becomes more significant, and this gives an additional influence on interaction and as consequence, the rheological behavior of DSS/SC systems. Second reason is a possible charge reversal that takes place during interaction of proteins with spherical polyelectrolyte [49, 50]. We assume, that similar to the systems described [49, 50], by suitable combination of the pH adjusted in the system and a low ionic strength, the local pH within the interacting layers of the protein may be lower than the iep of the protein. Hence, the net charge of the protein is reversed and a strong electrostatic attraction between unlike charged objects becomes operative. These points to the fact that the pH is an important but not a decisive parameter for complex formation in such systems. We named this phenomenon polyelectrolyte mediated protein association (PMPAS).

10.3.6 DISCUSSION ON THE STRUCTURE OF THE SC-DSS COMPLEXES AND SC ENRICHED PHASE OF THE DEMIXED W-SC-SA SYSTEM

The rheological behavior of the SC enriched phase of the W-SC-SA water-in-water emulsions at q = 0.14 and W-SC-DSS single phase systems show formation of a weak gel like structure which becomes more pronounced when the protein concentration increases (Figure 8, Figure 9, and Figure 12). Dynamic moduli as a function of frequency for such emulsions shows without doubt (Figure 8) the existence of a low frequency plateau that would be the signature of a weak (small G' values) network structure at the emulsion scale. This conclusion finds a good support in the data of ESEM for SC-DSS systems. Figure 13 ((a)–(d)) presents ESEM images of W-SC-DSS systems, obtained after quick freezing in liquid nitrogen, at concentrations of SC in the system 10 wt%

and 18 wt%. One can see that after addition of DSS in the SC solution, the system forms the pore network structure, the density of which increases at constant q values (and the size of the pores decreases) when the concentration of SC in the system is grows. Since the sizes of the SC-DSS network pores (2–4 μm) are much smaller than those of the droplets of the SA dispersed phase of W-SC-SA-DSS system (see Figure 3) we can assume that the disappearance of droplet morphology in the later system observed in Figure 3 is due to formation of the SC-DSS network, and partial inclusion of the droplets of the SA enriched phase into network of the SC enriched phase.

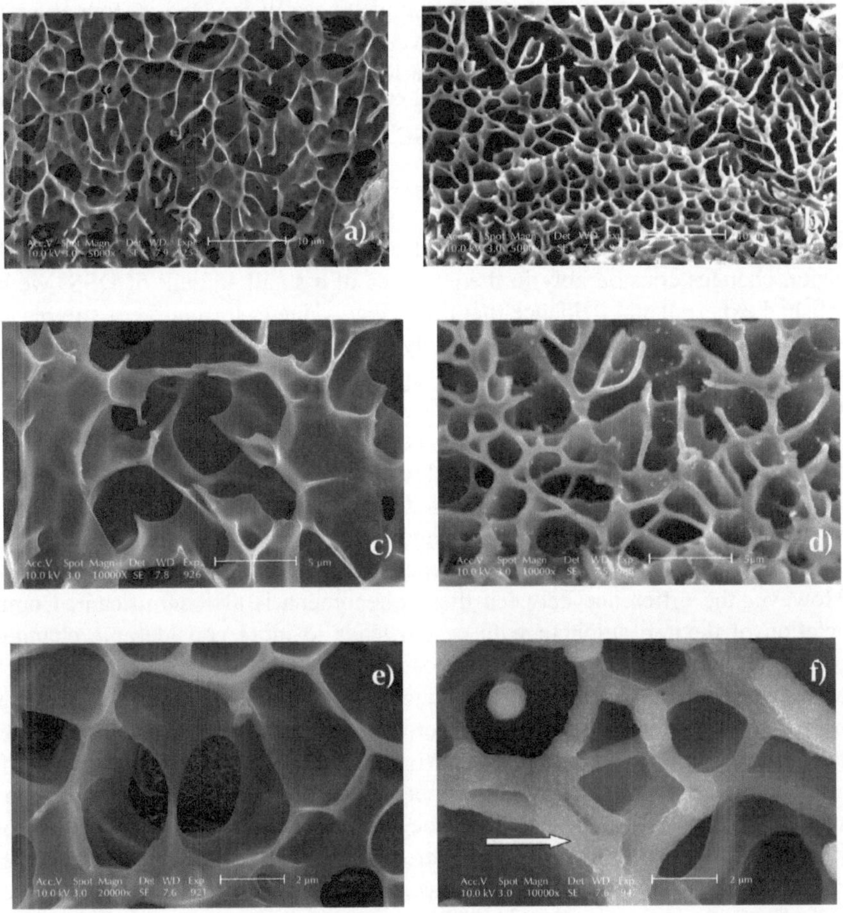

FIGURE 13 ESEM images of single phase W-SC-DSS systems (q = 0.14) obtained at different concentrations of SC in the system. ESEM images of two-phase W-SC-SA-DSS system at q = 0.14. (f) Concentration of SC—10 wt% (a,c,e) 18 wt% (b,d). Concentration of SC and SA in the W-SC-SA-DSS systems equal to 10 wt% and 0.29 wt% correspondingly. Arrow show possible places of SA absorption.Magnification—(a,b) – x5000; (c,d)- x10000. (e,f)-x20000. 23°C, pH 7.0, $I = 0.002$.

The phase behavior of the decompatibilized W-SC-SA systems, placed along the second time line (points B′-B′′′′ on Figure 1) is complicated. The solvent distribution between their coexisting phases changes dramatically (Figure 4). Approximately 15–20% vol. of the SA enriched phase becomes solubilized in the SC enriched phase (Figure 4). In order understand the reason of such phenomenon, two factors should be taken into account—formation of gel-like network in the presence of DSS at such conditions, and an appreciable cosolubility of SC in the SA enriched phase in the absence of DSS (0.5 wt%). We assume that in the presence of DSS, the SA enriched phase can absorbed partially into the SC enriched phase. This assumption find support in comparison of the ESEM images obtained for single phase W-SC-DSS system at $q = 0.14$ at concentrations of SC in the system 10 wt% (Figure 13 (e)) and SC (10.5 wt%)-SA (0.29 wt%) –DSS system at $q = 0.14$ (Figure 13 (f)). It can be seen that in the presence of SA, an appreciable reinforcement of the SC-DSS network becomes visible, which can be formed by absorption of SA in the SC enriched phase in the presence of DSS.

10.4 CONCLUSION

Concluding this work one can state that phase equilibrium in SC-SA water-in-water emulsion changes considerably in the presence of a small amount of DSS. We have established experimental evidence that this effect is due to formation of soluble complex associates between strong polyelectrolyte (dextran sulphate) and a net negatively charged ampholytic caseins. Our results suggest that at moderately high casein concentrations (≤ 6 wt%) electrostatic attraction is the main driving force for association. At higher protein concentration, the interaction becomes stronger, and the role of the secondary forces in stabilization of complex increases. This complex association and so called PMPA seems to be closely related phenomena—both can take place on the “wrong side” of the iep and the ionic strength within the system is one of the decisive factors. Both these phenomena can be traced back to the presence of positive patches. However the difference between these phenomena is also significant. Complex association of dextran sulphate with casein leads to more considerable changes in rheological behavior of casein-alginate system compared with PMPA, the interaction between casein and dextran sulphate strongly depends on the protein concentration in the system and it is less sensitive to increase in ionic strength at high protein concentration than PMPA. Accordingly, the driving forces of these phenomena are different, the Donnan pressure arising due to net release of numerous counter ions in the later one [48], and presumably, strong association of casein molecules and the protein charge reversal and in the former one. As the association of casein molecules at higher concentration is stronger, it can be the reason that the role of the secondary forces in complex formation becomes more significant.

The formation of the large SC/DSS associates leads to decrease their compatibility with alginate due to the effect of the excluded volume interactions between SA and DS + SC complex.

However, at high concentrations of these biopolymers (far from the CP) the presence of DSS results in the partial inclusions of the SA in the SC-DSS network, and as a consequence, increases in cosolubility of SC with SA. The interplay of segregative and associative processes, probably, determines the shape of the phase diagram of the system. At higher concentrations the contribution of a weak complexation becomes noticeable. In other words, the increase in the total protein + polysaccharide concentration is accompanied with the decrease of the parameter $\chi_{pr\text{-}ps}$.

The changes in phase equilibrium and rheological behavior of biopolymer water-in-water emulsions in the presence of sulfate polysaccharide can be used in protein processing in fiber-like materials, and phase separation technology. Protein molecules can be reversibly uptake by DSS at low ionic strength, and released by raising the amount of added salt. This allows to use systems based on sulphate polysaccharide and protein in biotechnology as a carrier of enzymes.

KEYWORDS

- **Biopolymer mixture**
- **Complex formation**
- **Decompatibilization**
- **Rheo-optics**
- **Structure formation**

REFERENCES

1. Kasapis, S., Morris, E. R., Norton, I. T., and Gidley, M. J. *Carbohyd. Polym*, **21**, 249–259 (1993).
2. Goff, H. D. Colloidal aspects of ice cream - a review, *Int. Dairy J.*, **7**, 363–373 (1997).
3. Bot, A., Mellema, M., and Reiffers-Magnani, C. K. *Ind. Proteins*, **11**, 11–13 (2003).
4. Tolstoguzov, V. B. Functional properties of protein–polysaccharide mixtures, S. E. Hill, D. A. Ledward, and J. R. Mitchel (Eds.), *Functional properties of food macromolecules*, Aspen Publishers Inc., Gaitherburg Maryland, pp. 252–277 (1998).
5. Swaisgood, H. E. Chemistry of milk protein, In P. F. Fox (Ed.), *Developments in dairy chemistry*, Appl. Sci. Publ., London, **1**, 1–59 (1982).
6. Piculell, L., Bergfeld, K., and Nilsson, S. In S. E, Harding, S. E. Hill, and J. R. Mitchell (Eds.), Factors determining phase behaviour of multicomponent polymer systems, *In Biopolymer Mixtures*, Nottingham University Press, Nottingham, pp. 13–36(1995).
7. Grinberg, V. Ya. and Tolstoguzov, V. B. *Food Hydrocolloids*, **11**, 145–158 (1997).
8. Antonov, Y. A. *Applied Biochemistry and Microbiology* (Russia), *Engl.Transl.*, **36**(4), 382–396 (2000).
9. Dickinson, E., and McClements, D. J. *Advances in food colloids* (chapter 3), Glasgow: Blackie (1995).
10. Harding, S. E., Hill, S. E., and Mitchell, J. R. (Eds.). *Biopolymer mixtures,* Nottingham University Press, Nottingham.
11. Antonov, Y. A. and Zubova O. M. *Int. J. Biol. Macromol.*, **29**, 67–71 (2001).
12. Dickinson, E., and Galazka, V. B. Emulsion stabilisation by protein±.polysaccharide complexes. In G. O. Philips, D. J. Wedlock, and. P. A. Williams (Eds.), *Gums and Stabilisers for the Food Industry*, Oxford: IRL Press., **4**, 351 (1992).
13. De Kruif, Cornelus G., Feinbrerk Fanny, de Vries R. *Opinion in Colloid and Interface science*, **9**, 340–349 (2004).

14. Kaibara K. T., Okazaki H., Bohidar, B., and Dubin, P. L. *Biomacromolecules*, **1**, 100–107 (2000).
15. Antonov, Y. A. and Wolf, B. *Biomacromolecules*, **6**, 2980–2989 (2005).
16. Antonov, Y. A., and Moldenaers, P. *Biomacromolecules*, ***10***(12), 3235–3245 (**2009**).
17. Gurov, A. N., Wajnerman, E. S., and Tolstoguzov, V. B. Starch/Stärke, **29**(6), 186–190 (1977).
18. Jourdain, L., Leser, M. E., Schmitt, C., Michel, M., and Dickinson, E. *Food Hydrocolloids,* **22**(4), 647–659 (2008).
19. Semenova, M. G., Belyakova, L. E., Polikarpov, Y. N., Antipova, A. S., and Dickinson, E. *Food Hydrocolloids*, **23**, 629–639 (2009).
20. Antonov, Y. A., Van Puyvelde, P., and Moldenaers, P. *Food Hydrocolloids*, **23**, 262–270 (2009).
21. Capron, I., Costeux, S., and Djabourov, M. Water in water emulsions: phase separation and rheology of biopolymer solutions. *Rheol. Acta*, **40**, 441–456 (2001).
22. Roberts, S. A., Kasapis, S., and Lopez, I. S. *Int. J. Food Sci. and Technology*, **35**, 227–34 (2000).
23. Whistler, R. L. *Industrial Gums, 2nd edn.*, New York: Academic Press (1973).
24. Farrer, D., and Lips, A. *Int. Dairy, J.*, **9**, 281–286 (1999).
25. Chu, B., Zhou, Z., Wu, G., and Farrell, H. M., *J. Colloid Interface Sci.*, **170**, 102–112 (1995).
26. Swaisgood, H. E. Chemistry of milk protein. In P. J. Fox (Ed.), *Developments in dairy chemistry*, Appl. Sci. Publ., London, **1**, 1–59 (1982).
27. Van Puyvelde, P., Antonov, Y. A., and Moldenaers, P. *Food Hydrocolloids*, **17**, 327–332 (2003).
28. Koningsveld, A. J. Staverman,. *Journal of Polym. Sci.* 1968, *A-2*, 305–323.
29. Koningsveld, A.J. Staverman, *Journal of Polym. Sci.* 1968, *A-2*, 305–323.
30. Polyakov, V. I., Grinberg, V. Ya., and Tolstoguzov, V. B. *Polymer Bulletin*, **2**, 760–767 (1980).
31. De Gennes, P. G. *Scaling Concepts in Polymer Physics*, Cornell University Press, Ithaca, NY, (1979).
32. Rowlinson, J. S. and Widom, B. *Molecular Theory of Capillarity*, Clarendon Press, Oxford, U.K., (1984).
33. Antonov, Y. A., Van Puyvelde, P., and Moldenaers, P. *Food Hydrocolloids*, **34**, 29–35 (2004).
34. Guido, S., Simeone, M., and Alfani, A. *Food Hydrocolloids*, **18**(3), 463–470 (2004).
35. Heinrich, M. and Wolf, B. A. *Polymer*, **33**, 1926–1931 (1992).
36. Scholten, E., Tuinier, R., Tromp, R. H., and Lekkerkerker, H. N. W. *Langmuir*, **18**, 2234–2238 (2002).
37. Simeone, M., Alfani, A., and Guido, S. *Food Hydrocolloids,* **18**, 463–470 (2004).
38. Lefebvre, J. and Doublier, J. L. *Chem. Inform*, **37**(9), 67 Pages - WILEY-VCH Verlag GmbH & Co. KGaA, Weinheim (**2006**).
39. Ruis, H. G. M. Structure-rheology relations in sodium caseinate containing systems PhD Thesis Wageningen University, The Netherlands ISBN: 978-90-8504-648–6 (2007).
40. Overbeek J. T. G. and Voorn M. J. J. *Cell Comp. Physiol.*, **49**, 7–39 (1957).
41. Sato, H., and Nakajima, A. *Colloid and Polymer Science,* **252**, 294–297 (1974).
42. Sato, H., Maeda, M., and Nakajima, A. *Journal of Applied Polymer Science*, **23**, 1759–1767 (1979).
43. Antonov, Y. A, Dmitrochenko, A. P, and Leontiev, A. L. *Int. J. Biol. Macromol.,* **38**(1), 18–24 (2006).

44. Antonov, Y. A. and Sato, T. *Food Hydrocolloids*, **23**(3), 996–1006 (2009).
45. Cooper, C. L., Dubin, P. L., Kaitmazer, A. B., and Turksen, S. *Curr. Opin. Colloid Interface Sci.*, **10**, 52–78 (2005).
46. Weinbreck, F., de Vries, R., Schrooyen, P., and. de Kruif, C. G. *Biomacromolecules*, **4**, 293–303 (2003).
47. Koh, G. L. and Tucker, I. G. *J. Pharm. Pharmacol.*, **40**, 233–236 (1988).
48. Plashchina, I. G., Zhuravleva, I. L., and Antonov, Y. A. *Polymer Bulletin*, **58**, 587–596 (2007).
49. Wittemann, A. and Ballauff, M. *Phys. Chem.Chem. Phys.*, **8**, 5269–5275 (2006).
50. Biesheuvel, P. M., and Wittemann, A. *J.Phys.Chem.* B. **109**, 4209–4214 (2005).
51. Biesheuvel, P. M., Leermakers, F. A. M., and Cohen-Stuart, M. A. *Phys. Rev.E.,* **73**, 011802, 1–9 (2006).

CHAPTER 11

STRONG POLYELECTROLYTE-INDUCING DEMIXING OF SEMIDILUTE AND HIGHLY COMPATIBLE BIOPOLYMER MIXTURES

Y. A. ANTONOV and P. MOLDENAERS

CONTENTS

11.1 INTRODUCTION

The weak intermacromolecular interactions caused by the presence of a complexing agent in two phase biopolymer mixture can affect its phase equilibrium and morphology. In this communication, the attempt was performed to induce demixing in semidilute and highly compatible sodium caseinate/sodium alginate (SC-SA) system mixtures in the presence of sodium salt of dextran sulfate at pH 7.0, (the isoelectrical point of caseins), and to characterize phase equilibrium, intermacromolecular interactions, and structure of such systems by rheo small angle light scattering (SALS), optical microscopy (OM), phase analysis, dynamic light scattering (DLS), fast protein liquid chromatography (FPLC), ESEM, and rheology. Addition of dextran sulfate sodium (DSS) salt to the semidilute single phase SC-SA system, even in trace concentrations (10^{-3} wt%), leads to segregative liquid-liquid phase separation, and a substantial increase in storage and loss modules of the system. The degree of the protein conversion in the complex grows, when the concentration of SC in the system increases from 1 to 2 wt%. It is also established here that demixing of semidilute biopolymer mixtures, induced by the minor presence of DSS is a rather common phenomenon, because it's also was observed here for other biopolymer pairs. At high shear rates SC becomes even less compatible with SA in the presence of DSS than at rest. Experimental observations suggest that the approach for inducing demixing of semidilute and highly compatible biopolymer mixtures by physical interactions of the constituents is a promising tool for regulation of biopolymer compatibility and achieving better predictions of phase behavior of aqueous protein charged polysaccharide systems.

The importance of the phase behavior in biopolymer mixtures is evident in many technological processes, such as isolation and fractionation of proteins (see, for example, [1-6]) and enzymes [7], enzyme immobilization [8, 9], encapsulation [10], and drug delivery [9, 11]. Aqueous, two phase systems are used in modern technological processes where clarification, concentration, and partial purification are integrated in one step [12]. Thermodynamic incompatibility, or, in other words, segregative phase separation, determines the structure and physical properties of biopolymers mixtures in quiescent state [13-15] and under flow [16-18] and plays an important role in protein processing in food products [14].

From a technological point of view, especially important are biopolymer systems which undergo liquid–liquid phase separation in a wide concentration range, starting from low concentrations [19]. But whether phase separation is desired or not, it is important for practical applications to understand the underlying mechanisms and molecular interactions governing the phase behavior of a given system [20]. Despite the considerable amount of research in the field of segregating polymer mixtures, the molecular interactions in the systems are inadequately understood, although, theoretical models have been proposed [21-28]. There have, as of yet, been comparatively few studies on phase separation in mixtures of similarly charged polyelectrolytes [29, 30]. Such systems may have advantages over uncharged systems in the separation of proteins due to the tunable charge in the system arising from the dissociated counter ions of the polyelectrolytes [29, 30]. Although, the majority of biopolymer mixtures show phase separation [14, 32], in most cases the phase separation takes places at critical total concentrations, which are much higher (7–12 wt%) [31, 32] compared with those of synthetic polymers (less than 1–2 wt%). Unlike synthetic polymers with flexible chains, many proteins are known to be relatively symmetric compact molecules and are usually able to form solutions that can still be considered dilute for concentrations 10-fold higher than for synthetic polymers of the same molecular weight [33].

The molecular weight, charge, and topography of the accessible surface of water soluble complexes of proteins with anionic polysaccharides are differ markedly from the "free" proteins. Therefore, it can be assumed that all these factors may affect the phase separation. In the present chapter, we focus study on the phase transitions in aqueous semidilute homogeneous SC-SA with the total concentration of biopolymers 1.5–2.5 wt%, that is much below the critical concentrations for phase separation [17]. The phase state of the SC-SA mixtures is not sensitive to changes in pH, ionic strength and temperature in the quiescent state [31, 32] and under of shear flow [17]. Therefore, the effect of demixing that can be reached for this system can be easily reproduced for other emulsions in which the phase equilibrium is more sensitive to physicochemical parameters. Here, it will be explored how far this strategy of demixing can be extended to other biopolymer pairs. For this reason gelatin-SA and gelatin-SC systems will be investigated to assess the generality of observations. In addition, the shear induced behavior of the decompatibilized semidilute SC-SA system will be presented and compared with that of the "native" SC-SA system.

Alginate is an anionic polysaccharide consisting of linear chains of (1–4)-linked ß-D-mannuronic and-α-L-guluronic acid residues. These residues are arranged in blocks of mannuronic or guluronic acid residues linked by blocks in which the sequence of the two acid residues is predominantly alternating [33, 34]. Casein is a protein composed of a heterogeneous group of phosphoproteins organized in micelles. These biopolymers are well known, widely used in industry for their textural and structuring properties [14, 31-33, 35] and the thermodynamic behavior of the ternary water–caseinate–alginate systems is known from literature [17, 31, 32, 35].

11.2 MATERIALS AND METHODS

The caseinate at neutral pH is negatively charged, like alginate and DSS. The SC sample (90% protein, 5.5% water content, 3.8% ash, and 0.02% calcium) was purchased from Sigma Chemical Co. The isoelectric point is around pH = 4.7–5.2 [36]. The weight average molecular mass of the SC in 0.15 M NaCl solutions is 320 kDa. The medium viscosity sodium alginate, extracted from brown seaweed (*Macrocystis pirifera*), was purchased from Sigma. The weight average molecular weight of the sample, M_w was 390 kDa [16]. Dextran sulfate, DSS (M_W = 500 kDa, M_n = 166 kDa, η (in 0.01 M NaCl) = 50 mL/g, 17% sulfate content, free SO_4 less than 0.5%) was produced by Fluka, Sweden (Reg. No. 61708061 A, Lot No. 438892/1). The gelatin sample used is an ossein gelatin type A 200 bloom produced by SBW Biosystems, France. The bloom number, weight average molecular mass, and the isoelectric point of the sample, reported by the manufacturer are respectively 207, 99.3 kDa, and 8–9.

11.2.1 PREPARATION OF THE PROTEIN/POLYSACCHARIDE MIXTURES

Most experiments were performed in the much diluted phosphate buffer (ionic strength, $I = 0.002$). To prepare molecularly dispersed solutions of SC, SA, and gelatin or DSS with the required concentrations, phosphate buffer (Na_2HPO_4/NaH_2PO_4, pH 7.0, $I = 0.002$, and 0.015) was gradually added to the weighed amount of biopolymer sample at 298K, and stirred, first for 1 hr at this temperature and then for 1 hr at 318K. The solutions of SC, SA, and DSS were then cooled to 296K and stirred again for 1 hr. The required pH value (7.0) was adjusted by addition of 0.1–0.5M NaOH or HCl. The resulting solutions were centrifuged at 60,000 *g* for 1 hr at 296K or 313 K (gelatin solutions) to remove insoluble particles. Concentrations of the solutions are determined by drying at 373K up to constant weight. The ternary water–SC–SA systems with required compositions were prepared by mixing solutions of each biopolymer at 296K. After mixing for 1 hr, the systems were centrifuged at 60,000 *g* for 1 hr at 296K to separate the phases using a temperature-controlled rotor.

11.2.2 DETERMINATION OF THE PHASE DIAGRAM

The effect of the presence of DSS on the isothermal phase diagrams of the SC-SA system was investigated using a methodology described elsewhere [36]. The procedure is adapted from Koningsveld and Staverman [37] and Polyakov et al [38]. The weight DSS/SC ratio in the system, (q) was kept at 0.14. The threshold point was determined from the plot as the point where the line with the slope −1 is tangent to the binodal. The critical point of the system was defined as the point where, the binodal intersects the rectilinear diameter, which is the line joining the centre of the tie lines.

11.2.3 RHEO-OPTICAL STUDY

A rheo-optical methodology based on SALS during flow, is applied to study *in-situ* and on a time-resolved basis the structure evolution. Light scattering experiments were conducted using a Linkam CSS450 flow cell with parallel plate geometry. A 5 mW HeNe laser (wavelength 633 nm) was used as light source. The 2D scattering patterns were collected on a screen by semi transparent paper with a beam stop and recorded with a 10-bit progressive scan digital camera (Pulnix TM-1300). Images were stored on a computer with the help of a digital frame grabber (Coreco Tci-Digital SE). The optical acquisition setup has been validated for scattering angles up to 18°. The gap between the plates has been set at 1 mm and the temperature was kept constant by means of a thermostatic water bath. In house developed software was used to obtain intensity profiles and contour plots of the images (New SALS SOFTWARE-K.U.L.). Turbidity measurements have been performed by means of a photo diode. *Microscopy observations* during flow have been performed on a Linkham shearing cell mounted on a Leitz Laborlux 12 Pol optical microscope using different magnifications.

11.2.4 RHEOLOGICAL MEASUREMENTS

Rheological measurements were performed using a Physical Rheometer, type CSL2 500 A/G H/R, with a cone-plate geometry CP50-1/Ti ~diameter 5 cm, angle 0.993°, Anton Paar. The temperature was controlled at 23°C by using a Peltier element. For each sample, flow curves were measured at increasing shear rate from 0.1 to 150 sec^{-1}. The ramp mode was logarithmic and the time between two measurements was 30 sec. Frequency sweeps ~0.1–200 rad/sec were carried out as well for a strain of 3.0%, which was in the linear response regime. During the rheological measurements, all samples were covered with paraffin oil to avoid drying.

11.2.5 DYNAMIC LIGHT SCATTERING

The determination of Intensity weighted distribution of hydrodynamic radii (R_H) of SC, SA, and DSS solutions and their mixtures was performed, using the Malvern ALV/CGS-3 goniometer. Concentration of the protein in protein-dextran sulfate mixtures was kept at 0.1 (w/w). For each sample the measurement was repeated three times. The samples were filtered before measurement through DISMIC-25cs (cellulose acetate) filters (sizes hole of 0.22 µm for the binary water-casein and water-dextran sulfate solutions and 0.80 µm for the protein polysaccharide mixtures). Subsequently, the samples were centrifuged for 30 sec at 4000 g to remove air bubbles, and placed in the cuvette housing which was kept at 23°C in a toluene bath. The detected scattering light intensity was processed by digital ALV-5000 Correlator software. The second order cumulant fit was used for the determination of the hydrodynamic radii. The asymmetry coefficient (Z) of the complex particles was estimated by Debye method based on determining the scattering intensity at two angles 45° and 135°, symmetrical to the angle 90°.

11.2.6 ZETA (ζ) POTENTIAL MEASUREMENT

The ζ potential measurements of SC and DSS solutions and their mixtures at different q values were performed at 23°C with a Malvern–Zetamaster S, model ZEM 5002 (England), using a rectangular quartz capillary cell. The concentration of the protein in solutions was 0.1 wt%, and the concentrations of DSS in the protein-polysaccharide solutions were variable. All solutions were prepared in phosphate buffer (Na_2HPO_4/ NaH_2PO_4, pH 7.0, $I = 0.002$).The ζ potential was determined at least three times for each sample. The ζ potential was calculated automatically from the measured electrophoretic mobility, by using the Henry equation:

$$U_e = \varepsilon z\rho f/6\rho\eta \quad (1)$$

where U_e is electrophoretic mobility, ε is the dielectric constant, ? is the viscosity, and $z\rho$ is the zeta potential. The Smoluchowski factor, $f = 1.5$ was used for the conversion of mobility into zeta potential.

11.2.7 ENVIRONMENT SCANNING ELECTRON MICROSCOPY (ESEM)

Micro structural investigation was performed with the environment scanning electron microscope Philips XL30 ESEM FEG. The instrument has the performance of a conventional scanning electron microscopy (SEM) but has the additional advantage that practically any material can be examined in its natural state. The samples were freeze-fractured in freon and immediately placed in the ESEM. Relative humidity in the ESEM chamber (100%) was maintained using a Peltier stage. Such conditions were applied to minimize solvent loss and condensation, and control etching of the sample. Images were obtained within less than 5 min of the sample reaching the chamber. The ESEM images were recorded multiple times and on multiple samples in order to ensure reproducibility.

11.2.8 FAST PROTEIN LIQUID CHROMATOGRAPHY (FPLC)

Solutions of SC, (0.5 wt %), dextran sulfate (0.5 wt%) and their mixtures, containing 0.5 wt% of the protein and variable amount of dextran sulfate were applied on a Superose 6 column (HR 10/30), Amersham Biosciences mounted on an FPLC apparatus (Pharmacia, Uppsala, Sweden). Elution was performed at room temperature with phosphate buffer (5mM Na_2HPO_4/NaH_2PO_4, pH 7.0) 2% (v/v) n-propanol (Riedel-de Haen, Seelze, Germany) and 0.015 M NaCI. The samples and the elution buffer were filtered through a 0.22 um sterile filter. The flow rate was 0.2 mL/min and the column was monitored by UV detection at 214 nm.

11.2.9 DETERMINATION OF DEXTRAN SULFATE CONTENT

The phenol-sulphuric acid method of Dubois et al [39] was applied. 50 uL. 80% (w/w) phenol in water and 5mL sulphuric acid were added to the measured samples of 0.5 mL. After 30 min at room temperature the absorbance at 485 nm was measured. A calibration plot was constructed with D-glucose (Riedel-de Haen).

11.3 DISCUSSION AND RESULTS

11.3.1 DSS-INDUCED DEMIXING

The experimental results shown in this chapter have been obtained on water (97.5 wt %)-SC (2.00 wt %)-SA (0.5 wt %) semidilute systems. This system is located in the one-phase region far from the binodal line. To study the effect DSS on the phase behavior, a flow history consisting of two shear zones is used. First, a pre-shear of 0.5 sec^{-1} is applied for 1000 sec (500 strain units) to ensure a reproducible initial morphology. Subsequently, this pre-shear is stopped, and the sample is allowed to relax for 30 sec leaving enough time for full relaxation of deformed droplets. Then SALS patterns are monitored.

The SALS patterns and the scattering intensity upon adding different amount of DSS are shown in Figures 1 ((a)–(f)) and 2, starting from a concentration of DSS as low as 2.08×10^{-3} wt%. In the absence of DSS no scattered light is observed (data are not presented). The presence of even only 2.08 × 10^{-3} wt% DSS in the homogeneous system led to appreciable increase the SALS pattern (Figure 1) and accordingly the light scattering intensity (Figure 2). It is important to note that that the SC-DSS system remains homogeneous in the DSS concentration range studied here. Centrifugation of the SC-DSS systems (120 min, 60.000 g, and 296K) prepared at the same conditions did not show phase separation.

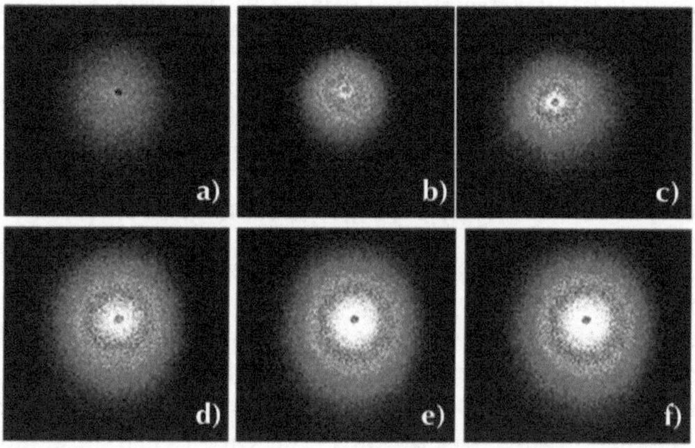

FIGURE 1 Effect of the concentration of DSS on the SALS patterns of water (97.5 wt %)-SC (2.00 wt %)-SA (0.5 wt %) single-phase systems. pH 7.0. I = 0.002 (phosphate buffer). Temperature 296K. Concentrations of DSS in mixture, wt%—(a) 2.08 × 10^{-3}, (b) 4.10 × 10^{-3}, (c) 1.61 × 10^{-2}, (d)7.50 × 10^{-2}, (e) 0.15, (f) 0.29, and resulting DSS/SC ratio—(a) 0.001, (b) 0.002, (c) 0.008, (d) 0.0375, (e) 0.075, and (f) 0.145.

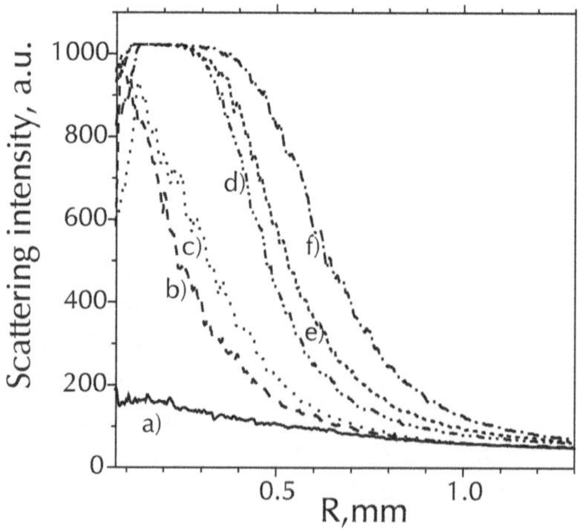

FIGURE 2 Effect of the concentration of DSS on the scattering intensity of water (97.5 wt %)-SC (2.00 wt %)-SA (0.5 wt %) single-phase systems as a function of the distance from the bean stop. The other parameters are the same as in Figure 1.

When the DSS concentration in the SC-SA system increases the SALS pattern (Figure 1) and the scattering intensity (Figure 2) of the system sharply grows. This indicates that the position of the system on the phase diagram changes deeply into the two phase range. The corresponding microscopy images for the same concentrations of DSS and the same flow conditions are shown in Figure 3. One can see that the phase separation led to formation of liquid-liquid emulsions. At the lowest DSS concentration (2.08×10^{-3} wt %), the system contains ultra small droplets of the dispersed phase having a size of 2–3 μm. At higher DSS concentrations the size of the droplets increases significantly in agreement with SALS data achieving more than 50 μm in diameter.

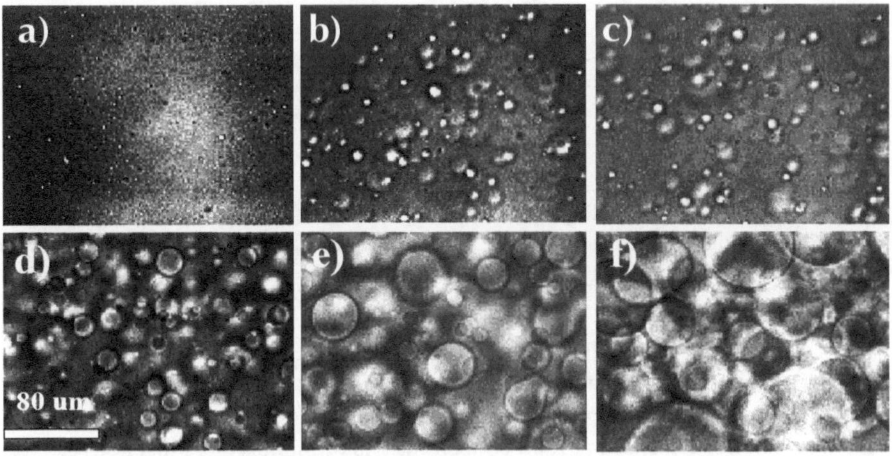

FIGURE 3 Microscopy images of water (97.5 wt %)-SC (2.00 wt %)-SA (0.5 wt %) system after addition of different amounts of DSS. pH 7.0, I = 0.002 (phosphate buffer). Temperature 296K. The other parameters are the same as in Figure 1.

In order to quantify the effect of DSS on phase equilibrium in semidiluted SC-SA system, the isothermal phase diagram of the system was determined in the presence of DSS, at DSS/SC weight ratio (q) = 0.14, plotted in the classical triangular representation, and compared with that obtained in the absence of DSS (Figure 4). The phase separation in the presence of DSS has a segregative character with preferential concentrating of SC and SA in different phases. The phase diagram of the initial system, without DSS, is characterized by a high total concentration of biopolymers at the critical point (C_c^t = 62.9 g/L), and a strong asymmetry (Ks = 15.5).

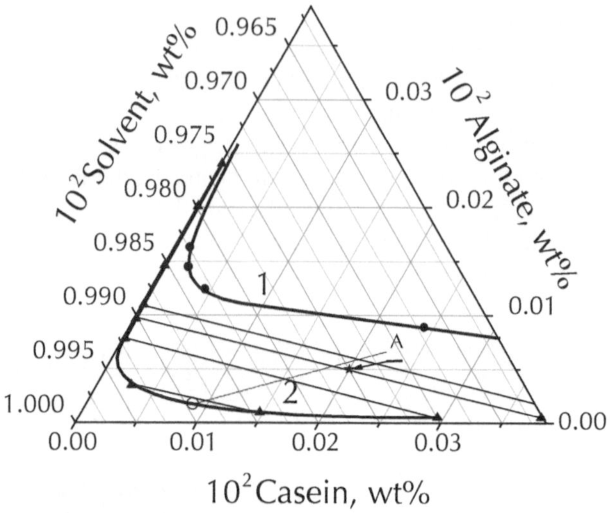

FIGURE 4 Isothermal phase diagrams of the W-SC-SA system. pH 7.0, $I = 0.002$ (phosphate buffer), 296 K. 1. In the absence of DSS. 2. In the presence of DSS, at DSS/SC weight ratio (q) = 0.14.

The presence of DSS affects dramatically the phase separation, significantly increasing the concentration range corresponding to two phase state of the system. The total concentrations of biopolymers at the critical point decreases to 10.6 g/L. The phase separation is observed at total concentrations of biopolymers just above 1 wt%, that is level of compatibility of the biopolymers after an addition of DSS seems to be one of the smallest known for biopolymer mixtures (see, for example, [40, 41]). The decrease in compatibility of casein and alginate is especially surprising when taking into account that the phase composition of this system is weakly dependent on many physicochemical factors, such as pH (in the pH range from 7 to 10), ionic strength, and temperature (from 5 to 60°C) [17, 32, 34].

11.3.2 RHEOLOGICAL BEHAVIOR OF THE DEMIXED SYSTEMS

For the rheological investigations, the homogeneous W-SC (2.0 wt%)-SA (0.5 wt %) system (point A on the phase diagram, Figure 4) was characterized before, and after addition of DSS at the DSS/SC weight ratio, q = 0.045, and q = 0.15 respectively. The latter two systems were two phase ones with the content of the casein enriched phase 15 w/w and 55% w/w accordingly. The experimental flow protocol applied was the same as the one used for rheo-SALS. The mechanical spectrum and flow curve were determined in order to characterize the state of the systems through their viscoelastic behaviors. It has been shown [42] that at moderately low shear rates, the biopolymer emulsions can be regarded as conventional emulsions and various structural models that are available in the literature for prediction of the morphology in these emulsions can also be used for prediction of the structure in aqueous biopolymer emulsions.

The evolution of the mechanical spectrum was investigated as a function of DSS concentration. These viscoelastic behaviors were monitored and compared with the behavior of the W-SC-SA system without DSS. The dynamic modulus *G*' (elastic) and *G*" (viscous) were measured with frequency sweep experiments at a constant strain of 3%, which was checked as being in the linear regime. The obtained data are presented in Figure 5.

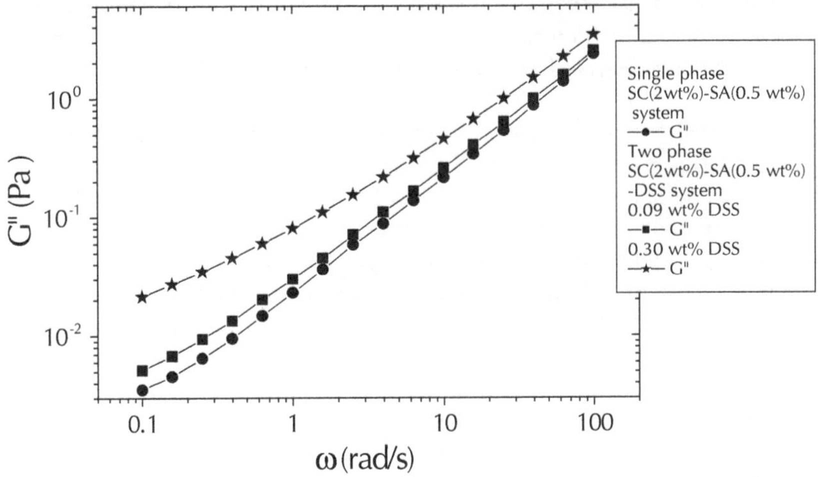

FIGURE 5 Dynamic spectra of single phase W-CS (2 wt%)-SA (0.5 wt %) system and two phase W-CS (2 wt%)-SA (0.5 wt%)-DSS systems. pH 7.0, *I* = 0.002 (phosphate buffer), 296K.

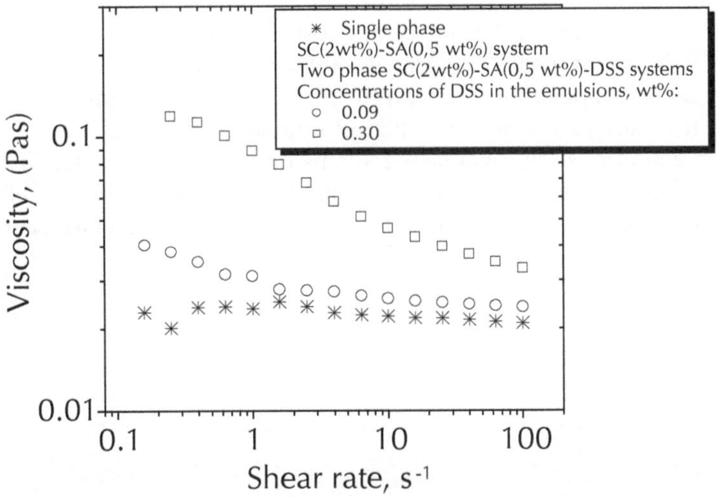

FIGURE 6 Flow viscosity of single phase W-CS (2 wt%)-SA (0.5 wt %) system and two phase W-CS (2 wt%)-SA (0.5 wt%)-DSS systems, after application of increasing shear rates. pH 7.0, *I* = 0.002 (phosphate buffer), 296K.

For the single phase system, and the system containing 0.09 wt% DSS, G' was too low to be measured accurately. Under these conditions the system behaves as purely viscous liquid with the curve of G" versus frequency displaying a slope of one on a double logarithmic graph. In the present of DSS the system undergoes phase separation and this transition leads to an appreciable increase of the modules. The elastic properties of the decompatibilized W-SC-SA system were mainly induced by the presence of the DSS. In the presence of high ionic strength (0.25 NaCl), when electrostatic interactions were suppressed the mechanical spectrum of the system (q = 0.14) becomes insensitive to the presence of DSS (data are not presented). Flow curves determined at the same concentrations show an increase in viscosity for the demixed systems, especially remarkable at a low shear rates (Figure 6).

More detailed experiments were then carried out on the single phase W-SC (4 wt%)-DSS (variable), and W-SA (0.5 wt%)-DSS (variable) systems to understand how DSS affects the mechanical spectrum of the casein and alginate solutions, and accordingly the coexisting phases. The behavior of these solutions in the presence of sulfated polysaccharide is clearly different (Figures 7 and 8); the casein-enriched phase is sensitive to the presence of DSS, while the viscoelastic properties of the alginate enriched phase in the presence of DSS remain almost unaltered. As reported in Figure 7 (a), the dependence of the G" on the DSS/casein ratio has an extreme character, with a maximum at a DSS/casein ratio around 0.14. In the presence of even small amounts of DSS (0.01–0.05 wt%), a dramatic increase of the G" of the emulsion takes place. Thus, in the presence of 0.5 wt% of DSS (at q = 0.14) and at a frequency 1 rad/sec, G" values is more than 1400 times, higher compared with those of the single phase system with almost the same composition. From theory we know that such dependences are typical for the formation of inter-polymer complexes [42]. Similar changes were observed for the viscosity (Figure 8 (a) and (b)). At q = 0.14 and a shear rate of 10 sec^{-1} the viscosity is more than 940 times higher compared with those of the single phase system with almost the same composition (Figure 8 (b)). It is important to note that in the shear rate range from 0.1 to 150 sec^{-1} we did not find any difference in the flow curves obtained in conditions with increasing versus decreasing shear rate (data are not presented). It can be assumed that the dramatic changes in rheological behavior of the casein-alginate system in the presence of DSS are due to interactions of the casein molecules with the DSS molecule.

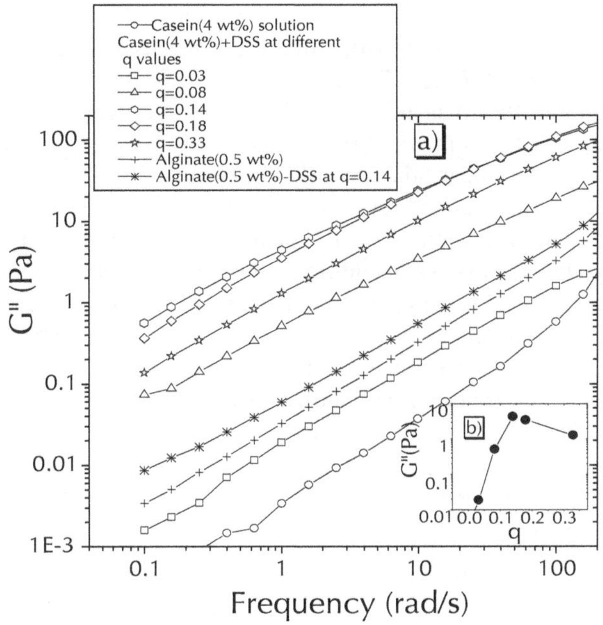

FIGURE 7 (a) G" of W-SA(0.5 wt%), W-SA(0.5 wt%)-DSS, and W-SC(4 wt%)–DSS, systems at different q values and (b) the dependence of G" on q values for W-SC(4 wt%)–DSS, system at frequency 1.0 rad/sec, pH 7.0, $I = 0.002$ (phosphate buffer), and 296K.

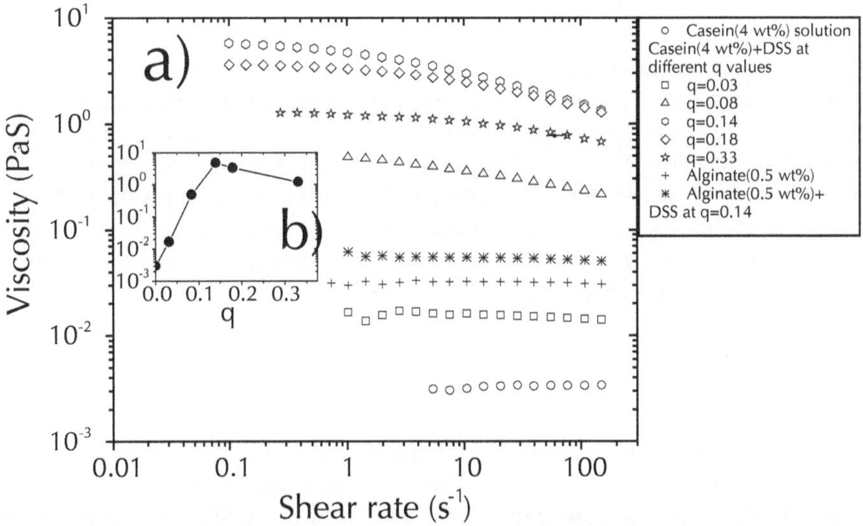

FIGURE 8 (a) Dependences of flow viscosity of W-SC(4 wt%), W-SA(0.5 wt%), and W-SC 4 wt%)–DSS (var) systems and (b) the dependence of flow viscosity of the W-SC-DSS system on q values at shear rate 1.0 sec^{-1}, pH 7.0, $I = 0.002$ (phosphate buffer), and 296K.

It can be suggested that casein interacts with DSS, and this interaction may have an effect on the phase separation. Note, that the viscosity of the demixed system in Figure 8 decreased from 5.72 to 1.74 Pa sec with increasing shear rate from 0.1 to150 sec^{-1}, which highlight the shear thinning behavior of the demixed system, indicating a structural change. The result was striking since most concentrated protein-polysaccharide mixtures can be shear thinning only due to the polysaccharide relaxations. In the absence of structure induced formation the rheological behavior of concentrated polysaccharide solutions is monotonically shear thinning; the viscosity varies between two extremes η_0 and η_∞. A possible additional mechanism would be the breakdown of structures due to the breakup of physical bonds at high shear. This structure was most probably due to the electrostatic interactions between SC and DSS. Indeed, in the presence of 0.25 M NaCl, when no attractive interaction took place, no shear thinning behavior was observed (data are not shown). More detailed experiments were then carried out to understand the mechanism of demixing.

11.3.3 INTERMACROMOLECULAR INTERACTIONS AND THE MECHANISM OF DEMIXING IN SC-SA-DSS SYSTEM

An important property of the demixed semidilute SC-SA systems described is their high stability against homogenization and low sensitivity to change in temperature. Thus, for the mixtures with different composition we observed constancy of absorption values at 500 nm during 6 hr storage, as well as in processes of their heating from + 5 to 70°C. The results obtained (Figure 7 and 8) show the presence the intermacromolecular interactions between SC and DSS. Usually coulomb protein-polysaccharide complexes are formed only in the vicinity of the isoelectric point of the protein [44], but for several systems formation of soluble protein–polysaccharide complexes has been registered even at pH 6.0–8.0 [45–47]. A beneficial consequence of complexation of sulfated polysaccharide with caseins at pH values IEP is the protection afforded against loss of solubility as a result of protein aggregation during heating or following high-pressure treatment [48, 49].The mechanism of this protection has been unclear until now. Snoeren, Payens, Jevnink, and Both, assumed [50] that there is a non-statistical distribution of positively charged amino acid residues along the polypeptide chain of kappa casein molecules and, as a consequence, the existence of a dipole interacting by its positive pole with sulfur polysaccharide is responsible for complex formation in such systems.

Many scientists suppose [51, 52] that nonelectrostatic forces, hydrophobic and (or) hydrogen bonds, play a determinant role in this process. In the case of sulfated polysaccharides this assumption is confirmed by experimental data showing the capacity of the sulfate groups to form hydrogen bonds with the protein cationic groups [53].

Introduction of NaCl in the initial buffer results in full insensitivity of the viscosity and the phase diagram of the SC-SA system to the presence of DSS in all the q range studied. On the other hand, an addition of 0.2 M NaCl in the SC-SA-DSS system at q = 0.14 after a 24 hr storage results in a sharp increase in the level of compatibility of SC with SA to that of SC-SA solution alone. This shows that the complexes are formed and stabilized *via* electrostatic interaction, rather than through hydrogen bonds formation or hydrophobic interaction. The role of salt is to "soften" the interactions, which is equivalent to making the electrostatic binding constant smaller.

To study intermacromolecular interactions in the process of demixing of the SC-SA system, at first, we focus attention to the interaction between SC and DSS in aqueous solutions within the region of pair interaction. To this aim, we have chosen SC and DSS concentrations low enough to exclude or considerably diminish effects of possible aggregation. This allows us to single out information on interaction processes between the two types of macromolecules, well separated from the subsequent aggregation process. The DLS can provide information about the hydrodynamic radius of proteins and polysaccharides and about the binding of ligands to these types of macromolecules. Figure 9 shows the intensity weighted distribution of hydrodynamic radius (R_H) of solutions of SC, dextran sulfate and their mixtures with the concentration of the protein equal to 0.1 (w/w),that is at the total concentrations below the critical concentration of phase separation of SC-SA system (see Figure 4). At 296 K, molecules of SC and DSS have R_H values 119 nm and 250 nm respectively.

An addition of DSS to SC solution at DSS/SC weight ratios ranging (q) from 0.025 to 0.05 leads to significant increase in the R_H toward the values R_H for DSS solution. At higher q values = 0.14, R_H of the mixed associates achieve the values R_H for DSS, and their size does not change with the further increase of q values. This is an indication of intermacromolecular interaction of the casein molecules with DSS and formation of complexes. At q = 0.14, function of the intensity weighted distribution of hydrodynamic radii (R_H) is placed completely outside that describing free SC.

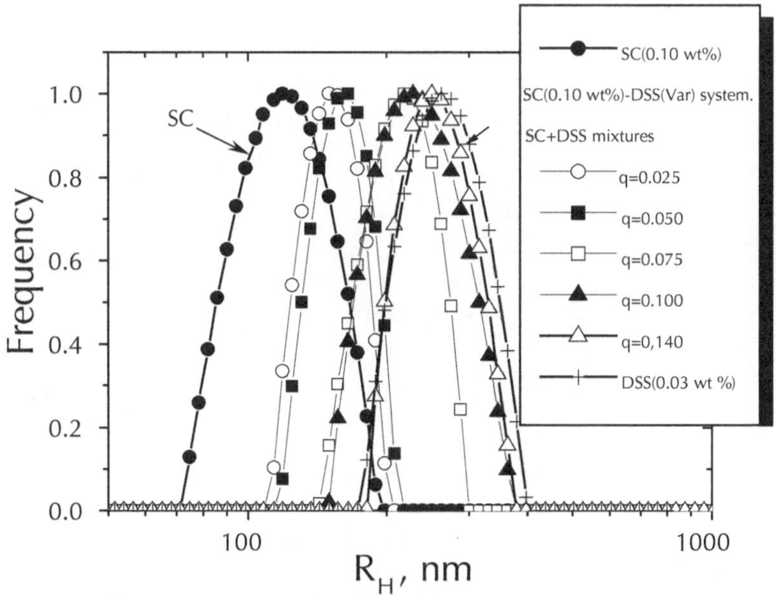

FIGURE 9 The intensity weighted distribution of hydrodynamic radius (R_H) of solutions of SC, dextran sulfate, and their mixtures. Concentration of SC is equal to 0.1 (w/w), pH 7.0, I = 0.002 (phosphate buffer), and 296K.

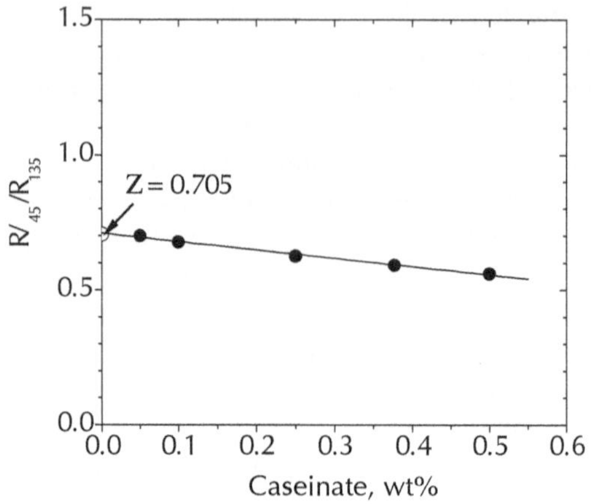

FIGURE 10 Dependence of the ratio of the scattering intensity, R at 45° and 135° on the concentration of casein in the SC-DSS mixture at q = 0.14, pH 7.0, I = 0.002 (phosphate buffer), and 296K.

The asymmetry coefficient (Z) of the complex associates was estimated by Debye method based on determination of the scattering intensity,(R) at two angles 45° and 135°, symmetrical to the angle 90° and subsequent extrapolation of the R45°/R/135° to zero concentration. The results obtained are presented in Figure 10. The complex associates are asymmetric with Z values equal to 0.7.

Figure 11 presents Zeta potential values and the total concentration of the biopolymer at the critical point C^t_{cr} as a function of the DSS/SC ratio, q after an addition of DSS the negative value of the zeta potential increases and C^t_{cr} decreases achieving correspondingly the maximal and minimal values at q = 0.14.

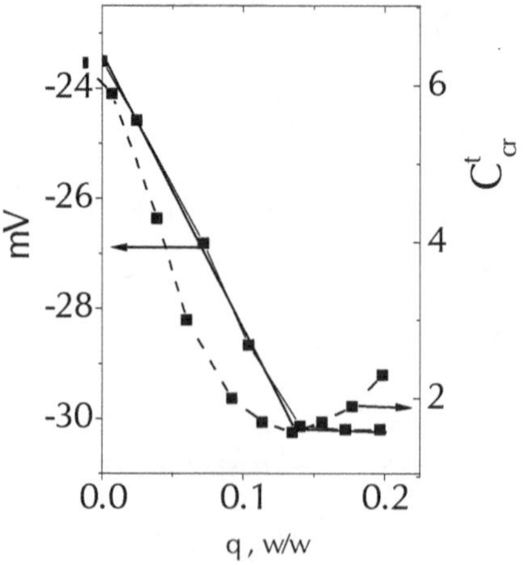

FIGURE 11 Dependence of Zeta potential and the total critical concentration of biopolymers corresponding to phase separation of W-SC-SA-DSS system on q. The SC/SA weight ratio is 4. pH 7.0, I = 0.002 (phosphate buffer), and 296K.

Once the negative charge of a protein becomes higher in the presence of DSS, interactions between casein molecules could be hindered by an overall effect of electrostatic repulsion. Thus, an increase in the net charge of casein due to DSS binding could lead to an enhancement in the extent of such repulsions, contributing to the suppression of the further association and aggregation. Obviously, at pH = 7.0, the total charge of the high molecular weight DSS molecule is higher than the total positive charge of the relatively small SC molecule. This gives the possibility to regard complex formation between these biopolymers (similarly to other weak-polyelectrolyte–strong-polyelectrolyte interactions [54, 55]) as a mononuclear association in which the DSS molecule is the nucleus and the casein molecule is a ligand. Therefore, the formation of casein–DSS complex can be regarded as the reaction of few casein molecules successively joining one molecule of DSS nucleus. Note that at pH = 7.0, (experimental conditions) all the cationic groups of casein, as well as all the sulfate groups of DSS are ionized. It easy to show that at q° = q* (0.135), the ratio of the total amount of sulfate groups in DSS molecule and cationic groups in casein molecule,

$$\left(\frac{[S^-]}{[Cat^+]}\right)$$

is close to unity. Actually, the total amount of cationic groups in casein molecule is 0.76 mmol/g [50, 56] and the content of sulfur groups in DSS molecule is equal to 5.43 mmol/g [57]. Therefore,

$$\frac{[S^-]}{[Cat^+]} = q\frac{5.43}{0.76}.$$

At $q^* = 0.135$ one can obtain $\frac{[S^-]}{[Cat^+]} = 0.964$.

Figure 12 presents the chromatograms of the initial solutions of SC (0.25 wt %) and DSS (0.25 wt %), and the SC-DSS system (q = 0.14, concentrations of SC =0.25 wt% and 1.0 wt %), showing distribution of the protein, polysaccharide, and complex associates in the chromatographic fractions.

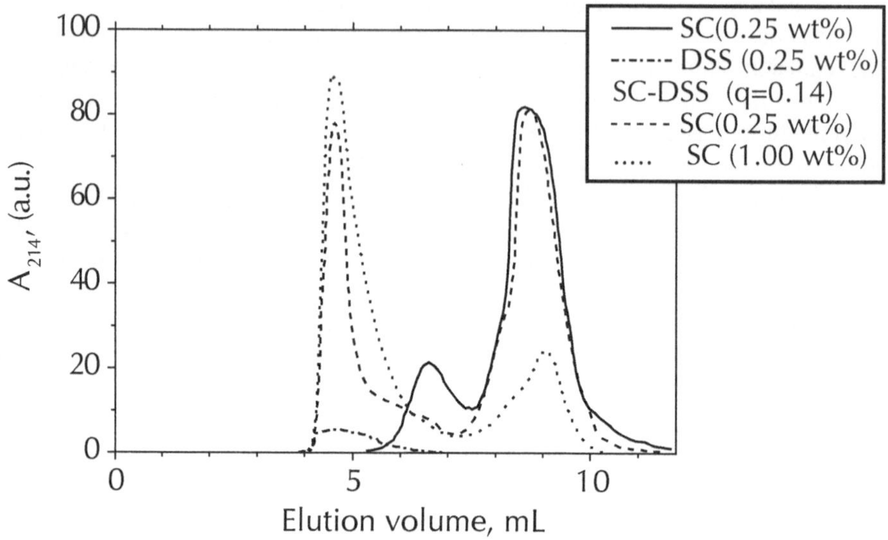

FIGURE 12 Chromatograms of the initial solutions of SC and DSS, and the SC- DSS system (q = 0.14), showing distribution of the protein, polysaccharide, and complex associates in the chromatographic fractions at concentrations of the protein 0.1 wt% and 1.0 wt%. pH 7.0, I = 0.015 (phosphate buffer), 296K.

Free SC exhibited at pH 7.0 two unequal peaks. The first peak (83% from the total square) presents SC molecules and the second one (17% from the total square) corresponds to the SC associates. Estimation of the molecular weights of these components on the basis of known molecular weights of alpha, beta, and gamma gelatins gave 260 kDa and 380 kDa accordingly. The weight average molecular weight of both fractions was about 300 kDa.

The DSS exhibited at the same conditions a weak wide signal in the excluded volume. The chromatograms of the SC (0.25 wt%) -DSS systems at q = 0.14 gave a new high molecular weight component corresponding to excluded volume and the peak corresponding to the elution volume of the free (unbounded SC). It is interesting to note that at concentration of SC below the critical concentration of the phase separation, the degree of conversion of the SC in water soluble complex with DSS is low (30%), and mainly the high molecular fraction of SC interact with DSS. The interaction becomes stronger when the concentration of the SC in the mixture increases up to 1.0 wt% (inside two-phase range of SC-SA system in the presence of DSS (q = 0.14).In such conditions 83% of SC form complex with DSS. Taking into account that the maximal yield of the complex takes place at q = 0.14, knowing the weight average molecular weights of SC and DSS and the degree of the protein conversion in protein-polysaccharide complex, we can roughly evaluate the SC/DSS molar ratio in the complex in the selected conditions corresponding to demixing of the mixed solutions of SC (2 wt%)-SA (0.5 wt%) in the presence of DSS (q = 0.14). Simple calculation showed that about 10 molecules of SC join to 1 DSS molecule, forming large associates with high molecular weight. Systematic experimental data concerning dependence of C^*_t upon the radius, or molecular weight of synthetic or natural polymers are unknown until now, although, it is generally accepted that thermodynamic compatibility of polymers decreases with increase in molecular weights. It has been shown recently [58] that the total concentrations of biopolymers at the threshold point (C^*_t) for casein-guar gum system changes in accordance to $C^*_t \infty {M^{cas}_w}^{-0.27}$, where M^{cas}_w is molecular weight of caseins. This dependence has been established in a wide range of M^{cas}_w (from 25 to 160.000 kDa). In that way, formation of large SC-DSS associates should decrease considerably compatibility of SA with bonded SC compared with that of "free" casein molecules that was observed in present work (Figure 4).

11.3.4 COMMONALITY OF THE DDS-INDUCED DEMIXING AT REST

The other question arising from the demixing phenomenon in diluted biopolymer systems in the presence of DSS is, what is the key factor determining complex formation between DSS and caseins at pH 7.0 (far from the pH value corresponding to iep of caseins)? Is the high local charge density of the positively charged kappa casein responsible for that, or its mainly determined by the structural features of DSS, such as the concentration of sulfate groups, charge density, and conformation of the polysaccharide. Specific interaction between k-casein and carrageenan has been ascribed by Snoeren et al [50] to an attraction between the negatively charged sulfate groups of carrageenan and a positively charged region of κ-casein, located between residues 97 and 112. It does not occur with the other casein types. Since the positive patch on κ-casein is believed to have a size of about 1.2 nm and is surrounded by predominantly negatively charged regions, the importance of the inter sulfate distances is unmistakable. To extend the Snoeren suggestion to system, containing more stronger polyelectrolyte than carrageenan, or to reject it, we investigated the effect of DSS on the phase equilibrium in semidilute single phase biopolymer systems containing the protein (gelatin) with the statistical distribution of the positively charged functional groups. Two systems were under consideration; gelatin type A-SA, and gelatin typeA-SC. The former is a single phase one in water over a wide concentration range, and it undergoes phase separation at ionic strength above 0.2 [59]. The latter system undergoes phase separation only at a very high ionic strength (above 0.5) [60] and is characterized by a very high total concentration of the biopolymer (>15–20 wt%) at the critical point [61].

The compatibility of these biopolymer pairs in water in the presence of DSS (at q = 0.14) was studies. The phase separation of both systems in the presence of DSS was established and the binodal lines for them were determined (Figure 13).The binodals for the systems without DSS are placed outside the concentration range studied. In both systems the phase separation leads to formation of water in water emulsions with liquid coexisting phases (Figure 13).Two important conclusions can be made from these data. First, the DSS induced phase separation in semidilute biopolymer solutions at rest is a rather general phenomenon not an exceptional case. Second, the structural features of DSS molecules is the more important factor determining complex formation of SC with DSS and subsequent demixing of the single phase semidilute systems, rather than the characteristics of the distribution of the positively charged groups in the protein molecules.

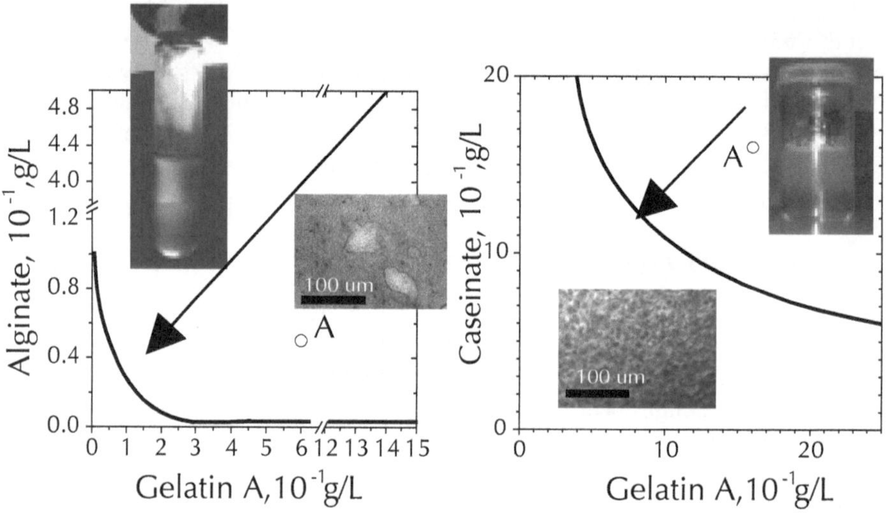

FIGURE 13 Shift of the bimodal line of the W-gelatin type A-SA and W-SC-gelatin type A systems in the presence of DSS at DSS/protein weight ratio 0.14; photo images and microscopy images of the demixed W-gelatin type A(6 wt%) –SA (0.5 wt%), and W-SC (16 wt%)-gelatin type A (16 wt%) systems (points A on phase diagrams). W-gelatin type A-SA system was prepared at pH 5.0 and W-SC-gelatin type A system was prepared at pH 7.0.

The last conclusion is in agreement with the FPLC data (Figure 12). As can be seen the degree of the protein conversion in complex achieves 80% whereas the content of kappa casein in SA is only 12–14% [62]. It is known that the sulfate groups of DSS are more closely packed than that of κ-carrageenan (0.5 nm for DSS and 1.2 nm for carrageenan [62, 63]. The later can allow for the attractive forces to overcome the repulsive forces acting outside the positive patch. Bowman, Rubinstein, and Tan, characterizing complex formation between negatively charged polyelectrolytes and a net negatively charged gelatin by light scattering, suggested [64] that the protein is polarized in the presence of strong polyelectrolyte. Junhwan and Dobrynin have recently presented the results of molecular dynamics simulations of complexation between protein and polyelectrolyte chains in solution [65]. They found that protein placed near polyelectrolyte chains is polarized in such a way that the oppositely charged groups on the protein are close to the polyelectrolyte, maximizing effective electrostatic attraction between the two while the similarly charged groups on the protein far away from the polyelectrolyte minimize effective electrostatic repulsion. In dilute and semidilute solutions, which are subjects of study, polyampholyte chains usually form a complex at the end of polyelectrolyte chains resulting from the polarization effect by polyelectrolyte. We believe that polarization-induced attraction is the main mechanism of complexation SC and DSS.

11.3.5 DISCUSSION ON THE STRUCTURE OF THE SC-DSS COMPLEXES AND SC ENRICHED PHASE OF THE DEMIXED SC-SA SYSTEM

From study of polyelectrolyte complexes we know that interaction between oppositely charged polyelectrolyte's leads to partial or complete neutralization of charges, complexes remain soluble or precipitate, and in some cases gel-like networks are formed. If neutralization of charges is significant, the so called "scrambled egg" compact structure will be formed. When neutralization of charges is far from complete, a "ladder" structure of complex can be formed [66].

The results of the Zeta potential measurements, DLS and flow experiments shown that the negative charge of the SC increases during interaction with DSS, and the maximal binding takes place at approx 0.14 DSS/SC weight ratio. Such features of the intermacromolecular interactions do not promote formation of the "scrambled egg" structure, because DSS molecule having many combined SC molecules and considerable negative charge cannot be fold. Therefore, the ladder structure is more preferable for the system (Figure 14 (a)). The overage size of the SC-DSS complex associates established from the DLS experiments is 0.2 um. Such a length scale would be in line with the fact that the SC/DSS solution is slightly turbid. This turbidity arises from a length scale in the micrometer range.

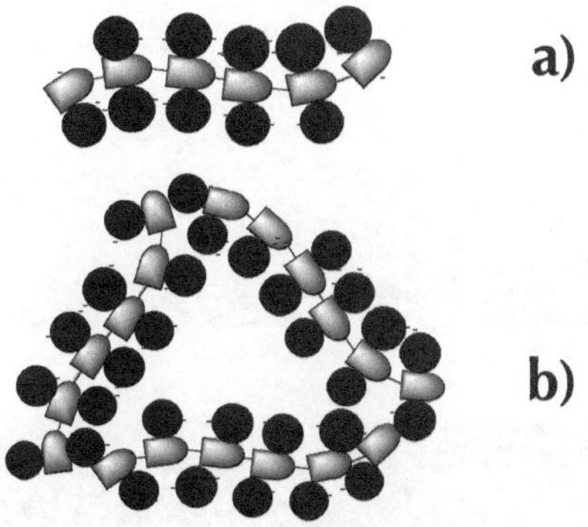

FIGURE 14 Schematic representation of the possible structures of (a) ladder-like and (b) gel-like. The long chain represent DSS molecule and the balls represent casein chains.

Obviously, heterogeneities on a micrometer scale were formed. If SC/DSS solution was made of a homogeneous structure of polymers on the nanometer scale, it would be transparent. In the presence of free polymer-SA, complex associates of SC and DSS undergo further association and the system becomes two phase. This suggestion finds confirmation in the flow experiments; viscosity of the demixed SC-SA system is considerably higher than that of undemixed SC-SA system having the same concentrations (Figure 6). This difference is even much higher in the case of higher protein concentration in the single phase SC-DSS system (Figure 8) ,this is a clear indication of association of the "ladder " structure of the complex associates, and formation of network (Figure 14 (b)). Figure 15 presents ESEM images obtained for the SC-DSS system at q = 0.14 (at maximal binding) and different concentrations of SC in the system. One can see that at concentration of SC equal to 2 wt% the formation of the regular structure is observed, which transfer to some "network" structure at higher concentration of SC (6 wt%) in the system.

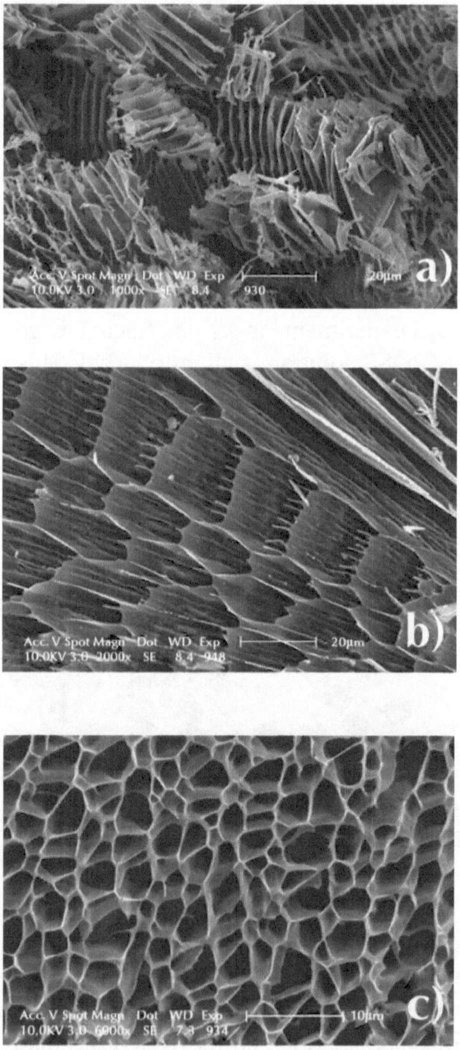

FIGURE 15 *(Continued)*

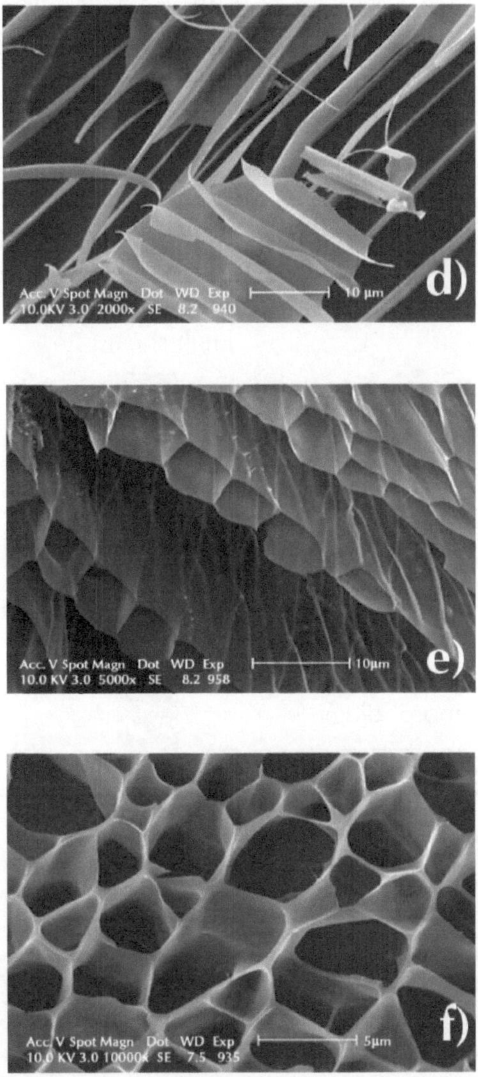

FIGURE 15 The ESEM images of the SC-DSS systems at q = 0.14. pH 7.0. Concentration of SC in the system, wt%; a,d- 1.0, b,e-2.0, c,f- 6.0.

11.3.6 SHEAR-INDUCED BEHAVIOR OF THE SC-SA SYSTEM IN THE PRESENCE OF DSS

The experimental results shown in this part have been obtained on a water (97.5 wt%)-SC (2.0 wt%)-SA (0.5 wt%)-DSS (2 × 10^{-3} wt%) system. It contains 99 w% of the SC enriched phase and 1 wt% of the SA enriched phase which have been mixed by hand, typically resulting in a very fine morphology. This emulsion is located in the two-phase region not far from the binodal line. The coexisting phases have Newtonian viscosities at 296K, of 0.03 Pa·sec and 0.02 Pa·sec for the SC enriched and the SA enriched phase, respectively.

To study the effect of flow on the phase behavior, a flow history consisting of three shear zones is used (Figure 16). First, a pre-shear of 0.5 sec^{-1} is applied for 1000 sec (500 strain units). It has been verified that this procedure leads to a reproducible initial morphology. Subsequently, this pre-shear is stopped and the slightly deformed droplets are allowed to retract to a spherical shape. The resulting droplet radius is of the order of 5 micron. Finally, the shear rate is suddenly increased to a high value for 80 sec, and after stopping flow the evolution of the SALS patterns are monitored.

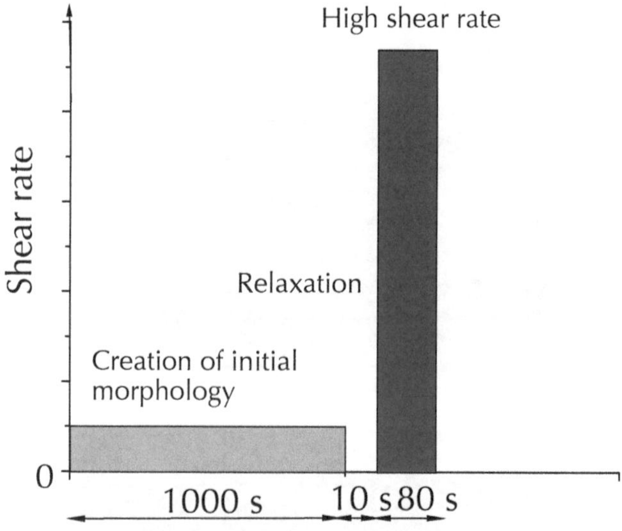

FIGURE 16 Schematic representation of the shear history.

The evolution of the SALS patterns after cessation of steady state shear flow at 60 sec^{-1}, 100 sec,$^{-1}$ and 150 sec^{-1} is shown in Figure 17. In each experiment, a freshly loaded sample has been used. As can be seen at all shear rates selected, the SALS patterns become more intensive just after cessation of flow. The higher the shear rate applied the more intensive the SALS pattern becomes. This is a clear indication of shear induced demixing in SC-SA system in the presence of DSS. After cessation of shear flow the light intensity is slowly decreasing (Figure 17) but the complete recovery of the initial SALS pattern takes place only after 1–2 hr (data are not presented). In Figure 18 microscopy images corresponding to the same emulsion as in SALS experiments are presented first, after pre-shear of the emulsion at 0.5 sec^{-1} for 1000 sec with subsequent cessation of steady state shear flow at 60 sec^{-1}(a), 100 sec^{-1} (b), and 150 sec^{-1}. One can see an appreciable increase of the droplet size after cessation of high shear rate flow, in accordance with SALS data.

In Figure 19 the light scattering intensity of semidilute demixed water (97.5 wt%)-SC (2.0 wt%)-SA (0.5 wt%)-DSS (2×10^{-3} wt%) system after pre-shear (curve 1) and just after cessation of flow at 60 sec^{-1} (curve 2) is compared with that of water (87.8 wt%)-SC (12.2 wt%)-SA (0.1 wt%) system, containing 1 wt% SA enriched phase at the same shear history (curves 3 and 4). It is seen that the increase in the light intensity after cessation of flow takes place for both systems, however, for the former system the light intensity increased much higher that for the latter one.

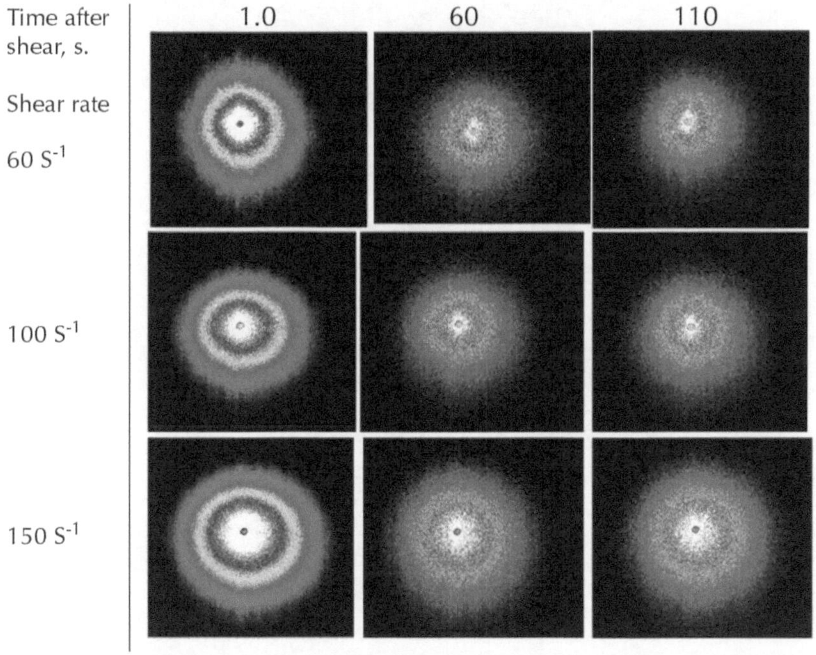

FIGURE 17 Evolution of the SALS patterns of water (97.5 wt%)-SC (2.00 wt%)-SA (0.5 wt%)-DSS (2 × 10^{-3} wt%) after cessation of a high shear rate flow. Shear rates and times of the shear as indicated on the figure. pH 7.0. I = 0.002 (phosphate buffer) and Temperature 296K. The SALS pattern of water (97.5 wt%)-SC (2.00 wt%)-SA (0.5 wt%)-DSS (2 × 10^{-3} wt%) system before high shear rate is shown in Figure1 (a).

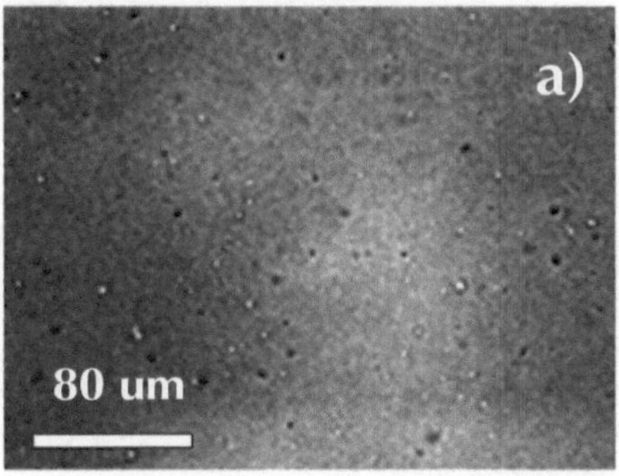

FIGURE 18 *(Continued)*

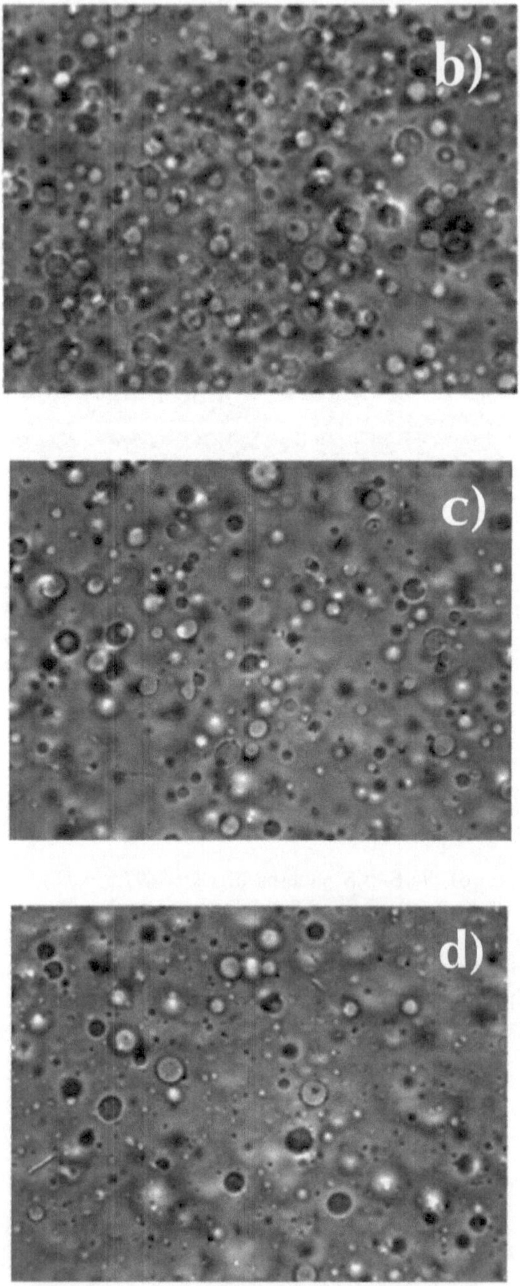

FIGURE 18 The evolution of microscopy images of water (97.5 wt%)-SC (2.00 wt%)-SA (0.5 wt%)-DSS (2×10^{-3} wt%) system before high shear rate flow (a) and just after cessation of a high shear rate flow. Shear rate—(b) 60 sec^{-1}, (c) 100 sec^{-1}, and (d) 150 sec^{-1}. pH 7.0. $I = 0.002$ (phosphate buffer) and Temperature 296K.

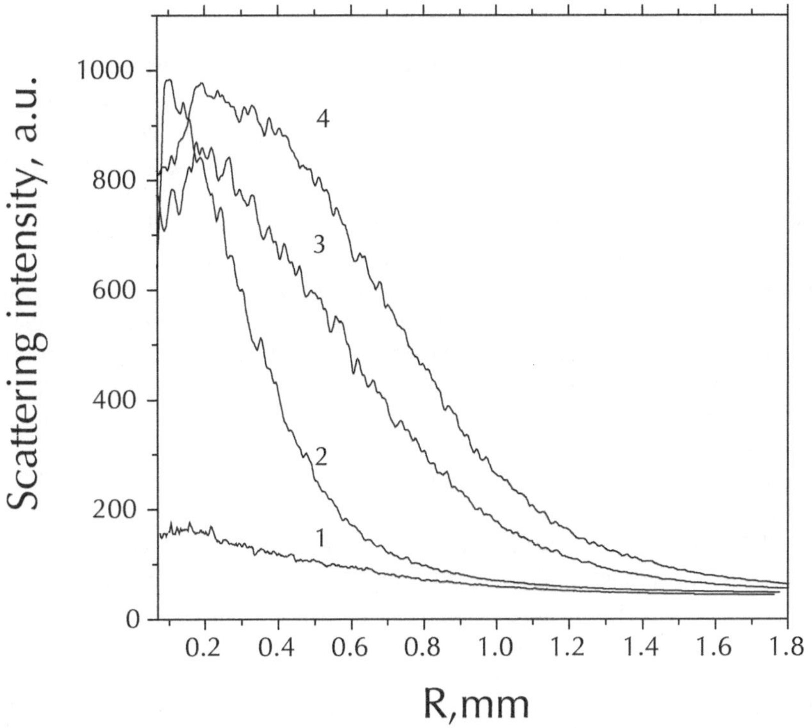

FIGURE 19 Dependence of the scattering intensity of the demixed water (97.5 wt%)-SC (2.00 wt%)-SA (0.5 wt%)-DSS (2 × 10^{-3} wt%) system after pre-shear (curve 1), and just after cessation of flow at 60 sec^{-1} (curve 2) and water (87.8 wt%)-SC (12.2 wt%)-SA (0.1 wt%) system after pre-shear (curves 3), and just after cessation of flow at 60 sec^{-1}, on the distance from the bean stop. Both systems contain 1.0 wt% SA enriched dispersed phase.

These observations can be explained on the basis of a comparison of the molecular weights of the "free" SC and SC, combined with DSS (see Figure 12). The molecular weight of the latter one is much higher than that of the former one. Note, that the second virial coefficients on the molar scale, related to pair interactions of similar SC macromolecules, A_{22} depends on the molecular weight inversely [67]. Therefore, according to conditions of the phase separation in biopolymer systems in flow [68]:

$$A_{23} > \sqrt{A_2\,A_3} \tag{2}$$

in which A_{ij} are the second virial coefficients on the molar scale, related to pair interactions of similar (2-protein, 3-polysaccharide) and dissimilar macromolecules, the protein-polysaccharide mixture containing macromolecules with lower values of A_{22} will be more predisposed to shear induced demixing.

11.4 CONCLUSION

It well known that phase equilibrium in aqueous system containing casein and linear acid polysaccharide is weakly sensitive to changes of the main physico-chemical parameters, such as pH, ionic strength, and temperature. This is the case both at rest [17, 31, 32, 34] and under shear flow [69]. It has been shown in many studies that proteins interact with acid polysaccharides forming intermacromolecular coulomb complexes mainly at pH values below the isoelectrical point of the protein (iep) when both biopolymers are oppositely charged, or at pH values slightly above iep.

In this work the attempt was performed to induce demixing in semidilute and highly compatible SC-SA mixtures in the presence of sodium salt of dextran sulfate at pH 7.0, (above the isoelectrical point of caseins) and to characterize phase equilibrium, intermacromolecular interactions, and structure of such systems by rheo-SALS, OM, phase analysis, DLS, FPLC, ESEM, and rheology. The results obtained in the present study can be summarized as follows:

The DSS is able to induce a deep segregative phase separation in semidilute SC-SA systems (at a SC concentration as low as 1 wt%) at a trace concentrations (10^{-3} wt%), DDS significantly increases the phase separation range, as well as viscosity and mechanical modules of the system. The phase separation observed is the result of formation at pH 7.0 (that is far away from the iep of the caseins/4.4–4.6/) of DSS/SC water soluble charged associates (1:10 mol/mol), having $R_H = 0.26$ um and electrostatic nature. The minimal compatibility of SC and SA was observed at the DSS/SC weight ratio of 0.14, which corresponds to an equality of the cationic groups in the protein molecules and sulfate groups in DSS. At a higher SC concentration (4 wt%) SC-DSS associates forms some kind of network.

Data of chromatography indicate that DSS interacts first with SC associates having higher molecular weight. The degree of the protein conversion in the complex increases from approx. 30 to 80% when the concentration of SC in the system grows from 1 to 2 wt%. Phase separation of semidilute ternary water-biopolymer 1-biopolymer 2 systems in the presence of DSS is observed here to be a rather common phenomenon, observed for different types of biopolymers, For example SC-gelatin type-A and gelatin type A-SA. Therefore, the use of DSS as a decompatibilizer for semidilute biopolymer systems can find applications in processes for concentrating biological materials in two phase systems, because DDS induced demixing allows decreasing the critical concentration of the phase separation significantly. Moreover, at a high shear rate flow (60–150 sec^{-1}), semidilute phase separated SC-SA-DSS systems undergo further segregative separation, decreasing the critical concentration of phase separation into the range of dilute solutions. Such peculiarities of thermodynamic and rheological behaviors allow us to consider sulfate polysaccharide interacting with protein in aqueous protein-polysaccharide mixture as a new type of decompatibilizer for biopolymer emulsions. Therefore the results obtained promote more thorough understanding of the relationships between intermacromolecular interactions in aqueous biopolymer systems and their thermodynamical properties.

KEYWORDS

- **Biopolymer mixture**
- **Demixing**
- **Complex formation**
- **Structure formation**
- **Rheo-optics**

REFERENCES

1. Hidalgo, J. and Hansen, P. M. T. *J. Dairy Sci.*, **54**, 1270–1274 (1971).
2. Wang, Y., Gao, J. Y.,and Dubin, P. L. *Biotechnol. Prog,* **12**, 356–362 (1996).
3. Dubin, P. L., Gao, J., and Mattison, K. *Sep. Purif. Methods*, **23**, 1–16 (1994).
4. Strege, M. A., Dubin, P. L., West, J. S.,and Flinta, C. D. Protein separation via polyelectrolyte complexation In *Symposium on Protein Purification: From Molecular Mechanisms to Large-scale Processes*, M. Ladisch, R. C. Willson, C. C. Paint, and S. E. Builder (Eds.), ACS Symposium Series 427, American Chemical Society, Washington, Chapter 5, p. 66. (1990).
5. Antonov, Yu. A., Grinberg, V. Ya., Zhuravskaya, N. A., and Tolstoguzov, V. B. *Carbohydrate polymers*, **2**, 81–90 (1982).
6. Antonov, Yu. A. Application of the Membraneless Osmosis Method for Protein Concentration from Molecular-dispersed and Colloidal dispersed Solutions. *Review Applied Biochemistry and Microbiology* (Russia) Engl.Transl., **36**, 382–396 (2000).
7. Kiknadze, E. V. and Antonov, Y. A. *Applied Biochemistry and Microbiology* (Russia) Engl. Transl., **34**, 462–465 (1998).
8. Xia, J. and Dubin, P. L. Protein-polyelectrolyte complexes In: *Macromolecular Complexes in Chemistry and Biology*, P. L. Dubin, R. M. Davis, D. Schultz, and C. Thies (Eds.), Springer-Verlag, Berlin, Chapter 15 (1994).
9. Ottenbrite, R. M. and Kaplan, A. M. *Ann. N. Y. Acad. Sci.*, **446**, 160–168 (1985).
10. Magdassi, S. and Vinetsky, Y. Microencapsulation: methods and industrial applications. In *Microencapsulation of oil-in-water emulsions by proteins*, S. Benita (Ed.) Marcel Dekker Inc., New York, pp. 21–33 (1997).
11. Regelson, W. *Journal of Bioactive and Compatible Polymers,* **6**, 178–216 (1991).
12. Albertsson, P. Å., Johansson, G., and Tjerneld, F. In *Separation Processes in Biotechnology*, J. A. Asenjo (Ed.), Marcel Dekker, New York, pp. 287–327 (1990).
13. Harding, S., Hill, S. E., and Mitchell, J. R. *Biopolymer Mixtures*, University Press, Nottingham, pp. 499 (1995).
14. Tolstoguzov, V. B Functional properties of protein-polysaccharide mixtures. In: *Functional properties of food macromolecules*, S. E. Hill, D. A. Ledward, and J. R. Mitchel, Aspen Publishers Inc., Maryland, Gaitherburg (1998).
15. Piculell, L., Bergfeld, K., and Nilsson, S. Factors determining phase behavior of multicomponent polymer systems. In *Biopolymer Mixtures*, S. E. Harding, S. E. Hill, and J. R. Mitchell, (Eds.), Nottingham University Press, Nottingham, pp. 13–36 (1995).
16. Antonov, Y. A., Van Puyvelde, P., and Moldenaers, P. *Biomacromolecules,* **5**(2), 276–283 (2004).
17. Antonov, Y. A., Van Puyvelde, P.,and Moldenaers, P. *IJBM*, **34**, 29–35 (2004).

18. Antonov, Y. A. and Wolf, B. *Biomacromolecules*, **7**(5), 1582–1567 (2006).
19. Walter, H., Brooks, D. E., and Fisher, D. In *Partitioning in Aqueous Two-Phase Systems: Theory, Methods, Uses, and Applications to Biotechnology*, Academic Press, London (1985).
20. Hellebust, S., Nilsson, S., and Blokhus, A. M. *Macromolecules*, **36**, 5372–5382 (2003).
21. Scott, R. *J. Chem. Phys*., **17**(3), 268–279 (1949).
22. Tompa, H. *Polymer Solutions*, Butterworth, London (1956).
23. Flory, P. J. *Principles of Polymer Chemistry*, Cornell University Press, Ithaca, New Yark (1953).
24. Prigogine, I. *The Molecular Theory of Solution*, North Holland, Amsterdam (1967).
25. Patterson, D. *Polym. Eng. Sci.*, **22**(2), 64–73 (1982).
26. Gottschalk, M., Linse, P., and Piculell, L. *Macromolecules*, **31**, 8407–8416 (1998).
27. Lindvig, T., Michelsen, M. L., and Kontogeorgis, G. M. *Fluid Phase Equilib.*, **203**, 247–260 (2002).
28. Piculell, L., Nilsson, S., Falck, L., and Tjerneld, F. *Polym. Commun.*, **32**, 158–160 (1991).
29. Bergfeldt, K., Piculell, L., and Tjerneld, F. *Macromolecules*, **28**, 3360–2270 (1995).
30. Bergfeldt, K. and Piculell, L. *J. Phys. Chem.*, **14**, 5935–5840 (1996).
31. Antonov, Y. A., Grinberg, V. Ya., and Tolstoguzov, V. B. *Starke*, **27**, 424–431 (1975).
32. Antonov, Y. A., Grinberg, V. Ya., Zhuravskaya, N. A., and Tolstoguzov, V. B. *J. Texture Studies*, **11**(3), 199–215 (1980).
33. Rha, C. K. and Pradipasena, P. In *Functional Properties of Food Macromolecules*, J. R. Mitchell and D. A. Ledward (Eds.), Elsevier Applied Science, London, pp. 79–119 (1986).
34. Whistler, R. L. *Industrial Gums 2nd edn.*, Academic Press, New York (1973).
35. Capron, I., Costeux, S., and Djabourov, M. *Rheol. Acta*, **40**, 441–456 (2001).
36. Van Puyvelde, P., Antonov, Y. A., and Moldenaers, P. *Food Hydrocolloids*, **17**, 327–332 (2003).
37. Koningsveld, R. and Staverman, A. J. *Journal of Polymer Science A-2*, pp. 305–323 (1968).
38. Polyakov, V. I, Grinberg, V. Ya., and Tolstoguzov, V. B. *Polymer Bulletin*, **2**, 760–767 (1980).
39. Dubois, M., Gilles, K. A., Hamilton, J. K., Revers, P. P., and Smith, T. *Anal.Chem.*, **18**, 350–356 (1956).
40. Zaslavsky, B. Y. *Aqueous two phase partitioning: Physical Chemistry and Bioanalytical Applications*, Marcel Dekker, New York (1995).
41. Grinberg, V. Ya. and Tolstoguzov, V. B. *Food Hydrocolloids*, **11**, 145–158 (1997).
42. Van Puyvelde, P., Antonov, Y. A., and Moldenaers, P. *Food Hydrocolloids*, **17**, 327–332 (2003).
43. Overbeek, J. T. G. and Voorn, M. J. *J. Cell Comp. Physiol.*, **49**, 7–39 (1957).
44. Kaibara, K., Okazaki, T., Bohidar, H. P., and Dubin, P. *Biomacromolecules*, **1**, 100–107 (2000).
45. Antonov, Yu. A., Dmitrochenko, A. P., and Leontiev, A. L. *Int. Journal Biol. Macromol.*, **38**(1), 18–24 (2006).
46. Antonov, Y. A. and Gonçalves, M. P. *Food Hydrocolloids*, **13**, 517–524 (1999).
47. Michon, C., Konate, K., Cuvelier, G., and Launay, B. *Food Hydrocolloids*, **16**, 613–618 (2002).
48. Galazka, V. B., Ledward, D. A., Sumner, I. G., and Dickinson, E. *Agric. Food Chem.*, **45**, 3465–3471 (1997).
49. Dickinson, E. *Trends in Food Science and Technology*, **9**, 347–354 (1988).

50. Snoeren, T. H. M., Payens, T. A. J., Jevnink, J., and Both, P. *Milchwissenschaft*, **30**, 393–395 (1975).
51. Kabanov, V. A., Evdakov, V. P., Mustafaev, M. I., and Antipina, A. D. *Mol. Biol.*, **11**, 582–597 (1977).
52. Noguchi, H. *Biochim. Biophys. Acta*, **22**, 459–462 (1956).
53. Kumihiko, G. and Noguchi, H. *Agriculture and Food Science,* **26**, 1409–1414 (1978).
54. Sato, H. and Nakajima, A. *Colloid and Polymer Science*, **252**, 294–297 (1974).
55. Sato, H., Maeda, M., and Nakajima, A. *Journal of Applied Polymer Science*, **23**, 1759–1767 (1979).
56. Bungenberg de Jong, H. G. Crystallisation-Coacervation-Flocculation. In *Colloid Science*, H. R. Kruyt (Ed.), Elsevier Publishing Company, Amsterdam, **2**, Chapter 8, pp. 232–258 (1949).
57. Gurov, A. N., Wajnerman, E. S., and Tolstoguzov, V. B. *Stärke*, **29**(6), 186–190 (1977).
58. Antonov, Y. A., Lefebvre, J., and Doublier, J. L. *Polym. Bull.*, **58**, 723–730 (2007).
59. Antonov, Y. A., Lashko, N. P., Glotova, Y. K., Malovikova, A., and Markovich, O. *Food Hydrocolloids*, **10**(l), 1–9 (1996).
60. Polyalov,V. I., Grinberg, V. Ya., Antonov, Y. A., and Tolstoguzov, V. B. *Polymer Bulletin*, **1**, 593–597 (1979).
61. Polyalov, V. I., Kireeva, O. K., Grinberg, V. Ya., and Tolstoguzov, V. B. *Nahrung*, **29**, 153–160 (1985).
62. Swaisgood, H. E. In *Advanced Dairy Chemistry-1:Proteins*, P. F. Fox (Ed.), Elsevier Applied Science, London (1992).
63. Langendorff, V., Cuvelier, G., Launay, B., Michon, C., Parker, A., and De Kruif, C. G. *Food Hydrocolloids*, **13**, 211–218 (1999).
64. Bowman, W. A., Rubinstein, M., and Tan, J. S. *Macromolecules,* **30**(11), 3262–3270 (1997).
65. Junhwan, J. and Dobrynin, A. *Macromolecules*, **38**(12), 5300–5312 (2005).
66. Smid, J. and Fish, D. In: *Encyclopedia of polymer science and engineering, 2nd edn. Polyelectrolyte complexes*, H. F. Mark, N. M. Bikales, C. G. Overberger, and G. Menges (Eds.), New York, Wiley/Interscience, **11**, 720 (1988).
67. Striolo, A., Ward, J., Prausnitz, J. M., Parak, W. J., Zanchet, D., Gerion, D., Milliron, D., and Alivisatos, A. P. *J.Phys. Chem. B*, **106**(21), 5500–5505 (2002).
68. Antonov, Y. A., Van Puyvelde, P., and Moldenaers, P. *Biomacromolecules*, **5**(2), 276–283 (2004).
69. Antonov, Y. A.,Van Puyvelde, P., and Moldenaers, P. *Food Hydrocolloids*, **23**, 262–270 (2009).

CHAPTER 12

COMPARISON OF TWO BIOREMEDIATION TECHNOLOGIES FOR OIL POLLUTED SOILS (RUSSIA)

V. P. MURYGINA, S. N. GAIDAMAKA, and S. YA. TROFIMOV

CONTENTS

12.1 INTRODUCTION

This chapter deals with two bioremediation technologies of bogs, accidentally polluted with oil, which are applied in the Northern part of Russia in the Komi Republic and the Western Siberia. One of the technologies is a typical *ex-situ* and the other is *in-situ* one without gathering of oil out of the surface of the bog and milling of the moss. So, different results were obtained after bioremediation of bogs with an oil-oxidizing preparation Rhoder there.

The basic oil production places in Russia are situated in the Northern parts of the Komi Republic and in the Western Siberia. Vast scale territories polluted by oil often are located in difficultly passable bogs. Penetration depth of oil in such bogs does not exceed of 0.3–0.6 m and oil is usually propped up with a permafrost layer. Areas contaminated with oil are huge: from 1–2 to 10 hectares or more. Climate of the regions is similar with severe prolonged winter and cool short summer. In these two regions different remediation technologies are applied for restoration of bogs polluted with oil. This is due to many factors: economic, geologic, availability of contaminated sites for facilities, volume of oil spills, sizes of areas contaminated with oil and so on.

In the Komi Republic before the bioremediation a lot of oil is carefully gathered from the surface of polluted bogs. The surfaces of the bogs are often washed from the residual oil by water with pumps, sometimes with addition of surfactants [1, 2]. Gathered oily water and previous collected oil are transferred to refinery plant. In winter or in early spring until bogs thaw out, the top layer of 5–7 cm of highly contaminated soil is usually cut off and excavated, washed off on special devices from oil and all hydrocarbons (HC) are transferred to refinery plant. However, it is impossible to wash off the soil completely till the most probably concentration of oil. And the soil with the residual oil in it is returned to the same original location. The bioremediation with or without oil-oxidizing microorganisms and fertilizers and following phytoremediation are usually performed on this soil.

In the Western Siberia *in-situ* bioremediation of the soil is preferred because the excavation of a top layer of the soil polluted by crude oil from huge areas of impassable bogs is technically difficult and economically inefficient. Besides, the sheet of water settles down under a moss layer of thickness about 3-10 m on the most part of impassable bogs. With the best case such bogs are once milled at the very beginning of summer, until the permafrost layer completely thawed. At once a large amount of fertilizers, lime, seeds of oats and any oil-oxidizing preparation is brought into the moss [3]. At worst for example behind the Polar Circle of the Western Siberia the polluted bogs are left without any treatments.

In the present study, the efficiency of two remediation technologies for bogs polluted with oil which were applied in Russia are compared. This study presents an approach for development of a new bioremediation technology for impassable bogs polluted with oil in the Western Siberia without application of consecutive stages of a classical remediation of technical, agrochemical, and biological ones.

12.2 MATERIALS

12.2.1 THE OIL-OXIDIZING PREPARATION RHODER

The Rhoder consists of two bacterial strains belonged to the genus Rhodococcus, (*R. ruber* Ac-1513 D and *R. erythropolis* Ac-1514 D), isolated from soils polluted with crude oil. The strains are non-pathogenic and non-mutagenic to humans, animals, plants, and bacteria. The Rhoder is approved for widespread using in nature and it has been successfully used for bioremediation of oil refinery sludge, soils, wetlands, and water surfaces polluted with oil [4-11] and the Rhoder is used in these described field-scale tests.

12.2.2 BACKGROUND

THE KOMI REPUBLIC, USINSK TOWN

In 2008 soil polluted with a residual oil was taken from a special device of washing off the oil sludge and returned to the same place from which previously was excavated. This place was an area with the size of 136 × 82 m^2 near Usinsk town. The bioremediation of soil was started in early July, 2008. The weather at the beginning of July, 2008, was dry and hot for two weeks, the temperature varied from 17°C to 33°C and then it went down. Rains were rare that was atypical for this region at that time of a year. In August the temperature went down till 15–11°C and rains began.

The subsoil of the site for the washed off soil was presented by genetic types of a lake and marine alluvial precipitations. The alluvial sediments situated into surfaces were presented also with brown peat clay sand and light gray-brown sequence of semi loam. Marine deposits were presented by loam of gray-blue clay with inclusions of gravel and pebble of the large size.

BIOREMEDIATION

Bioremediation of the soil was carried out with addition of 100 kg of dry fertilizer into the soil and treated it with a disk harrow. Then a working solution of the Rhoder with the most probably number (MPN) of active hydrocarbon oxidizing (HCO) cells of 2.5 × 10^6 cells/ml with addition of 0.1% fertilizer was sprinkled with a water cart on the soil surface and then the soil was milled again. Three such treatments with the interval of 2–3 weeks were performed and the Rhoder in a total quantity of 30 kg of a dry powder with HCO bacteria cells of 1.0 × 10^{10} in 1g of powder was used for the bioremediation of this area. The soil was milled after each introduction of the Rhoder.

SAMPLING FOR ANALYSIS

Soil samples for analysis were selected from five points of the site before the bioremediation and before each next treatment with the Rhoder and a half month after the completion of treatments. Samples taken from one point of the site were mixed carefully and homogenized samples (about 0.5 kg) were passed to analyze. Every time five samples were taken for analysis.

THE WESTERN SIBERIA, MURAVLENKO TOWN

In 2011, an impassable bog with a size about 0.8 hectares polluted with spring accidental oil spill and halved by high knolls was offered for the bioremediation with using the Rhoder only. This bog located near the Muravlenko town. Typical marsh plants (moss, cloudberry, wild rosemary) existed on the knolls, which were practically not affected by the oil spill. Large spots of the oil were situated on swampy impassable depressions. The vegetation (moss, sedge) on these depressions perished almost completely. A layer of the oil with a thickness about 1 cm and more was presented on the water surfaces on these depressions. The penetration of the oil into the moss was about 40-45 cm. The oil contamination of the bog was unequal. The bog had a slight bias towards a sand bank which had been made to prevent spreading of the oil pollution and, in fact, turned into the road. Two previously digged pits to collect oil with the pump were presented on the bog. However, oil was gathered poorly, and these pits still had much of oil. The thickness of the oil on the water surfaces of the pits were more than 1 cm. The oil under an air temperature below 10°C became viscous on the surface of the water of the depressions and pits. The oil from the surface of the polluted bog was not collected additionally. The soil was not mixed by a disk harrow or other devices. An attempt to perform the bioaugmentation with the Rhoder was undertaken without additional gathering of the oil and without application of milling because of technical complication of doing classical *ex-situ* remediation on the impassable bog polluted with oil. It was needed to minimize expenses on the bioremediation.

The weather during the bioremediation of this bog was not favorable, the air temperature did not exceed 10–14°C, and it was raining from time to time.

BIOREMEDIATION

The oil polluted bog was treated three times with intervals for three weeks by the working solution of the Rhoder with the MPN of hydrocarbon oxidizing cells of 1.0×10^8 per 1 ml by the sprinkling from the fire-engine vehicle, previously washed with water. The Rhoder was used in total quantity of 120 kg as a liquid concentrate with HCO bacteria cells of 1.0×10^{11} cells/mL.

CONTROL OF THE SOIL TOXICITY

After the third application of the Rhoder the seeds of oats and perennial grasses were sown on the cleaning area to determine a toxicity of the soil and perform the phytoremediation. Half of seeds were previously treated with a solution of the hamates "Extra" (Russia) to identify an impact of the hamates on a stability of herbs germination on the bog after bioaugmentation with the Rhoder. The oil contaminated bog was divided in two parts with using landmarks. One part of the site was sowed with seeds previously treated with the solution of hamates (right), second (left) one was sowed of the seeds without any treatment. On the right and left halves of the bog two plots (size of 1.5×5.0 m^2) were done and covered with a non-woven material to assess the impact of it on the bog restoration. This non-woven material usually is recommended in the farming and gardening sectors to protect sprouts of plants from adverse environmental conditions.

SAMPLING FOR ANALYSIS

Soil samples were collected before and after finishing of the bioremediation from 12 points of the bog contaminated with oil from the depths of 0–10 cm and 10–25 cm, and 25–40 cm (by using GPS) for microbiological, chemical and agrochemical analysis. Each sample had weight about 150 g.

12.3 METHODS

12.3.1 CHEMICAL AND AGROCHEMICAL ANALYSES.

Several samples (8 samples) of moss (No. 1, 2, 3, 12, 18, 22, 24, 26) from the bog in the Western Siberia, Muravlenko town, were excessively polluted with crude oil. The oil from these samples were at first extracted by chloroform (150 mL) in chemical flasks (each flask with a capacity of 400 ml), which were shaken for 30 min at room temperature. Received solutions of the oil were transferred to the other flasks through waterless sodium sulfate (Na_2SO_4) to remove remains of water. The chloroform was evaporated at 75°C. Each sample of the moss excessively polluted with oil was then extracted three times as described above. The chloroform extracts in flasks were heating at 105°C till a constant weight. Samples of the moss after oil extraction were dried at 75°C, weighed and the oil was calculated per 1 kg of a dry moss. Chemical analyses of HC in all other samples and in the dried samples of the moss after previous chloroform extractions at room temperature were carried out by a gravimetric method with using of Soxhlet's apparatus and column chromatography with Silica gel, gas chromatograph (GC), and high-performance liquid chromatography (HPLC) methods were used too [12].

Saturated HCs from each sample after column chromatography with Silica gel (1 μL of hexane fraction) were analyzed on GC Cristallux 4000 m by company Meta

Chrom (Russia), column was OV-101, 50 m × 0.22 mm × 0.50 mkm, detector was flame ionization detector (FID), gas-carrier was nitrogen. Detector temperature was 300°C, initial column temperature was 80°C, velocity of heating was 12° per minute till the temperature 270°C. Time of analysis was 40 min. Mixture of Undecane, Dodecane, Tetradecane, Hexadecane, and Squalane were used as external standards in concentrations of 5 μg/μL for each substance.

The HPLC analyses were carried out on Knauer HPLC with the ultra-violet detector, on the reversed-phase column of Diasfer-110-C18 for HPLC, length of the column was 250 mm, the diameter was 4 mm, and the grains were 5 microns. Samples for analyses on HPLC were prepared after drying of hexane fractions and following extraction each sample with 1 mL of acetonitrile during 20 min under shaking and then analyzed. Phenantrene, pyrene, and benzo(e)pyrene were used as external standards in concentrations of 10 μg/mL for each substance in acetonitrile [13].

The humidity and pH in samples of the moss were determined with standard agrochemical methods and the soluble nitrogen and phosphorus were determined calorimetrically [14].

12.3.2 MICROBIOLOGICAL ANALYSES

The MPN of microorganisms in the treated soils and in the Rhoder were estimated by method of ten-fold dilutions with using the meat-peptone agar for heterotrophic bacteria (HT) and selective agar for detection of actinomicetes, pseudomonas, nitrifying, ammonifying, oligotrophic microorganisms, and fungi [15].

Modified liquid Raymond media with crude oil was used to determine of hydrocarbon oxidizing (HCO) bacteria [16] (g/L): Na_2CO_3 [0.1]; $CaCl_2 \times 6\ H_2O$ [0.01]; $MnSO_4 \times 7\ H_2O$ [0.02]; $FeSO_4$ [0.01]; $Na_2HPO_4 \times 12H_2O$ [1.0]; KH_2PO_4 [1.0]; $MgSO_4 \times 7\ H_2O$ [0.2]; NH_4Cl [2.0]; NaCl [5.0]; pH = 7.0. Raymond media was prepared on the distilled water with consistently bringing the components. Then 4.5 ml of media was placed in each test tube. The 50 mg of crude oil were added to each test tube and all test tubes were sterilized under 121°C during 30 min.

Two drops of 0.05% solution of Twin 80 were added in the first test tubes with the soil samples (about 1 g) and carefully stirred up to wash away most fully cells of HCO bacteria from soil particles to prepare ten-fold dilutions. After preparing ten-fold dilutions of the investigated soil samples (or the Rhoder) 0.5 ml of dilutions was passed into test tubes with Raymond media. Test tubes were incubated at 28°C for two weeks and results were considered on a dispersion of the oil or disappearance of the oil film, or by a turbidity or oil inflation that testified an ability of microorganisms from samples of soil to utilize oil. Total number of HCO microorganisms in samples was determined by the last number of the test tube in which dispersion of oil or inflation or turbidity in the liquid media or disappearance of oil was observed.

Microorganisms were identified by microscopic observation: cells morphology, motility, gram-coloring, capsule- and spore-forming, acid-fast, and conidial stage. Morphology examination involved the shape and color of colonies, mycelia, and pseudo mycelia, growth on the selective agar media and a biochemical characterization.

12.4 RESULTS AND DISCUSSION

12.4.1 *BIOREMEDIATION OF WASHED OFF SOIL POLLUTED WITH RESIDUAL OIL IN THE KOMI REPUBLIC*

MICROBIOLOGICAL MONITORING

Preliminary microbiological analysis of samples of the soil polluted with residual oil after washing off (from 14.05.2008) had showed that the MPN of different groups of microorganisms varied from 10^3 to 10^4 CFU/g of soil (Table 1). Composition species of soil microorganisms was presented of genus Bacillus, Pseudomonas, Rhodococcus, Nitrosomonas, Nitrosococcus, Nitrobacter, Azotobacter, Aspergillus, Penicillium, and Fusarium. In soil samples, which were selected before directly starting of bioremediation (03.07.2008), the MPN of the same groups of microorganisms was higher by 2-4 orders, caused by a positive effect of a warm weather. However, the MPN of HCO microorganisms (7.2×10^5 CFU/g of soil) was still insufficient for effective degradation of HC in soil polluted with residual oil and an introduction of the oil-oxidizing preparation Rhoder was justified.

TABLE 1 MPN of different types microorganisms which presented in the washed soil with residual oil before and during bioremediation with the Rhoder

Sample	MPN, CFU/g of soil						
	HT	Ammo-nifying	Nitrify-ing	Oligo-nitro-philic	Pseudo-monas	Mold	HCO
14.05.2008	5.2×10^4	1.4×10^4	5.2×10^4	1.7×10^4	1.2×10^4	280	1.0×10^3
03.07.2008	1.5×10^8	1.7×10^8	2.9×10^8	2.2×10^7	1.3×10^6	2.5×10^6	2.7×10^5
16.07.2008	6.2×10^8	1.9×10^8	8.9×10^8	2.2×10^7	1.0×10^7	4.1×10^6	8.1×10^7
04.08.2008	6.0×10^7	3.9×10^7	5.0×10^7	1.3×10^6	1.1×10^8	4.1×10^6	2.1×10^6
29.09.2008	5.7×10^7	2.9×10^7	5.0×10^7	6.9×10^6	1.2×10^7	1.4×10^6	7.8×10^6

Note: the averages are the total number of microorganisms identified in samples taken from five points of the plot

Introduction of Rhodococcus from the Rhoder into the soil during the bioremediation (three times) had increased the MPN of HCO bacteria responsible for degradation of oil (Table 1), and the process of oil decontamination of the soil was activated. In addition, the introduction of the Rhoder did not adversely affect on the indigenous microorganisms that was evident in the obtained results (Table 1). The peak of MPN of all analyzed groups of microorganisms was observed in the middle of July till early August 2008, and the MPN of microorganisms was remained relatively high for a long time.

AGROCHEMICAL MONITORING

Preliminary agrochemical analysis of the washed off soil (May 2008) showed a relatively high content of nitrogen (as ammonia) and phosphorus which were available to plants and microorganisms (Table 2). Additional introduction of the fertilizer (100 kg/ha) before the bioremediation of the soil had optimized concentration of the fertilizer for microorganisms and plants [14]. Soil moisture was sufficient for the life of microorganisms responsible for the biodegradation of HC and restoration of the biological activity of the soil. The pH of the soil was close to neutral and was supported by the introduction of the lime before the bioremediation and before the third treatment with the Rhoder and subsequent phytoremediation on the bog (Table 2).

TABLE 2 Monitoring of agrochemical parameters of the soil before and in the process of bioremediation with the Rhoder

Date of sampling	**pH**	**Humidity, %**	**Nitrogen $N\text{-}NH_4^+$ mg/kg of soil**	**Phosphorus PO_4^{3-} mg/kg of soil**	**Salinity of soil, %**	**Contamination, g/kg**
14.05.08	6.73 ± 0.24	65.80 ± 7.13	12.92 ± 5.59	56.39 ± 10.11	-	126.0
03.07.08	6.62 ± 0.33	76.70 ± 9.00	13.13 ± 3.97	56.39 ± 10.11	10.9	101.0
04.08.08	6.48 ± 0.06	73.47 ± 10.24	31.02 ± 2.23	89.45 ± 27.74	-	73.0
29.09.08	6.84 ± 0.21	66.46 ± 4.55	16.40 ± 2.42	75.58 ± 23.22	6.6	69.8

Note: not determined

HC DEGRADATION DURING THE BIOREMEDIATION

The initial concentration of the oil in the washed off soil before bioremediation was 126.1 g/kg of dry matter (DM). Group composition of HC (average value) in this soil before application the Rhoder and during the bioremediation and after the phytoremediation of the soil is shown on Figure 1(a). Three times application of the Rhoder decreased the concentration of the pollutant in the soil an average by 43.6%. The concentration of the saturated HC was decreased by 61.5%, the aromatic HC by 30.4%, resins and asphaltenes only by 3.4% (Figure 1(a)).

The GC analysis of the soil from the bog before and after bioremediation confirmed the obtained results. According to results of GC analysis the degradation of oil products in soil was by 74–84 % (Figure 1(B-C)).

Analysis of aromatic HC on HPLC in the washed off soil before and after the bioremediation showed a significant reduction of these compounds, though a small peak similar to benzo(e)pyrene was found in samples of the soil after the bioremediation with the Rhoder (Figure 1(d)). An appearance of such peak looked like the benzo(e) pyrene associated with the formation of intermediate compounds by microorganisms, which would be eventually degraded by cells of the Rhoder. Previously in laboratory experiments sometimes the appearance and then disappearance of a like benzo(e)pyrene substance in soils was observed during the bioremediation with the Rhoder.

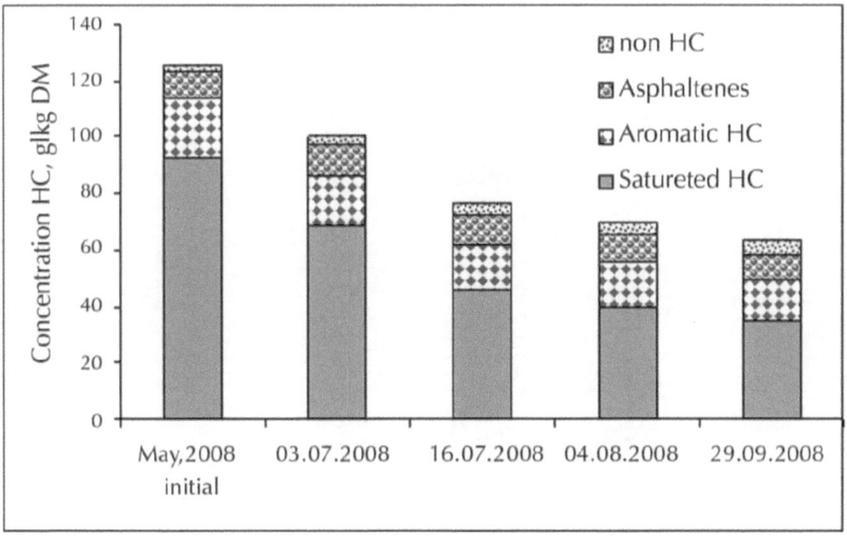

FIGURE 1 *(Continued)*

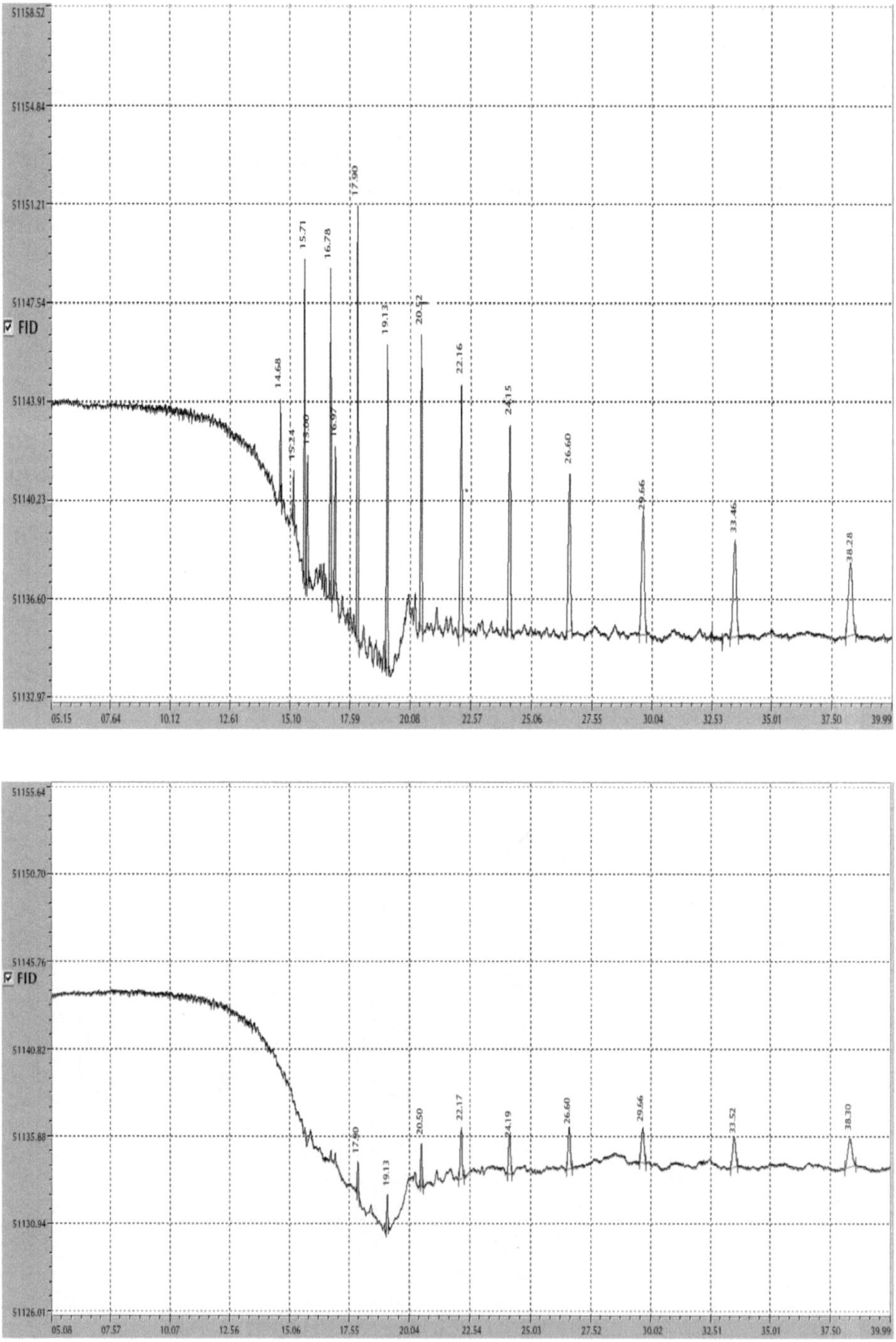

FIGURE 1 *(Continued)*

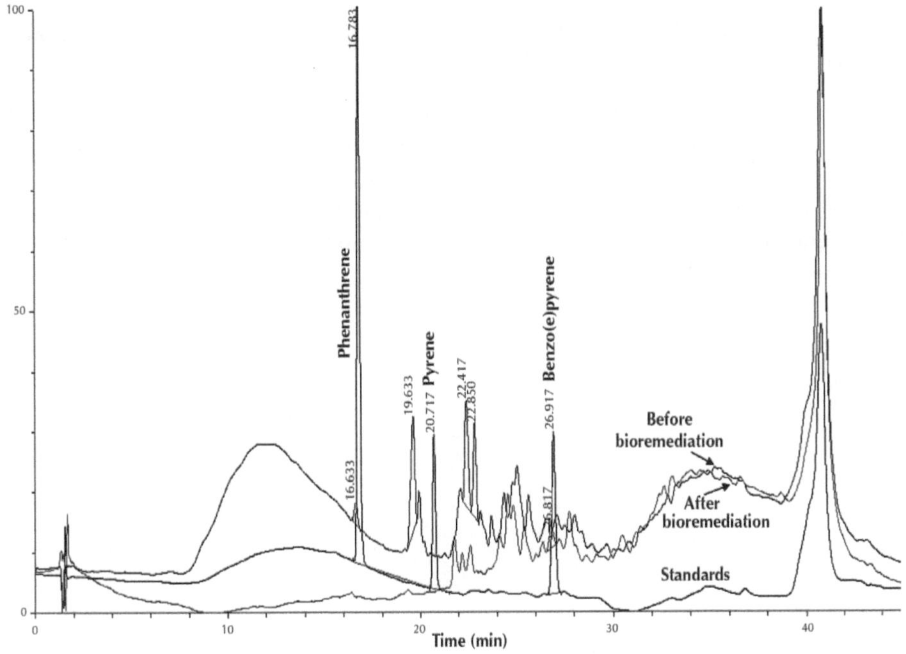

FIGURE 1 (a) Bioremediation of washed off soil with the Rhoder, 2008; (b) GC analysis of washed off soil before bioremediation with the Rhoder, 2008; (c) GC analysis of the washed off soil after bioremediation with the Rhoder, 2008; (d) HPLC analysis of the washed off soil before and after bioremediation with the Rhoder, 2008

PHYTOREMEDIATION OF THE WASHED OFF SOIL WITH RESIDUAL OIL AFTER BIOREMEDIATION

The phytoremediation of the soil after the previous bioremediation with the Rhoder in September, 2008, was not successful because of rain during all month. A lot of plantlets rotted through it. Salinity of the soil under these circumstances was decreased almost by 40% on this site (Table 2).

Thus, this technology, including the excavation of the soil polluted with oil out of the bog, and more complete extraction of the oil out of this soil and subsequent processing this oil and bringing it into a commodity form and then a realization of it, allow partially to offset outlays on the equipment service and energy expense. Subsequent the bioremediation of the washed off soil from residual oil (or the bioremediation + the phytoremediation) allows reducing a restoration period of oil polluted areas till 1–2 years and improving the quality of the remediation. This remediation technology of any soil contaminated by accidental oil spill in the Komi Republic can be really regarded as a comprehensive and non-waste one.

12.4.2 BIOREMEDIATION OF THE IMPASSIBLE BOG IN THE WESTERN SIBERIA, MURAVLENKO TOWN

The allocated object was very strongly polluted with oil (Figure 2), and it was difficult to expect a big success in such situation. Nevertheless, it was made a decision to test oil-oxidizing ability of the Rhoder in such extreme conditions. It was on the one hand; on the other hand it was necessary to be convinced that bioaugmentation with the Rhoder can initiate of self-restoration process though it may be not such effective as the *ex-situ* technology.

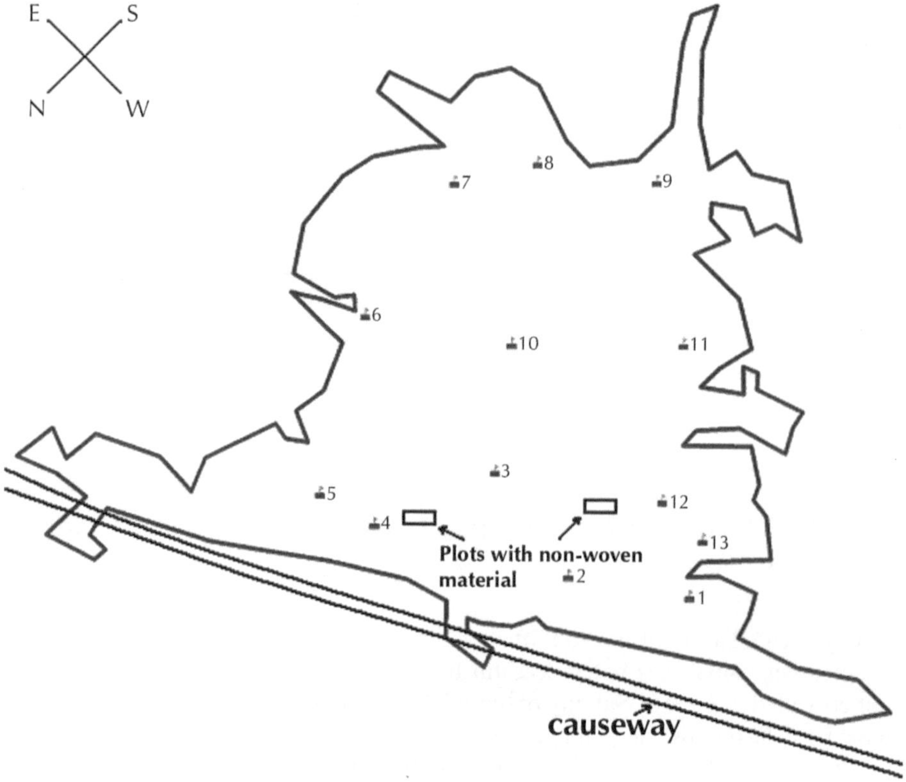

FIGURE 2 Scheme of the oil polluted bog with points of sampling, Western Siberia, Muravlenko, 2011.

MICROBIOLOGICAL MONITORING

Preliminary microbiological analysis of soil samples showed (Table 3) that a lot of microorganisms were presented in layers of 0–10cm. In these upper layers of the soil the MPN of heterotrophic bacteria (HT) varied from 1.1×10^7 to 6.1×10^8 CFU/g of the soil. In these points the level of the oil contamination varied from 60.3 g/kg DM to 903.6 g/kg DM. The MPN of HCO bacteria varied from 1.2×10^6 cells/g to 1.1×10^8cells/g of the soil. In samples with a very high oil pollution the MPN of HCO cells was only 1.0×10^3cells/g of the soil. In other samples taken from different depths of the bog the MPN of HT and HCO microorganisms was lower (Table 3). After three times introduction of the Rhoder the total number of HT microorganisms as a whole did not decrease and even was increased in some samples by 1 order. The MPN of HCO bacteria increased by about 2 orders and more in the majority of the samples (Table 4). The negative influence of the oil-oxidizing preparation Rhoder on indigenous microorganisms was not observed.

TABLE 3 Microbiological and agrochemical characteristics of soil samples selected from different depths in the oil polluted area before the augmentation with the Rhoder

Samp-les	Depth of samples selec-tion	Soil pH	HT CFU/g of soil	HCO, cells g of soil	Nitrogen N-NH_4^+ mg/kg of soil	Phosphorus PO_4^{3-} mg/kg of soil
1	(0–10)	-	-	-	-	-
	(10–25)					
2	(0–10)	5.2	2.8×10^7	3.6×10^4	5.08	-
	(10–25)	4.9	2.5×10^7	4.3×10^4	2.99	33.11
3	(0–10)	4.9	6.1×10^8	8.1×10^7	11.58	31.80
	(10–25)	5.0	3.8×10^8	6.0×10^4	9.18	22.28
4	(0–10)	5.4	2.8×10^8	1.1×10^8	17.40	33.53
	(0–25)	5.1	6.1×10^7	3.8×10^7	7.56	-
5	(0–10)	5.2	1.1×10^7	4.9×10^7	21.02	20.81
	(10–25)	5.0	5.1×10^7	8.0×10^5	15.91	-
	(25–40)	4.9	1.8×10^7	7.9×10^5	9.67	-

TABLE 3 *(Continued)*

6	(0–10)	-	-	-	-	-
	(10–25)	5.	2.5×10^6	6.0×10^4	6.2	-
7	(0–10)	5.3	1.9×10^7	8.4×10^4	10.72	16.64
	(10–25)	4.9	2.6×10^7	5.0×10^4	16.01	-
8	(0–10)	4.9	7.6×10^7	8.0×10^4	11.14	19.10
	(10–25)	5.0	6.4×10^6	7.7×10^3	7.07	-
9	(0–10)	5.0	1.1×10^8	7.1×10^3	4.83	-
	(10–25)	4.9	8.9×10^5	8.1×10^5	7.56	-
10	(0–10)	5.1	7.3×10^7	1.0×10^4	6.35	-
	(10–25)	5.2	2.8×10^6	1.0×10^7	6.54	-
11	(0–10)	-	-	-	-	-
	(10–25)	5.0	5.1×10^6	7.7×10^5	3.30	-
12	(0–15)	-	-	-	-	-
	(15–30)	4.9	5.9×10^7	1.2×10^6	18.51*	-
13	(0–10)	-	-	-	-	-
	(10–25)	5.0	6.1×10^7	9.6×10^4	8.35	-

Note: - not detected because samples were unable to determine due to their high oil content or an insufficient amount of it

AGROCHEMICAL ANALYSIS

The boggy soil had initial pH from 4.9 to 5.4 and a content of nitrogen and phosphorus compounds in the soil was low (Table 3). Unfortunately, lime in amount of 400 kg and fertilizers (600 kg) were added manually on the oil polluted surface of the bog before the third application of the Rhoder due to circumstances beyond our control. Nearly one and a half months later it was observed that introduction of the lime had led to some increase of value pH in the soil and more favorable conditions for activity of the oil-oxidizing bacteria of the Rhoder (Table 4).

TABLE 4 Microbiological and agrochemical characteristics of soil samples selected from different depths on the oil polluted area after augmentation with the Rhoder

Samples	Depth Samples selec-tion	Soil pH	MPN of HT CFU/g of soil	MPN of HCO cells/g of soil	Nitrogen $N\text{-}NH_4^+$ mg/kg of soil	Phosphorus PO_4^{3-} mg/kg of soil
2	(0-10)	5.2	2.8×10^7	4.4×10^6	172.5	5.17
	(10-25)	6.3	2.5×10^7	4.5×10^6	99.6	8.84
3	(0-10)	6.4	2.2×10^7	4.3×10^6	284.9	9.98
	(10-25)	6.3	3.8×10^8	5.1×10^6	66.0	9.46
4	(0-10)	5.7	1.7×10^7	5.9×10^7	275.8	12.46
	(0-25)	5.8	7.1×10^6	8.9×10^6	163.3	2.60
5	(0-10)	-	-	-	-	-
	(10-25)	6.0	2.4×10^6	3.7×10^6	88.8	3.55
7	(0-10)	-	-	-	-	-
	(10-25)	6.0	3.2×10^7	1.1×10^6	139.9	3.58
8	(0-10)	6.3	2.4×10^7	6.6×10^6	240.7	2.15
	(10-25)	4.9	1.8×10^7	1.0×10^6	71.0	1.83

Note: Not analyzed

HC DEGRADATION

Some samples of the moss selected for the preliminary examination of the bog were visually represented by oil slightly contaminated with moss. Several samples looked relatively non-polluted, the others were moderately polluted. Twenty seven samples were selected from the different depth of the bog and analyzed before the bioremediation of this bog with the Rhoder.

On the right side of the bog in some places the preliminary concentration of the crude oil in the moss layers of 0–10 cm was from 35.13 to 14.35 kg/kg DM and residual concentration of HC in the same samples after extraction of the crude oil at the room temperature became from 290.6 to 66.9 g/kg DM. The concentration of HC on the right side in two samples (0–10 cm) varied from 543.1 to 522.99 g/kg DM. In the soil layers of 10–25 cm the concentration of HC varied from 516.6 to 43.6 g/kg DM. In soil layer of 15–30 cm the concentration of HC was about 300.0 g/kg DM. This part of the bog was heavily polluted with the oil (Table 5, samples with a letter R). On the left side of the bog in one place in the moss layer of 0–10 cm the crude oil concentration was 29.0 kg/kg DM and after extraction of this crude oil under room temperature the residual HC concentration became 173.3 g/kg DM. In the other samples the concentration of HC varied from 567.2 to 508.1 g/kg DM. In the depth of 10–25 cm the concentration of HC varied from 9.3 to 82.3 g/kg DM. In the soil layer of 25–40 cm the concentration of HC was about 27 g/kg DM. This part of the bog visually seemed a little bit purer than the right one (Table 5, samples with letter L).

TABLE 5 Content of crude oil and saturated hydrocarbons in soil samples before and after augmentation of oil polluted soil with the Rhoder

Samp-les	Depth of samples selection	Free crude oil, kg/kg DM	Saturated HC, g/kg **	Free crude oil, kg/kg DM	Saturated HC, g/kg **	Degradation,%
1_R	(0-10)	15.24	66.9	5.39	105.5	0
	(10-25)	2.94	51.3	2.94	59.8	0
2_R	(0-10)	35.13	73.7	6.37	234.1	0
	(10-25)	*	516.6	*	470.9	8.8
3_R	(0-10)	*	543.1	*	312.4	40.8
	(10-25)	*	84.7	*	45.3	31.2
4_L	(0-10)	*	567.2	*	567.8	0
	(10-25)	*	38.1	*	24.4	35.9
5_L	(0-10)	*	546.7	15.64	230.7	57.8
	(10-25)	*	11.8	*	5.1	56.8
	(25-40)	*	27.1	*	207.6	0
6_L	(0-10)	29.03	173.3	*	47.8	72.4
	(10-25)	*	82.3	*	433.9	0
7_L	(0-10)	*	515.0	*	11.3	97.8
	(10-25)	*	38.8	*	339.7	0
8_R	(0-10)	*	522.9	*	27.6	94.7
	(10-25)	*	43.6	11.37	217.1	0
9_R	(0-10)	25.84	77.8	*	26.5	65.9
	(10-25)	*	76.9	*	260.9	0
10_L	(0-10)	*	508.1	*	8.02	98.4
	(10-25)	*	9.3	7.18	330.2	0
11_R	(0-10)	14.35	280.0	*	190.7	32.1
	(10-25)	*	196.7	4.96	314.1	0
12_R	(0-10)	25.04	187.6	*	123.5	34.2
	(10-25)	*	53.5	8.47	301.5	0

TABLE 5 *(Continued)*

13_R	(0-15)	14.52	290.6	*	318.0	0
	(15-30)	*	308.0	*	384.1	0

Note: R – right side of area, L – left side of area, * - free oil is absent; ** - residual saturated HC in the samples after separated the crude oil

The oil in the samples of the moss, which were severe contaminated of the real crude oil (35.1–14.5 kg/kg DM), contained the saturated HC of 62.5 ± 1.7%, the aromatic HC of 19.3 ± 1.4%, resins and asphaltenes of 11.8 ± 0.8% and from 5 to 7% of non HC (oxidized substances). Such composition of the oil is a typical for any high quality oil and such oil should be gathered and directed to a refinery plant. Oil contaminating samples of the moss with concentration HC of 850–460 g/kg DM contained the saturated HC of 61.8 ± 1.3%, the aromatic HC of 16.7 ± 0.3%, resins and asphaltenes of 8.7 ± 1.9%. Such contamination also represents the high oil quality and such oil should be gathered too. The oil in moss samples from the layers of 10–25 cm, 15–30 cm, and 25–40 cm contained saturated HC of 49.4 ± 1.12%, aromatic HC of 19.6 ± 2.3%, resins and asphaltenes of 13.4 ± 5.2% and 18% of non HC (oxidized substances). Such HC composition of the pollution indicated that the processes of the oil biodegradation with indigenous anaerobic microorganisms had begun inside of these layers. Obtained results showed that the initial huge amount of the crude oil in some places was decreased after bioaugmentation with the Rhoder (Table 5 and Figure 3), but the oil had appeared in other places, where previously it was absent. Content of the total saturated HC increased in these places. Probably, such changes in the amount of the crude oil and more impregnation of the top layers (0–10 cm) of the moss with the oil could be due to movement and displacement of the oil because of the small bias to a bulk of the sandy road.

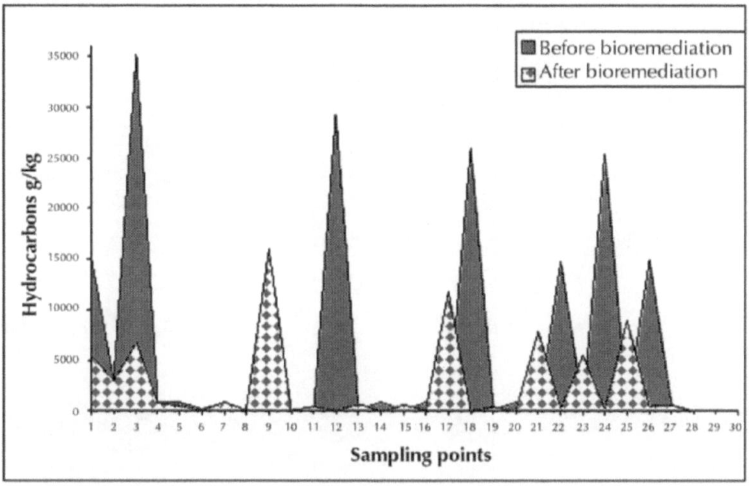

FIGURE 3 Bioremediation of the oil polluted bog with the Rhoder, Muravlenko, 2011.

Chromatograms of HC of the contaminated moss with extremely high and medium levels of the oil pollution before and after bioaugmentation are presented on Figure 4(a-c) and Figure 4(a1-c1) and confirm results that are described as given.

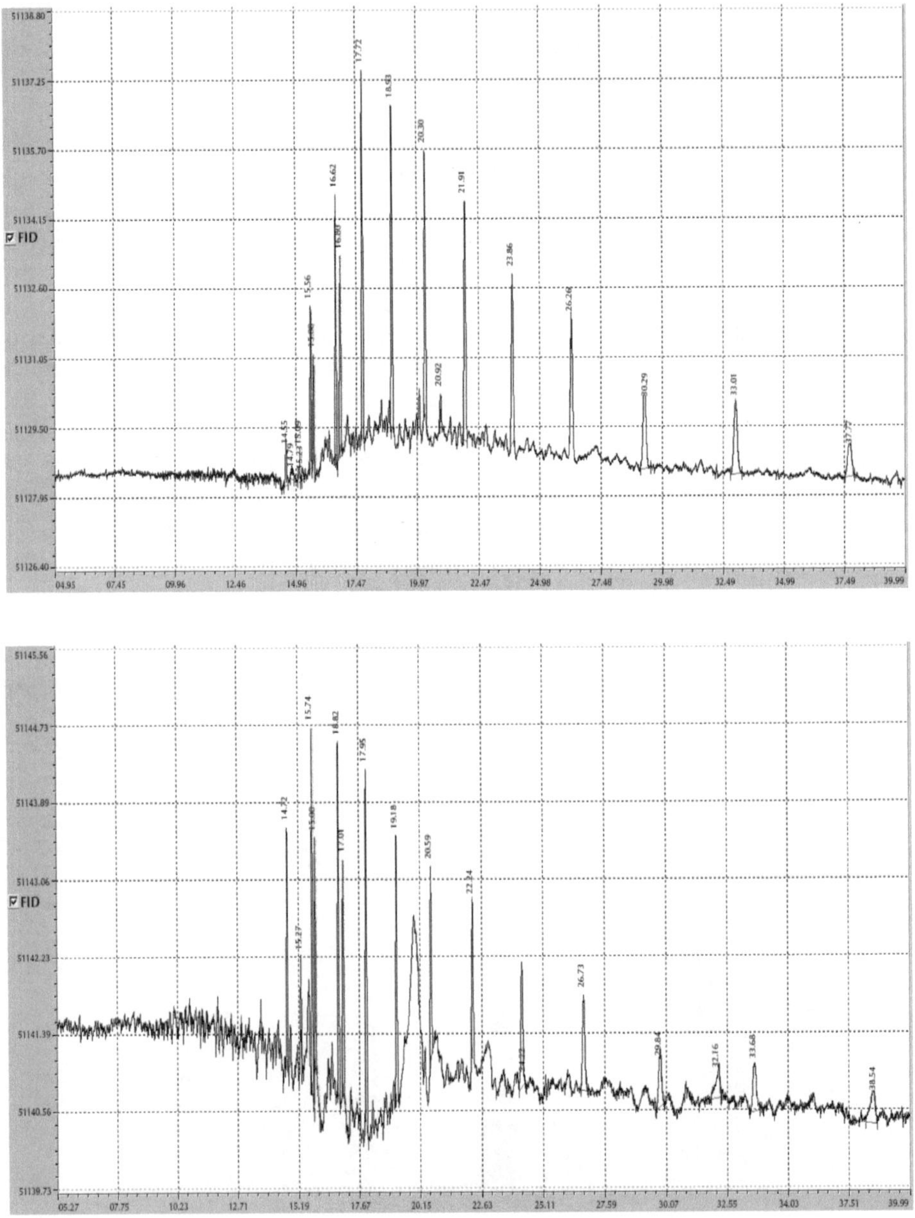

FIGURE 4 *(Continued)*

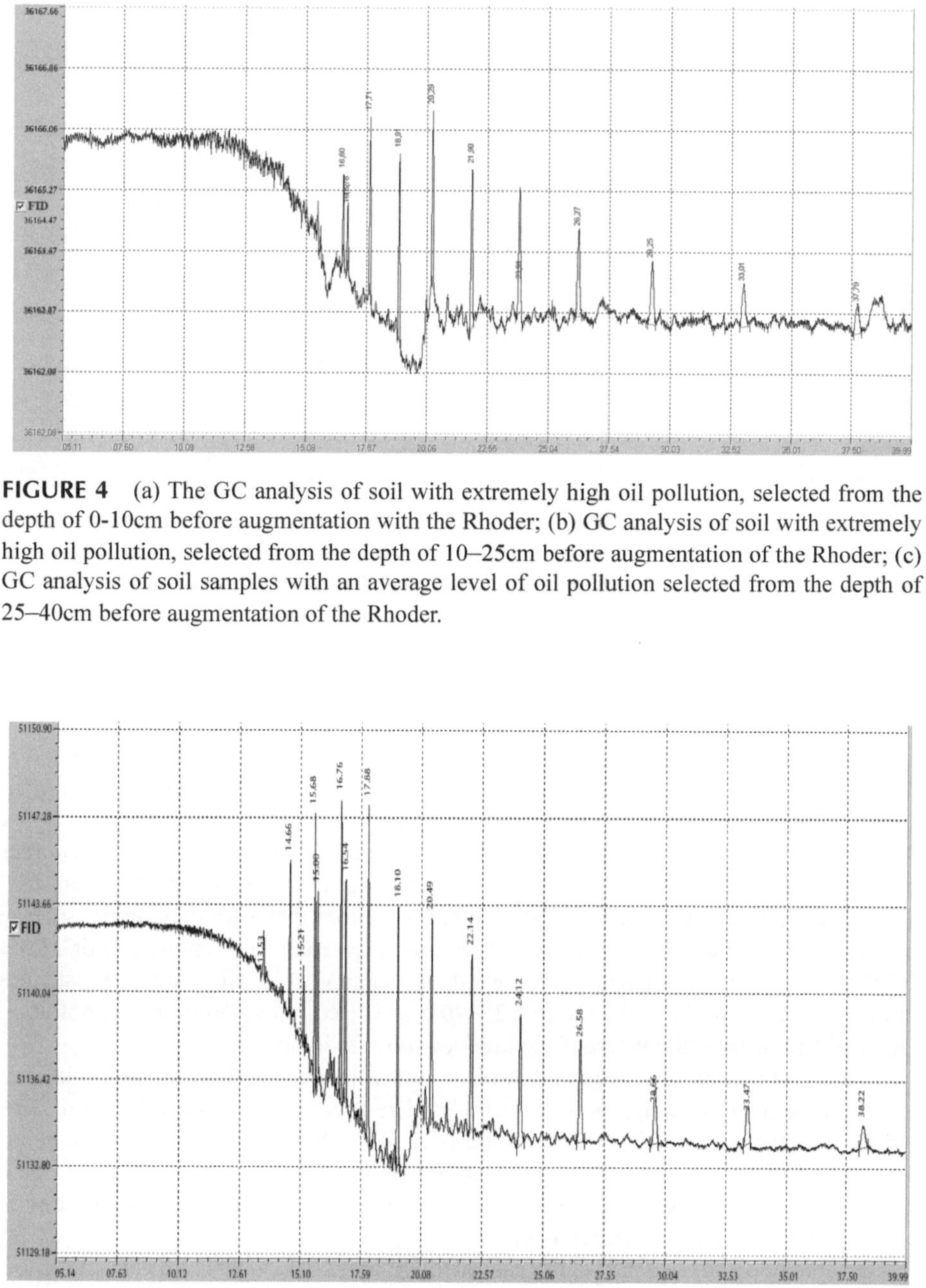

FIGURE 4 (a) The GC analysis of soil with extremely high oil pollution, selected from the depth of 0-10cm before augmentation with the Rhoder; (b) GC analysis of soil with extremely high oil pollution, selected from the depth of 10–25cm before augmentation of the Rhoder; (c) GC analysis of soil samples with an average level of oil pollution selected from the depth of 25–40cm before augmentation of the Rhoder.

FIGURE 4 *(Continued)*

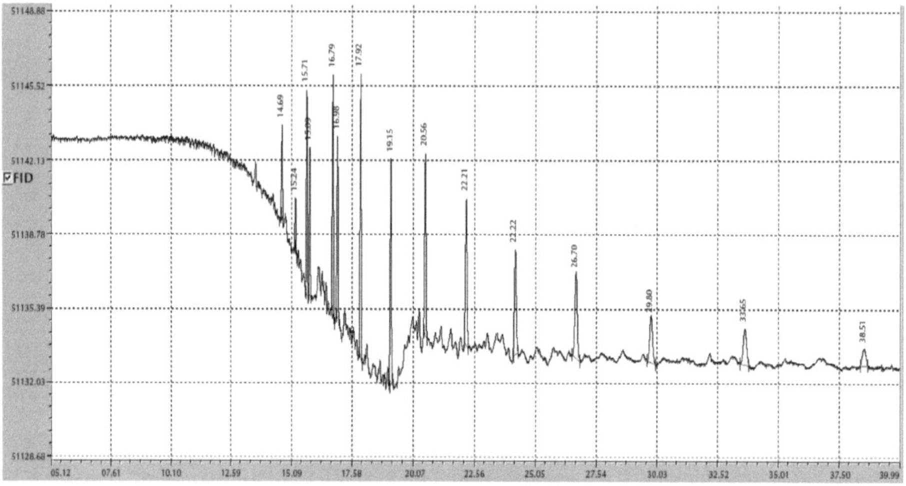

FIGURE 4 (a1) The GC analysis of soil selected from the depth of 0-10cm after augmentation with the Rhoder; (b1) GC analysis of soil from the depth of 10–25cm after augmentation with the Rhoder; (c1) GC analysis of soil from the depth of 25–40, after augmentation with the Rhoder.

After ending of the bioaugmentation with the Rhoder on the right side of the bog the saturated HC of 60.5 ± 0.7% and the aromatic HC of 21.5 ± 0.7%, and resins and asphaltenes of 10.0 ± 0.01% and about 8% of non HC (oxidized substances) were found in samples of moss from the depth of 0–10 cm, which initially contained a lot of crude oil. The oil contained saturated HC of 54.0 ± 0.01% and aromatic HC of 19.5 ± 8.5% and resins and asphaltenes of 6.8 ± 0.4% and about 20% of non HC in samples of the soil from layers of 10–25 cm. It is interesting, in the depth of 10–25 cm (anaerobic conditions) degradation process often was more intensive than on the surface of soil. Oil contained of 53.5 ± 0.01% of the saturated HC, the aromatic HC of 23.5 ± 0.01%, resins and asphaltenes of 11.5 ± 0.01% and about 13% of oxidized substances in the samples from the moss layer of 25–40 cm. The composition of the oil pollution changed and became the worse if the layer of soil was lower.

Another situation was observed in oil samples from the soil on the left side after ending the bioaugmentation with the Rhoder. The saturated HCs were found of 32.9 ± 5.8% and the aromatic HCs were found of 23.3 ± 1.8% and resins and asphaltenes were found of 29.3 ± 5.9% and oxidized substances were more than 14% in the depth of the moss layers of 0–10 cm, that indicated on significant oil oxidizing processes which caused by using of the Rhoder.

Oil contained 60.0 ± 1.6% of the saturated HC and 21.0 ± 0.9% of the aromatic HC and 10.8 ± 1.3% of resins and asphaltenes and about 8% of oxygenated compounds in the samples from the moss layers of 10–25 cm (the left side). The quality of oil in 10–25 cm of soil layers was better than in the upper layers.

The analysis of the residual HC contamination by HPLC method in the moss after bioaugmentation with the Rhoder showed the oil degradation (Figure 5(a–c)) in the layers of 0–10 cm and 10–25 cm of the moss (except the layer of 25–40 cm) and confirmed that the degradation of the aromatic HC was observed in these layers of soil. Tentatively the average efficiency of the Rhoder application can be estimated as 55.2 ± 26.2% for not so favorable weather conditions if an average percentage of oil degradation be calculated (Table 5). It is significant that the oil spill on the bog was the fresh (in spring), and the Rhoder was prepared as a liquid concentrate of cells with a high hydrocarbon oxidizing activity (1.0 × 10^{11} cells per 1 mL of the concentrated product).

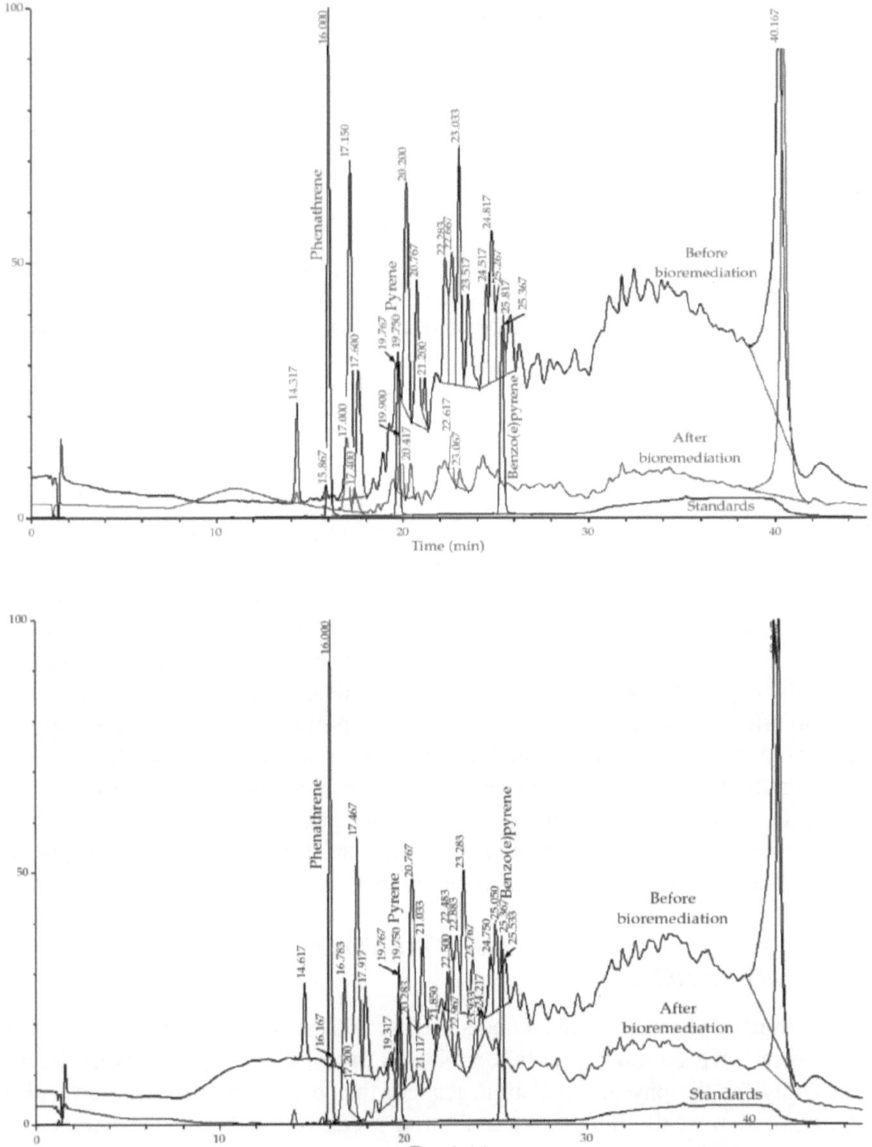

FIGURE 5 *(Continued)*

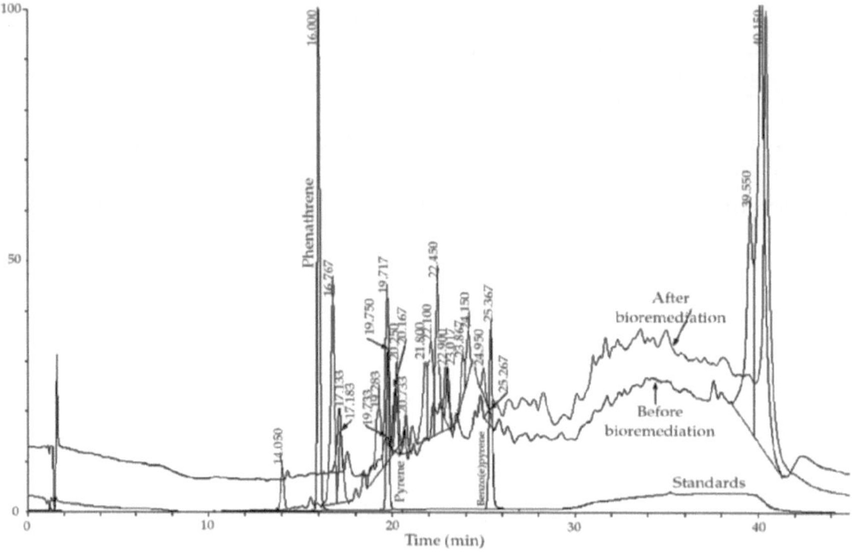

FIGURE 5 (a) HPLC analysis of the soil before and after augmentation with the Rhoder from the depth of 0-10cm; (b) HPLC analysis of the soil before and after augmentation with the Rhoder from the depth of 10–25 cm; (c). HPLC analysis of the soil before and after augmentation with the Rhoder, sample from the depth of 25–40 cm.

Thus, the obtained results have shown, on the one hand, that the Rhoder is able to operate in extreme conditions, such as a super high level of the oil pollution under unfavorable weather conditions without milling of moss that useful for the bioremediation at all. On the other hand, despite of the results described above, there was still a lot of oil on the surface of the bog. Multiple repetition of the bioaugmentation with the Rhoder on the bog heavily polluted with oil will be required for several years to fully restore this bog. The bioaugmentation technology described above cannot be considered as an effective one for the restoration of bogs polluted with oil in severe climatic conditions in the northern part of the Western Siberia. It is necessary to develop a new bioremediation technology, may be with using aerobic-anaerobic process of oil biodegradation. Besides, the valuable energy feedstock has been irretrievably lost, and the negative influence on the environment will keep for a long time because of spreading the oil contamination far from the oil polluted sites and in pour into the groundwater.

12.4.3 ANALYSIS OF THE MOSS TOXICITY AFTER THE BIOAUGMENTATION

After the third application of the Rhoder the seeds of oats and a mixture of perennial grasses were sown on the bog polluted with oil to test the phytotoxicity of the moss and possibly for the phytoremediation. Half of the seeds had been watered with the working solution of the humate “Extra”. The seeds watered with the humate were sowed on the right side of the bog more polluted with oil. On the left side of the bog the seeds were sowed without watering with the humate. Two small plots were isolated on the right and on the left sides of the bog after ending application of the Rhoder (Figure 2) and covered these plots with non-woven material to protect seedlings against adverse weather conditions. This material passes oxygen and rain moisture to the soil.

Seeds watered with the humate and without it did not grow up at all after six weeks in places with the very high level of the oil pollution on the whole bog. In the left side of the bog where the concentration of oil previously was below 800–900 g/kg DM seeds of oats without humate grew up (length of seedlings was about 10 cm) but seeds of perennial grasses mixtures did not grow up at all. On the right side of the bog where seeds watered with the humate were sown, the seeds of oats and especially perennial grasses germinated. Grown seedlings mixtures of perennial grasses were about 7 cm in length and possessed strong roots. In plots under non-woven material it was observed the same situation: on the left half of the oil contaminated bog (seeds without the humate) mainly oats seedlings grew. On the right side the grass mixture and a little bit seedlings of oat grew too. And oat, and grass mixture well grew in places of both plots where the level of the oil contamination was below 100 g/kg DM.

These results showed that the humate, containing humic and fulvic acids, had a positive effect on the germination and growth of roots of the perennial grasses mixture and practically had no effect on the germination and growth of oats (*Avena sativa*), which was more resistant to the oil pollution and used for the phytoremediation mainly in the northern part of Russia. Similar results were obtained earlier in laboratory experiments on bioremediation of oil polluted soil with the Rhoder and addition of Pawhumus (Germany) [17].

12.5 CONCLUSION

Comparison of two remediation technologies for oil-polluted bogs in the Northern part of Russia has shown that the most effective technology is *ex-situ*. It allows to do full remediation of soil polluted with oil in a short time, but a cost of this technology is significantly high. In addition *ex-situ* bioremediation technology allows refining gathered oil and partially offsetting outlays on equipment service and energy expense. The second bioremediation technology (*in-situ*) cannot be considered as the effective one for impassable bogs polluted with oil behind the Polar Circle in the Western Siberia, because it will really require of 3–4 years or more to restore bogs in severe climatic conditions there. Obtained results showed that the processes of oil biodegradation had begun inside of the bottommost layers of the bog due to indigenous anaerobic microorganisms. So, it is necessary to develop in future a new option of bioremediation technology with using aerobic anaerobic biodegradation of oil for such contaminated bogs which would be more favorable for environment and attractive for economic. Nevertheless, the oil-oxidizing preparation Rhoder during *in-situ* bioremediation is able to degrade oil (55.2 ± 26.2%) in extreme conditions: a super high level of oil pollution (from 14.4–35.1 kg of oil per kg of dry moss to 516.6–43.6 g/kg of dry moss) for unfavorable weather conditions without milling which is favorable for any bioremediation.

The humate "Extra" (Russia) containing humic and fulvic acids had a positive effect on the germination and roots growth of perennial grasses mixture and practically had no effect on the germination and growth of oats (*Avena sativa*) on the bog contaminated with oil.

ACKNOWLEDGMENTS

The authors express their gratitude to Mr. A. B. Kurchenko (Director-General of Joint Stock Co SPASF "Priroda", the Komi Republic) and Mr. I. I. Zhukov (Deputy Director-General of the AIE "Ecoterra", Moscow) for financial support and opportunity to use the oil-oxidizing preparation Rhoder for the bioremediation of natural objects polluted with oil.

KEYWORDS

- **Bioremediation**
- **Bogs**
- **Degradation**
- **Moss**
- **Oil pollution**
- **Oil-oxidizing preparation**
- **Soil**

REFERENCES

1. Kurchenko, A.: In Proc. 496 Intren. Oil Spill Confer, Seattle, p. 231 (1999).
2. Kurchenko, A. B.: In Proc. of the fifth scientific and practical conference: *Ecology works on oil fields of the Timano-Pechorsky province. Current state and prospects*, Syktyvkar, p. 132 (2008).
3. Murygina, V. P., Arinbasarov, M. U., and Kalyuzhnyi S. V.: *Ecology and Industry of Russia*, (in Russian), (8), p. 16 (1999).
4. Murygina, V. P., Markarova, M. Y., and Kalyuzhnyi, S. V.: *Environmental International*, **31**(2), 163 (2005).
5. Murygina, V. and Markarova, M. Sergey Kalyuzhnyi: in Proc. of IPY-OSC Symp., Norway, Oslo, (2010). http://www.ipy-osc.no/
6. Murygina, V., Gaidamaka, S., Iankevich, M., and Tumasyanz, A. *Progress in Environmental Science and Technology*, **3**, 791 (2011).
7. Gally, P., Murygina, V. P, and Kalyuzhnyi, S. V.: Proc. of the 8th International In Situ and On-Site Bioremediation Symposium, Baltimore, MD., p. 421 (2005).
8. Kamentschikov, F. A., Chernikh, L. N., and Murygina, V. P. *Oil Economy*, (in Russian), (3), p. 80 (1998).
9. Ouyang, W., Yu, Y., Liu, H., Murygina, V., Kalyuzhnyi, S., and Xiu, Z.: *Process Biochemistry*, **40**(12), 3763 (2005).
10. Ouyang, W. Liu, H. Yu, Y. Y. Murygina, V. Kalyuzhnyi, S. Xiu, and Z. D.: *Huanjing Kexue/Environmental Science*. **27**(1), 160 (2006).

11. De-Qing, S., Jian, Z., Zhao-Long, G., Jian, D., Tian-Li, W., Murygina, V., and Kalyuzhnyi, S.: *Water, Air, and Soil Pollution*, **185**(1–4), 177 (2007).
12. Drugov, Yu. S., Zenkevich, I. G., and Rodin, A. A. *Gaschromathography Identification of Air, Water and Soil and Bio-nutrients Pollutants* (in Russian). Binom, Moscow, p. 752 (2005).
13. Aksenyuk, D. A., Gerasimova, C. A., and Mikhailik, Yu. V.: (2009). lab@ecocity.ru
14. V. G. Mineev (Ed.) *Practical handbook on Agro chemistry* (in Russian). Moscow State University, Moscow, Russia, p. 688 (2001).
15. A. I. Netrusov, (Ed.) *Practical handbook on microbiology* (in Russian). Academia, Moscow, Russia (2005).
16. Nazina, T., Rozanova, Ye., Belyayev, S., and Ivanov, M.: *Chemical and microbiological research methods for reservoir liquids and cores of oil fields* (in Russian). Preprint Biological Centre Press, Pushchino, p. 35 (1988) .
17. Salem, K. M., Perminova, I. V., Grechischeva, N. Yu., Murygina, V. P, and Mescheryakov, S. V. *Ecology and Industry of Russia* (in Russian). (4), p. 19 (2003).

CHAPTER 13

UNSATURATED RUBBER MODIFICATION BY OZONATION

V. F. KABLOV, N. A. KEIBAL, S. N. BONDARENKO,
D. A. PROVOTOROVA, and G. E. ZAIKOV

CONTENTS

13.1 INTRODUCTION

The chapter considers the investigation related to modification of unsaturated rubbers by means of ozonation for improving their adhesion properties. The mechanism of macroradicals formation in the ozonation process as well as the influence of the contact time in the modification on adhesion properties of the rubbers has been studied. It has been shown that the best characteristics comparing with the initial ones can be obtained at 1 hr ozonation. The gluing strength increases by 10–70% on average at that.

Today, despite the existence of a large number of adhesives which differ not only in the composition and properties but also in manufacturing technology, formulations, and intended purpose, the problem of creating new adhesives with a certain set of properties is still relevant. This is due to fact that glue compositions are imposed to ever more high demands related to operation conditions of construction materials and products.

The problem may be solved by applying the targeted modification of a film-forming polymer which is a base component of any glue composition. The modification is of a priority than creating completely new adhesive formulations. This process is more advantageous from both an economic and technological points of view and allows not only to improve the performance of rubber but also to maintain a basic set of properties.

As it is known, there are several methods of film-forming polymer modification. They are physical, chemical, photochemical modification, modification with biologically active systems, and combinations of these methods.

The epoxidation, being one of the variants of chemical modification, represents a process of introduction of epoxy groups to a polymer structure that improves properties of this polymer. Materials based on epoxidized polymers show high physical and mechanical characteristics, meet the requirements to the strength and dielectric parameters, and manifest good adhesion to metals which is achieved due to high adhesive activity of epoxy groups. Thanks to these properties they find application as coatings for metals and plastics, adhesives, mastics, and potting compounds in electrotechnics, microelectronics, and other areas of engineering [1].

Chlorinated natural rubber (CNR) is applied as an additive in glue compositions based on chloroprene rubber that, in its turn, is widely used in industry for gluing of different rubbers together with metals or each other [2]. As an individual brand glues based on CNR are rarely produced.

Isoprene rubber is an analogue of natural rubber used in the majority of rubber adhesives but due to its low cohesive strength applied in their formulations much more rarely.

Thereby, the investigations concerned with the development of glue compositions based on these rubbers with improved adhesion characteristics to materials of different nature are of particular interest. It can be achieved by modification [3].

13.2 METHODS

In this chapter, it is investigated a possibility of CNR and isoprene rubber epoxidation by means of ozonation with the aim of improving adhesion characteristics of glues based on these rubbers as it is known that epoxy compounds are good film-formers in glue compositions and increase the overall viscosity of the last ones.

The introduction of epoxy groups in a polymer structure was carried out by means of ozonation as ozone is highly reactive towards double bonds, aromatic structures and C-H groups of a macro-chain.

The contact time (0.5–2 hr) was varied in the ozonation process. The rest parameters which are ozone concentration (5×10^{-5}%vol.) and temperature (23°C) were kept constant. Further, glue compositions based on the ozonized rubbers were prepared.

Glue compositions based on the ozonized CNR were 20% solutions of the rubber in an organic solvent which was ethyl acetate. The compositions based on the isoprene rubber were 5% solutions of the ozonized rubber in petroleum solvent.

The gluing process was conducted at 18–25°C with a double-step deposition of glue and storage of the glued samples under a load of 2 kg for 24 hr. Glue bonding of vulcanizates was tested in 24 (± 0.5) hr after constructing of joint by method called "Shear strength determination" (State Standard 14759–69), in quality of samples there were used polyisoprene (SKI-3), ethylene propylen (SKEPT-40), butadiene-nitrile (SKN-18), and chloroprene (Neoprene) vulcanized rubbers.

13.3 DISCUSSION AND RESULTS

In ozonation, a partial double bonds breakage in the rubber macromolecules that leads to the macroradicals formation occurs (Figure 1). Ozone molecules are attached to the point of the rubber double bonds breakage with the formation of epoxy groups [4]:

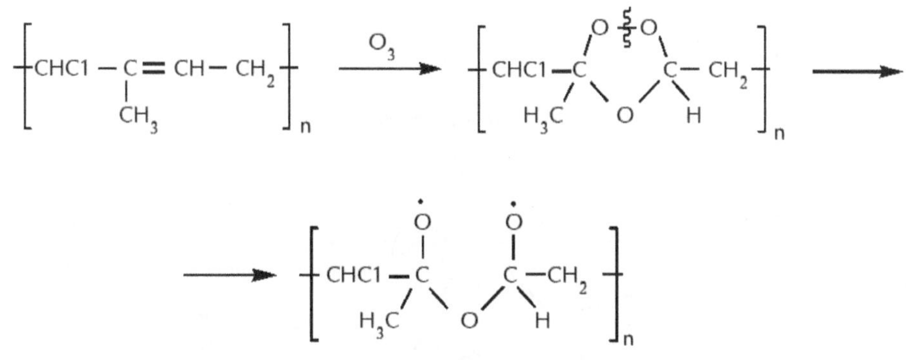

FIGURE 1 Reaction of the macroradicals formation (on the example of CNR).

The formed macroradicals are probably interacting with macromolecules of the rubber which is a substrate material, thereby, higher adhesion strength is provided.

At first, the CNR of three brands was investigated, they are as follows—S-20, CR-10, and CR-20. The results obtained in ozonation of three rubber brands are shown on the Figure 2.

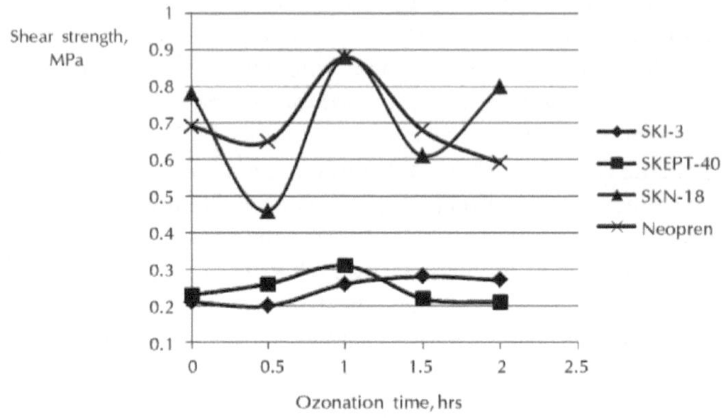

(a)

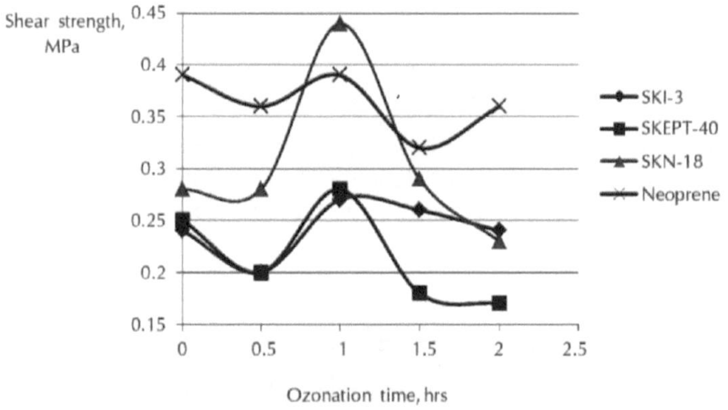

(b)

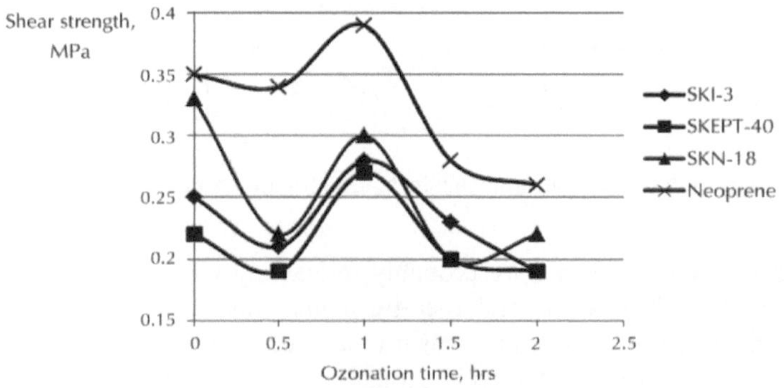

(c)

FIGURE 2 Influence of ozonation time on shear strength at gluing of vulcanizates with glue compositions based on CNR of the S-20 (a), CR-10 (b), and CR-20 (c) brands accordingly.

A decrease in the adhesion strength at $\tau = 0.5$ hr may be connected with preliminary destruction of macromolecules under action of reactive ozone.

From the Figure 2 we can see that the maximal figures correspond to 1 hr ozonation. The adhesion strength for rubbers based on different caoutchoucs increases by 10–40% at that. The results of the shear strength change depending on the rubber brand and adherend type are shown on the Figure 3.

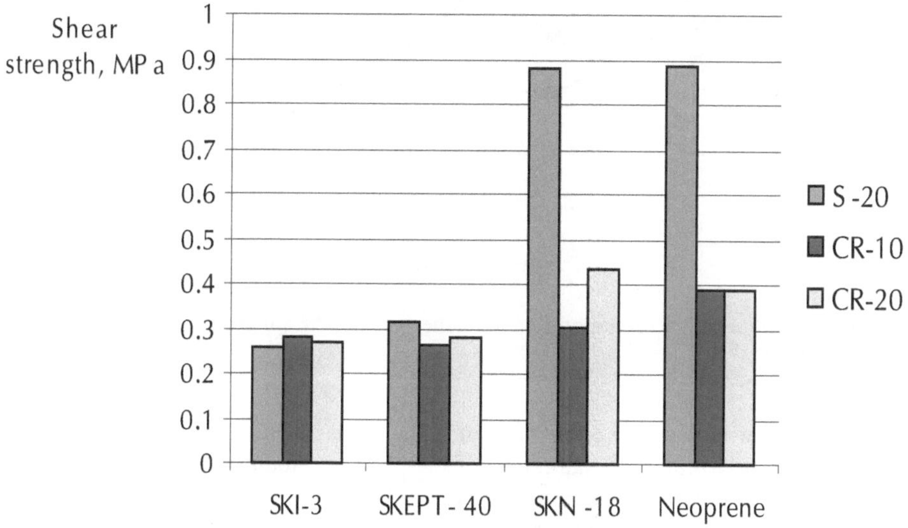

FIGURE 3 A change in shear strength for different CNR brands depending on adherend type (ozonation time $\tau = 1$ hr).

It should be noticed that the extreme nature of the dependences can be explained by the diffusion nature of the interaction between adhesive and substrate. As it is shown from the figures, with the increasing content of functional groups the strength began to reduce, having reached of certain limit in the adhesive. In this case, only adhesive molecules have ability to diffusion [5].

Isoprene rubber was also treated with ozone at the same parameters maintained for CNR. The results are on the Figure 4.

The data in Figure 4 confirm the ambiguity of the ozonation process. When the contact time is equal to 15 min (as in the case of CNR epoxidation [6]), possible preliminary destruction of the rubber macromolecules takes place, which on the graphs is proved by almost simultaneous reduction in the shear strength values. Concurrently, formation and subsequent growth of macroradicals go on, as evidenced by the increase in adhesive characteristics at ozonation time 0.5 and 1 hr. Here, the shear strength at gluing of different vulcanized rubbers increases by 10–70% on average, and then it starts to reduce again.

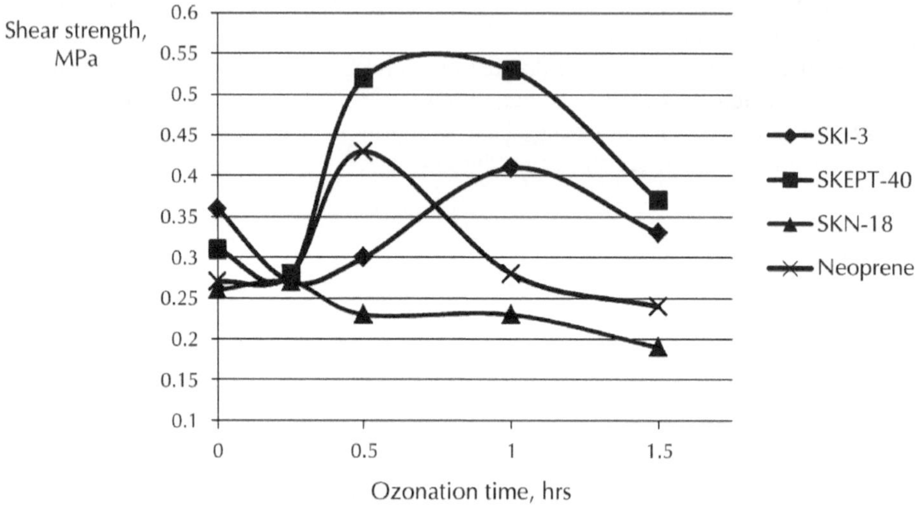

FIGURE 4 Influence of ozonation time on adhesive strength for the compositions based on isoprene rubber.

With further increase in ozonation time, the values of adhesion strength decline. That is apparently related to saturation of the polymer chain with epoxy groups and decrease in mobility of the macromolecules, and, consequently, the degree of interaction of the substrate with the adhesive composition as well as with destruction of the polymer chains.

13.4 CONCLUSION

Thus, ozonation can be applied as an effective method of enhancing adhesion properties of rubbers at the modification of film-forming polymers that are a main component in glues. Changing one of the parameters during the ozonation process we can obtain such content of epoxy groups at which the characteristics of adhesion strength will be maximal.

KEYWORDS

- **Glue compositions**
- **Gluing**
- **Modification**
- **Ozonation**
- **Unsaturated rubbers**
- **Vulcanizates**

REFERENCES

1. Solovyev, M. M. Local dynamics of oligobutadienes of different microstructure and their modification products. Thesis of Ph. D. in Chemistry Sciences: 02.00.06. Mikhail Mikhailovich Solovyov, Yaroslavl, p. 201 (2009).
2. Dontsov, A. A., Lozovick, G. Ya., and Novitskaya, S. P. *Chlorinated polymers*. Khymiya, Moscow, p. 232 (1979).
3. Kablov, V. F., Bondarenko, S. N., and Keibal, N. A. *Modification of elastic glue compositions and coatings with element containing adhesion promoters*, monograph. IUNL VSTU, Volgograd, p. 238 (2010).
4. Zaikov, G. E. Why do polymers age, *Soros Educational Journal*, **6**(12), 52 (2000).
5. Berlin, A. A. and Basin, V. E. *Basics of Polymer Adhesion*, Khymiya, Moscow, p. 320 (1969).
6. Keibal, N. A. Bondarenko, S. N., Kablov, V. F., and Provotorova, D. A. Ozonation of chlorinated natural rubber and Studying its Adhesion Characteristics. *Rubber: Types, Properties and Uses.* Gabriel A. Popa (Ed.) Nova Publishers, New York, pp. 275–280 (2012).

CHAPTER 14

BURN DRESSINGS SORPTION AND DESORPTION KINETICS AND MECHANISM

K. Z. GUMARGALIEVA and G. E. ZAIKOV

CONTENTS

14.1 INTRODUCTION

The principal medical treatment of burns is the use of dressings, which often worsen the effects of the injury. It is difficult to estimate the effectiveness of the new burn dressings. Their physicochemical properties are not usually presented in literature. This chapter firstly discusses this subject. The authors address the complexity of physicochemical methods of analysis in order to create criteria for an efficient dressing for a burner wound surface.

The experimental methods to estimate the main physicochemical properties were worked out. Based on theoretical and experimental data we found that maximal sorptional ability of burn dressing equals the free volume of the dressing material, calculated from the value of the material density. We found that the study sorption ability water can be used as a model liquid instead of the blood plasma medium. The kinetic parameters were determined from the sorption curves. These parameters showed that the first aid burn dressings markedly differ in the value of the rate or liquid media sorption at stages close to the sorption limits. We established that the air penetrability parameter in wet state decreases abruptly by 2–3 orders of magnitude for the majority of tested dressings. This is due to the filling of porous space by liquid medium. We recommended that the air penetrability parameter be determined in wet state which represents the common condition of action for the first air burn dressing. The value of adhesive strength after the end of its action on the wound must not exceed 20 N/m. From the data obtained in this study we formulated the following criteria to estimate the efficiency of the first aid bum dressing—maximum sorptional ability for water must be at least 10 g/g, optimal thickness of dressings, fulfilling this value of sorptional capacity, must be ~ $\times 10^{-3}$ m, the adhesive strength must not exceed 20 N/min, and the average diameter of open (connected) pores must not exceed 5×10^{-7} m.

14.2 MODERN SURGICAL BURN DRESSINGS

The dressings for wounds and burns must primarily be protective, surptional, and atroumatic. In currently used dressings these properties are provided by a multilayered structure or structural modifications. The different classifications of the dressings can be found in literature—by material, by construction or by functions [1-3].The dressings applied in the modern treatment of wounds and burns are subdivided into three groups according to material of the layer, sorbing the exudate of wounds, that is, material of animal origin, synthetic, foamy polyurethane, and material of vegetable origin.

14.2.1 DRESSINGS BASED ON MATERIALS OF ANIMAL ORIGIN

Typical dressings for this group are collagen sponges, which are porous. Beside hydrophilic properties, collagen sponger provides higher sorption of liquid (in the range of 40–90 g/g) [1, 4-9]. The patent literature describes in detail the methods of obtaining collagen dressings for wounds and burns. In the form of sponges and felt [10-13] based on materials of animal origin also include these made from biological artificial leathers based on lyophilized goby's and swine cutis, produced as plates 0.5–0.7 mm thick. However, these materials possess lower sorptional capacity than collagen dressings. The dressings called "cultivated cutis" are also obtained from the epithelia of cells of the patient himself [13]. The shortcoming of biological artificial leathers or biodressings their expense and their inability to retain, as a rule, their properties on storage.

14.2.2 DRESSINGS BASED ON SYNTHETIC MATERIALS

The demands inexpensive raw materials for the production of dressings for wounds and burns led to the production based on synthetic polymeric materials, particularly cellular polyurethane [14-19]. The cellular urethanes, intended for medical purposes, are synthesized on the basis of toluene diisocyanate and polyoxipropylene glycole [20].

The dressings based on polyurethanes have a pores distribution of about 200–300 pores/cm^2, and allow the regulation of the number and size of pores in layers [21]. On the whole, dressings from this group are prepared double layered, the outside layer being of higher density in order to prevent liquid evaporation and penetration of microorganisms. In rare case, these dressings are made homogeneous by all the thickness.

The influence of the pores sizes on sorptional properties of polyurethane sponges is described in the chapter [22] macroporous sponge with the pore size from 200 μm up to 2000 μm is completely nourished by exudate under pressure only. In this case through pores should possess the sizes of several micrometers.

Other polymeric materials (polyvinyl chloride, nylon, and so on) are applied as the sorbing layer apart from polyurethanes [23-26]. This group should be added by a compositional burn dressing, based on a silicon film, polyamide network, and hydrophilic admixture produced by Hall Woodroof Co. (USA) [13]. The polyurethane covers with atraumatic lower layer from polyglycolic acid may be considered as variety of compositional dressings [27]. It is characteristic of dressings from this group that they preserve their high strength properties at the absorption of the wounds exudate.

A two-component protective dressing "Hydron" was recently applied in the treatment of burns. It is a film, formed on the wound, a powder of poly-2-hydro-oxyethylmetacrylate dissolves in polyethylene glycol 400 [2, 28]. Although they possess good protective properties, "Hydron" dressings have low strength and sorptional capacity.

14.2.3 DRESSINGS BASED ON THE MATERIALS OF VEGETABLE ORIGIN

A large number of burn dressings are so-called "cotton-batting" ones, based on cellulose, viscose or a combination of the two [29-33]. These dressings differ from each other by structure and composition of the upper and lower layers. Most often the sorptional layer from cellulose is used in complex dressings.

TABLE 1 Characteristics of dressings in brief

No	Name	Company	State	Structure	Composition
			Group 1		
1.	Collagen burn dressing	"Helitrex"	USA	Dressings uniform by thickness, dense, pores of 0.01 mm size. It has gauze cover	Collagen
2.	Collagen sponge	"Helitrex"	USA	Similar to No. 1, differs by big radius of pores, formed by fibril weaving of cylindric form preferably	Collagen
3.	Collagen dressing	"Bayer"	Germany	Friable dressing, possesses rough porous structure with pores-holes sizes from 1.5 to 0.1 mm	Collagen
4.	Burn curative dressing	"Combutec-K 11"	USA	Dressing of large-porous structure with pores size from 1 to 0.05 mm. Pores are of cylindric form preferably formed by fibrils weaving of collagen	Collagen
5.	Biological dressing	"Corretium-2"	USA	Dense, pressed plate. Fibrillar structure is observed in dense layers	Collagen
6.	Biological dressing	"Corretium-3"	USA	The same	Collagen
			Group 2		
7.	Compositional burn dressing	"Biobrant"	USA	Double-layered elastic, porous dressing, consists of the upper layer of 0.01–0.005 mm and flexible fabric nylon base. It represents combination of hydrophilic components with elastic silicon films	Silicon, the main layer - from polyamide

TABLE 1 *(Continued)*

8.	Synthetic dressing	"Epigard"	USA	Double-layered elastic porous dressing. Upper layer is dense, non-porous 0.2 mm thick	The main layer - from polyurethane, the upper one - polypropylene
9.	Synthetic burn dressing	"Syncrite"	Czeck	Single-layered dressing on gauze base with through large pores, medium flexible	Polyurethane
10.	Synthetic wound dressing	"Syspurderm"	Germany	Dressings of homogeneous composition, with different pores distribution: upper layer is 0.1 mm thick, possesses small porous structure with pores of 0.01 mm size; lower layer, adjoining wound possesses large pores of 0.05 mm size. The dressing is "elastic", accepts a form badly	Polyurethane
11.	Synthetic wound dressing	"Farmexplant"	Poland	Antiseptic double-layered dressing. Main polyurethane layer possesses pores of 0.1–1.5 mm. Upper layer is 0.1 mm thick, more dense, non-porous	Polyurethane
12.	Atraumatic caproic dressing		Russia	Large-cellular dressing on basis of weaved nylon	Polyamide
Group 3					
13.	Cover for wound, burns	VNIImedpolymer"Algipor"	Russia	Wound large-cellular dressing, homogeneous by its composition	Alginic acid salts
14.	Needle-pierced fabric		Russia	Porous cotton balling dressing with atraumatic layer	Cellulose

TABLE 1 *(Continued)*

15.	Wound non-adhering dressing	"Bayersdorf	Germany	Three-layered dressing of plaster type with tricot lower layer. Dressing is of the sandwich type: upper layer-crepe paper, main part is cotton balling, lower layer-tricot network. Atraumatic action is provided by the effect of dressing "bending" (tunneling effect)	Main and upper layers - from cellulose, lower one is a film of Dacron or nylon
16.	Wound absorbing dressing	"Johnson-Johnson"	USA	Three-layered dressing with perforated lower and upper layers 0.01 mm thick, main part is cotton balling, porous	Main layer - from cellulose, external layers - from polypropylene
17.	Haemostatic		Sweden	Double-layerd dressing with perforated lower layer, sewed to the main layer	Viscose main layer, atraumatic one - from polyethylene
18.	Wound dressing	"Mesorb"	France	Cotton balling or viscose dressing, crepe paper - lower and upper layers	Cellulose
19.	Surgical dressing	"Kendall"	USA	Similar to No. 16 with cellulose base and atraumatic synthetic lower layer	Cellulose
20.	Dressing with perforated metallized layer		Germany	First aid dressing with hydrophobic layer and lower metallized layer. Internal layers represent non-fabric pressed layer of crepe paper	Min layer - from cellulose, lower layer is aluminum spray-coated

TABLE 1 *(Continued)*

21.	Dressing. lower layer is not metal-lized		Germany	Similar to No. 20	Cellulose with spray-coated lower layer
22.	Non-adhesing dressing	"Switin"	Czeck	Dense cotton balling dressing, lower and upper layers are non-fixed nylon networks	Cellulose and poly-amide
23.	Series of experimen-tal dressings with various quantitative viscose-cotton com-position		Russia	Cellulose or viscose dressings with atraumatic layer	

Such dressings are usually layered, with the separate layers being produced from either the same material or from different ones, and may be a fixed mechanically or by using thermoplastic material. To decrease their adhesion to the wound surface, the lower layer is produced from various fabric and non-fabric materials (perforated Da-cron, polypropylene, pressed paper, metallized fabric material, and so on). The total sorptional ability of these dressings is defined by hydrophilicity and porosity of the basic material and is usually equal 15–25 g/g.

This chapter [1, 34, 35] shows the dressing action data on wounds and burns of dressings based on another vegetable material-derivatives of alginic acid. Typical "Algipor" specimens are used based on mixed sodium-calcium salt of alginic acid as spongy plates of about 10 mm thick with high absorption ability (Table 1).

14.3 SELECTION OF PROPERTIES OF BURN DRESSING BEING TESTED

The literature data showed that burn dressings, particularly the first aid ones, must perform three main functions [1, 2, 36, 37]:

1. Absorb the wound exudate, which contains metabolism products and toxins,
2. Provide optimum water, air, and heat exchange between the wound and the atmosphere,
3. Protect the wound from the penetration of microorganisms from the air.

Moreover, the burn dressing must be removable from the wound without further injury to the patient. Therefore, the following properties of burn dressings are studied to determine their efficiency.

14.3.1 SORPTION-DIFFUSIONAL PROPERTIES

The sorptional-diffusional properties of dressings are extremely important, because they determine the performance of the three main functions of dressings.

WATER ABSORPTION

Water is the main component of exudate of wounds. At present, there is no opinion on how fast and in what degree the dressing must absorb the exudate enough to clean the wound from toxins and metabolism products while, the same time, keeping wound wet, enough to present the removal of water from healthy tissue [1, 2, 36, 37].

AIR PENETRABILITY

Sufficient air must be allowed to penetrate the dressing since increased oxygen concentration helps the healing process.

VAPOR PENETRABILITY

Vapor penetrability of skin of a healthy man is 0.5 mg×·······×$cm^{-2}hr^{-1}$ [38]. In the absence of technical data it may be concluded that high vapor penetrability will lead to "drying" of dressing with a corresponding change of surface energy on the dressing-wound interface. On one hand this will promote undesirable removal of water from the tissues. On the other hand, it may cause the dressing to come off the wound. Low vapor penetrability of dressing will lead to the accumulation of liquid under the dressing that may cause edema.

PENETRABILITY WITH REFERENCE TO MICROORGANISMS

The penetration of microorganisms through the dressing must be blocked to prevent infection.

14.3.2 ADHESIVE PROPERTIES

The adhesive properties of dressings determine their ability to stay attached to the wound. Thus, the surface energy of the dressing's sole facing the wound must always be lower than that of the wound surface.

14.3.3 MECHANICAL PROPERTIES

The following two mechanical properties are important for dressings—flexural rigidity and strength at break.

The former defines the ability of the dressing to accept wound profile, the latter is important since it allows the dressing to be removed from the wound completely without breaking.

14.4 METHODS OF INVESTIGATION OF PHYSLCO-CHEMICAL PROPERTIES OF BURN DRESSINGS

14.4.1 DETERMINATION OF THE MATERIALS POROSITY

The porosity of materials (the relation of pore space volume to total volume) is determined according to two methods:

1. Measuring density

$$Q = 1 - \frac{\rho}{\rho_0} \tag{1}$$

   Where Q is the material porosity, ρ is the observable density, and ρ_0 is density of the material forming porous medium.
   The value of ρ is determined by weighting of a sample with the known geometrical sizes. The value of ρ_0 is determined similarly for the samples, pressed by 500 GPa.

2. From photos obtained by microscope we get

$$Q = \left[\frac{S_{\text{pores}}}{S_0}\right]^{3/2} \tag{2}$$

   where S_{pores} and S_0 are total surface of pores and general surface of the material in the field of vision of the microscope, respectively.

14.4.2 DETERMINATION OF SIZES AND THE AMOUNT OF PORES

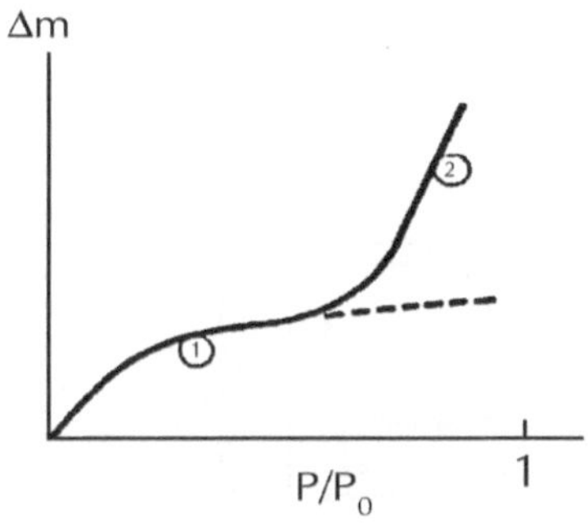

FIGURE 1 Sorption isotherm of low-molecular liquids by microporous materials—(1) real dissolving of liquid by the material, (2) condensation of liquid in the material microporous, and m–sorped liquid mass. P/P_0 is relative pressure of liquid in thermostating vessel (P_0 is pressure of saturated vapor of liquid in the present conditions).

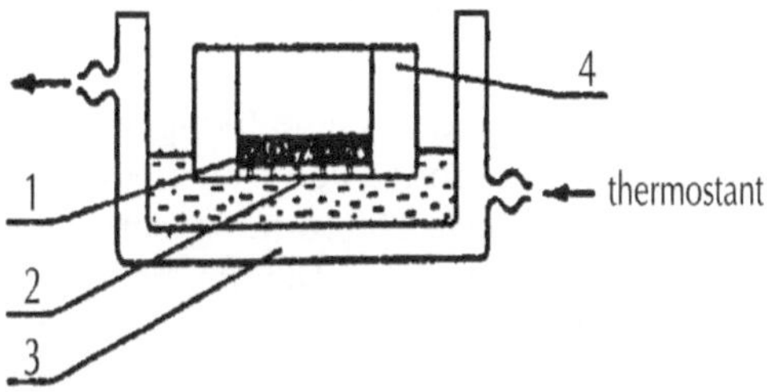

FIGURE 2 The scheme of device for determination of absorbing ability of the porous materials—(1) sample, (2) perforated plate, (3) thermostated camera, and (4) float.

The amount and sizes of pores are determined with the help of MIN-10 microscope in reflected light. The curve of pores distribution by radii is calculated.

14.4.3 ESTIMATION OF SURFACE ENERGY ON THE MATERIAL-MEDIUM INTERFACE

The surface energy on the material–medium interface is estimated using the wetting angle of the material surface by the medium. A drop of liquid is applied to the surface of the material, and the angle is measured between the tangent at the base of the drop and the material surface. The wetting angle is determined using horizontal microscope HM. The accuracy of angle measurement does not exceed ± 1°.

14.4.4 DETERMINATION OF SORPTIONAL ABILITY OF MATERIALS

The total amount of liquid sorped by tile material, includes the liquid in macropores with the size over 0.1 μm, in micropores with the size smaller than 0.1 μm and in the material matrix (dissolved liquid).

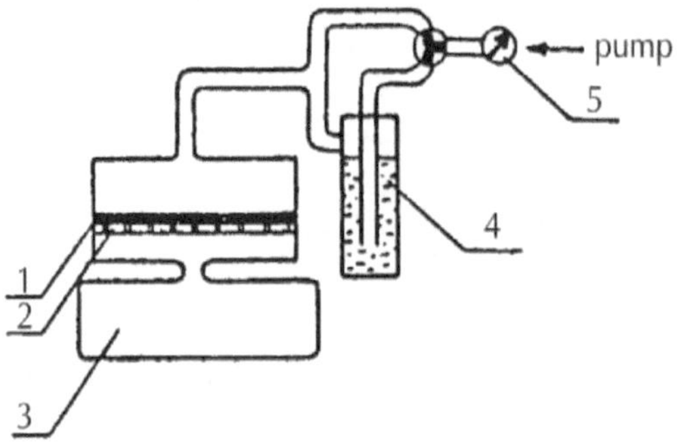

FIGURE 3 Device for determination of air penetrability of materials—(1) sample, (2) perforated plate, (3) polyethylene sack, (4) manometer, and (5) pressure controller.

The amount of dissolved liquid, and of liquid filling the micropores, is calculated from the dependence of the amount of sorted liquid on pressure of vapors of that liquid over the sample (isotherms of sorption). The sorption isoterms for the materials with micropores possess S-type forms (Figure 1). The first part of the curve is connected with the real desolving of liquid, the second one-with the condensation of liquid in micropores.

The maximal sorption (the amount of liquid, really solved and filling micro- and macropores) is determined with the help of the device, shown in Figure 2. The device represents a vessel with liquid medium, in which a float of special perforated square construction is placed. The float construction is calculated to prevent its sinking. This requires that the liquid medium docs not penetrate through the perforation of the square, but forming meniscus on the side of the square facing the porous interlayer. The change of the mass of porous material is determined from the immersion of the float with the sample. It is performed with tile help of a horizontal microscope HM.

14.4.5 DETERMINATION OF AIR PENETRABILITY OF BURN DRESSINGS

Air penetrability (the volume of the air, passed through the specific surface during the specific time) was determined with the help of device specially designed for this purpose. The device is a cylindric cell with perforated plate supporting the sample (Figure 3). The air was passed through the cell with the help of air compressor, equipped with manometer and pressure controller. We measured the time required to fill with air a polyethylene 45 volume sack.

The sample of a round form was cut with the help of mould. Then the sample was placed on the perforated plate of the cell. The sample was pressed by screwing the air was compressed through the cell. The time of polyethylene sack filling was measured with the help of stop-watch. The method allows determining the air pen etrability of dry or wet materials.

14.4.6 DETERMINATION OF ADHESION OF BURN DRESSINGS

The adhesion of burn dressings was investigated on the device, representing the modification of the device, described in the chapter [39].

A 1 mm thick liber glass plate, covered with three layers of medical gauze, was placed into the fiber glass cell having a working surface of 3 × 10 mm. The cell was filled of 5 ml of the whole blood added by 1 ml of 2% thrombine. The dressing tested was placed on the plate surface 1 min after. The cell with the sample was placed into thermostat at 37°C for 24 hr. The sample removal was performed at 90° angle with respect to the surface of the tested material.

14.4.7 DETERMINATION OF VAPOR PENETRABILITY OF BURN DRESSINGS

The vapor penetrability (the mass of water, passed through the specific surface during the specific time) was determined with the help of the device described in [40]. Glass vessel was filled of a definite amount of liquid, for example, water or water solution of sulfuric acid. This amount provided the definite relative humidity. The investigated sample was placed on the vessel surface, metal ring was set and pressed to the vessel by special clamp. The vessel with the contents was weighted and placed into desiccator with dryer at 37°C. After definite period of time the vessel was taken out from the desiccator, weighted and then put back to the desiccator. The amount of water, passed through the sample, was determined by the mass losses of the vessel contents. The vessel of 40 mm diameter and 20 mm high was used in the experiments.

TABLE 2 Composition of edema liquid and blood plasma, g/ml

Components	Edema liquid	Blood plasma
Urea	5.1×10^{-4}	5.5×10^{-4}
Sugar	5.8×10^{-6}	11.0×10^{-6}
Protein	3.4×10^{-2}	7.2×10^{-2}
Salts	1.0×10^{-2}	1.0×10^{-2}
Albumin/Globulin	3.9	1.5

14.5 EXPERIMENTAL DISCUSSION AND RESULTS

14.5.1 DETERMINATION OF SORPTION ABILITY OF BURN DRESSINGS

At applying the dressings on the burn wound there occurs first the wetting of the surface layer of the material and then sorption of the wound exudates into the dressing volume. In this connection it is necessary to answer the following two questions:

- What are the components of the exudates of wounds and burns, being able to sorp by the material, and what is the way of sorption?
- What is the maximum sorption of the separate components of the exudates by the dressing material?

The first question had not yet been addressed in the published literature. In regard to the second question, the maximum water sorption of different materials as follows was previously determined [41]. The sample was immersed into the water, dried fast by filter paper and then weighted. Such method did not allow one to measure the sorption kinetics, and its accuracy of the maximum sorption was low.

That is the reason why we have worked out the device for continuous measurement of sorption, which solved the above mentioned shortcomings.

The exudate of the wounds contains water, salts, proteins, cells the damaged cells, and various low- and high-molecular substances in relatively lower amounts. Table 2 shows the approximate composition of edema liquid in the burn wound. The composition of edema liquid changes in dependence on the burn degree: The worse the bum is, the higher the content of protein and the lower albumine to globulin ratio [3]. Similar data for the blood plasma are shown for comparison in Table 2.

The sorption of wounds exudate may proceed via filling of micro- and macropores, or dissolving in the material matrix. Let us consider, sorption of different components of the wound exudate by the dressing material.

Water fills pores and dissolves in the material matrix. Water solubility is defined by the material hydrophilicity. The solubility of water, salts and other low molecular substances in polymers is the subject to the following rules:

- Solubility in hydrophilic polymers is defined by size and charge of low molec ular substance,
- In hydrophobic polymers it is defined by vapor elasticity (the higher vapor elasticity is, the higher is the solubility) [42].

Protein fills the pores of the size up to 10^{-2} m and may dissolve only in hydrogels of "Hydron" type with water content over 30% by mass.

Cells fill only open pores of the size over 0.1–0.2 μm.

SOLUBILITY OF WATER IN POLYMERS

As mentioned modern burn dressings represent heterogeneous materials, usually consisting of several layers. The upper one having the air is, as a rule, more hydrophobic and less porous than the others. The solubility of water in this layer will define its evaporation from the dressing surface and the heat exchange between the wound and the surrounding. The information about solubility of water in various polymers (Table 3) is shown in the chapter [43].

The solubility of water was determined by the sorption method. Extreme values of sorption at the definite pressures of water vapors were calculated from the sorption curves, and then the sorption isotherms were constructed using the method described in the paper [44]. Extreme values of solution $\varphi^{\infty}_{H_2O}$ at the saturation pressure were determined by extrapolation of φ_{H_2O} to $P/P_s = 1$. The value $\varphi^{\infty}_{H_2O}$ equals to the solubility of water in polymer.

TABLE 3 Solubility of water in various polymers

Polymer	Solubility, 10^2 g/g	T, K
Cellophane	40	303
Viscose fiber	46	303
Cotton	23	303
Cellulose diacetate	18	303
Cellulose triacetate	11.5	303
Polycaproamide	8.5	303
Polyethyleneterphthalate	0.3	303
Polydimethylsiloxane	0.07	308
Poly(2-oxyethylmethacrylate)	40*	310
Polypropylene	0.007	298
Polytetrafluoroethylene	0.01	293
Polyethylene (= 0.923)	0.006	298
Polyurethane	1*	298
Polyvinylchloride	1.5	307

Note: * is measured by the authors.

MAXIMUM SORPTION ABILITY OF BURN DRESSINGS

The modern burn dressings represent large-porous or fibrillar heterogeneous materials, possessing high free volume. At the contact with the wound the exudate will fill the free volume of the dressing, filling degree being defined by hydrophilicity of the material, the size, and geometry of the free volume fraction.

THEORETICAL

Consider, the filling process of the entire free volume of the material by liquid medium. The maximum sorption of the medium by the dressing will be calculated in the following way. Mass of the medium, sorped by the dressing material:

$$m_c = V_\infty \rho_c - V_0 \rho_0 \quad (1)$$

Where V_∞ and V_0 are the volumes of the dressing after and before sorption of the liquid medium, respectively; ρ_c and ρ_0 are the density of the liquid medium and the dressing material, respectively.

The maximum sorption of the liquid medium in this case is;

$$C_c^\infty = \frac{m_c^\infty}{m_0} = \frac{V_\infty \rho_c}{V_0 \rho_0} - 1 \quad (2)$$

where m_0 is the initial material mass.

The two particular cases are possible:

1. There is no swelling of the polymer during sorption, that is liquid medium fills the free volume space only. In this case $V_\infty = V_0$ and

$$C_c^\infty = \frac{\rho_c}{\rho_0} \quad (3)$$

2. For materials possessing low density, $\rho_0 < 0.1$ g/cm^3

$$C_c^\infty = \frac{\rho_c}{\rho_0} \quad (4)$$

MAXIMAL SORPTION OF WATER BY BURN DRESSINGS

The sorption of water by burn dressings is measured using a device developed for this purpose by the authors. The experiments were performed in the following way. First, the weights of different mass were placed on the perforated plate of the device and the relative device immersing to the water was measured with the help of a horizontal microscope HM. Calibrating curve was represented in "weights masses–depth of device immersing" coordinates in the units of eyepiece of microscope. Angle coefficient equals 0.70 ± 0.02 g/unit.

Subsequently, the sample of a dressing was placed into the device, and the depth of the device immersing during time "h" was measured. The mass of the medium sorted by the material was calculated from the correlation:

$$m_c - 0.70h \tag{5}$$

The extreme value of the sorted medium mass was determined at $t \rightarrow \infty$. Maximal sorption of the medium by the material was calculated from Equation (3) and Equation (4). Table 4 shows experimental and theoretical (calculated from Equation (5)) values of $C^{\infty}_{H_2O}$ and values of ρ_0, determined experimentally and used for theoretical calculation.

A good correlation was observed between the experimental and theoretical values of $C^{\infty}_{H_2O}$ for the majority of dressings. This shows that practically the entire free volume is filled by liquid medium at the contact of dressings with water. The exception is the "Algipore" dressing, the large pores of which become denser on filling of water entering because of collapse of the pore walls. In the end this leads to the decrease of the total volume of the dressing. The liquid medium may fill not the whole volume of dressings, if the material is sufficiently hydrophobic and poorly wetted with water.

TABLE 4 Experimental and theoretical data of the maximum sorption of water by burn dressings

Covering name (material)	$C^{\infty}_{H_2O}$, **g/g**		ρ_0, **g/cm^3**
	Experimental	**Theoretical**	
Helitrex (collagen)	32 ± 2	33 ± 2	0.030 ± 0.007
Helitrex (collagen sponge)	58 ± 3	55 ± 3	0.018 ± 0.005
Collagen dressing	1.8 ± 0.1	2.8 ± 0.3	0.350 ± 0.07
Corretium-2 (collagen)	3.5 ± 0.3	3.3 ± 0.3	0.300 ± 0.07
Corretium-3 (collagen)	2.1 ± 0.2	3.0 ± 0.1	0.330 ± 0.05
Combutec-2 (collagen)	77.0 ± 5.0	66.0 ± 3.0	0.015 ± 0.005
Epigard (foamy polyurethane	10.0 ± 0.3	15.0 ± 1.0	0.067 ± 0.005
Silicon-nylon composite	7.5 ± 0.2	7.7 ± 0.5	0.130 ± 0.03

TABLE 4 *(Continued)*

Syspurderm (foamy polyurethane)	6.2 ± 0.2	7.1 ± 0.5	0.140 ± 0.03
Syncrite (foamy polyurethane)	20.0 ± 2.0	22.0 ± 1.5	0.050 ± 0.01
Farmexplant (foamy polyurethane)	12.0 ± 0.5	15.6 ± 3.0	0.064 ± 0.007
Johnson-Johnson (cellulose)	11.4 ± 0.5	10.0 ± 1.5	0.100 ± 0.03
Blood stopping (cellulose)	15.7 ± 0.9	12.5 ± 0.7	0.080 ± 0.006
Tunneling (cellulose)	4.3 ± 0.2	5.0 ± 0.4	0.200 ± 0.005
Switin (cellulose)	18.0 ± 2.0	20.0 ± 1.0	0.050 ± 0.005
Metallized (cellulose-paper)	12.4 ± 0.7	10.0 ± 0.5	0.100 ± 0.04
Needle-perforated (cellulose-viscose)	28.0 ± 2.5	30.0 ± 2.0	0.033 ± 0.007
Viscose 100%	25.0 ± 2.0	30.0 ± 2.0	0.033 ± 0.007
70% of cotton + 30% of viscose	31.0 ± 3.0	33.3 ± 3.0	0.030 ± 0.007
50% of cotton + 50% of viscose	25.0 ± 2.0	28.5 ± 2.0	0.035 ± 0.007
30% of cotton + 70% of viscose	28.0 ± 2.0	27.7 ± 2.0	0.036 ± 0.007
Algipore (vegetable)	30.0 ± 3.0	90.0 ± 5.0	0.011 ± 0.0002

TABLE 5 Density, maximal water sorption, wetting angle, and heat effect of sorption of water by different collagens

ρ_0, g/cm^3	$C^{\infty}_{H_2O}$, g/g		φ^{o}	ΔH, cal/g
	Experimental	**Theoretical**		
0.011	74	91	170	34.6
0.016	53	62.5	70	25.4
0.013	49	77	90	30.2
0.013	47	77	110	31.9
0.013	8	77	120	31.2
0.014	4	71.4	110	29.8
0.014	30	71.4	50	27.2

To test this assumption 7 collagen materials were investigated, which differed in the production method. We investigated—density, maximal water sorption, wetting angle, and heat effect of water sorption by the material. The latter was determined using the microcalorimeter LKB 2107 as follows. The sample of definite mass was exposed to vacuum in the Butch type cell, thermostated, and then the excess amount of water was in troduced into the cell causing a forced filling of the material volume. The obtained results are presented in the Table 5 and on the Figure 4. The following conclusions can be made on the basis of tile data presented in the Table 5:

The experimental value of $C^{\infty}_{H_2O}$, is lower than the "theoretical" one. This may be explained by two causes—the decrease of the total volume (as in the case of "Algipore") and non-filling of a part of the material free volume by water.

A satisfactory correlation exists between the theoretical values of $C^{\infty}_{H_2O}$ and ΔH. Thus, the main reason of the difference between experimental and theoretical values of $C^{\infty}_{H_2O}$ is evidently the non-filling of a part of the material free volume by water.

The absence of correlation between maximal water sorption and wetting angle, defined on the external surfaces of the material, shows that the values, obtained as mentioned above, do not reflect real interaction of water with internal surface of collagen.

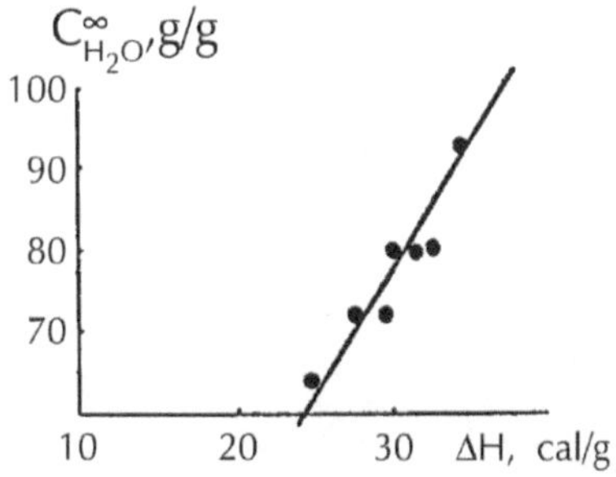

FIGURE 4 The dependence of maximum sorption of water by collagen materials on heat effect sorption.

Thus, it may be concluded that for the most number of burn dressings from hydrophilic materials, the maximum sorptional capacity with reference to water may be predicted satisfactorily. To fulfill this it is sufficient to use the Equation (4). For example, the experimental values of $C^{\infty}_{H_2O}$, correlate well with the free volume part of the materials (Figure 5). The correlation coefficient is 0.96.

MAXIMAL SORPTION OF PLASMA BY BURN DRESSINGS

The sorption of the blood plasma by burn dressings was determined by a similar method. Plasma was obtained by centrifugation of the conserved blood. The treatment of the experimental results was carried out similarly to the case of the investigation of maximal sorption of water. The value of C^{∞}_{plasma}, differs from $C^{\infty}_{H_2O}$. The difference is not higher than 10%, that is why the data for C^{∞}_{plasma} are not shown in Table 5.

14.5.2 STUDY OF THE KINETICS OF THE SORPTION OF LIQUID MEDIA BY BURN DRESSINGS

The study of the kinetics of the sorption of the wound exudate by burn dressings is of great importance for the estimation of their efficiency. The difficulty occurs at the mathematical description of the kinetics of the sorption process, connected with the absence of strictly quantitative description of the dressing's structure.

STRUCTURE OF BURN DRESSINGS

The burn dressings are heterogeneous systems, consisting of several component-phases. As the general attention in dressings must be paid to the material possessing the maximum penetrability with reference to liquid medium, it is necessary to classify the types of heterogeneous systems. For example, the penetrable parts of the material are placed under the layer of another weakly penetrable material in the way that diffusing flow is perpendicular to the surface layer. This is the case of double-layered dressings with a dense external layer. The penetrable parts of the material can be dispersed in a continuous weakly penetrable phase.

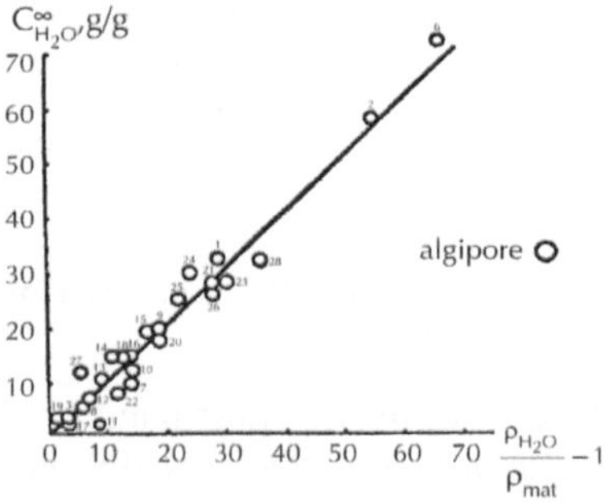

FIGURE 5 The dependence of $C^{\infty}_{H_2O}$ on free volume of the material various burn dressings—(1) collagen dressing Helitrex, (2) collagen sponge neutron, (3) collagen dressing Braun, (4), (5) artificial leather Corretium 2 and 3, (6) Combutek-11, (7) synthetic dressing Epigard, (8) foamy polyurethane dressing Syspurderm, (9) synthetic dressing Syncrite, (10) foamy polyurohane dressing Famlexplant, (11) burn face mask, (12) compositional dressing Biobrant, (13) cellulose dressing Johnson-Johnson, (14) cellulose dressing Candall, (15) cellulose non-adhesive dressing Torcatee, (16) blood-stopping cellulose dressing, (17) cotton balling dressing Mesorb, (18) dressing with tunneling effect, (19) cellulose dressing with nonadhesive synthetic layer, (20) cotton balling dressing Svutyn, (21), (22) metallized dressings, (23), (24) needle perforated fabric with atraumatic layer, (25) viscose 100%, (26–29) viscose + cotton balling.

The dressings, based on collagen and cellulose, possess fibrillar structure and the fibers are randomly placed. In some cases spatial orientation of fibers is present. The sufficient amount of open pores in the dressings of such type is large but the open pores possess irregular form and great tortuosity in the direction of mass transfer. Modern burn dressings are multilayered with denser external layer. Table 6 shows the mean-radius of macropores and their amount per unit of square for dressings, based on polyurethane. All dressings possess a mean-radius of macropores in the range of $(2.0 \div 3.0) \times 10^{-2}$ cm and a sufficiently narrow distribution (Figure 6).

THEORETICAL

The detailed analysis of a number of mathematical models and results of experimental investigations of heterogeneous systems was performed by Barrer [45].

TABLE 6 Mean-radius of macropores and their amount per unit of square N for dressings, based on polyurethane

Dressing name	R, cm $\times 10^{-2}$	N, pore/cm^2
Epigard	2.2 ± 0.2	370 ± 10
Syspurderm	1.8 ± 0.2	266 ± 5
Syncrite	2.8 ± 0.2	275 ± 5
Farmexplant	2.2 ± 0.2	300 ± 10

Dressings are of a membrane form. If a membrane is in contact with the solution in a way that the concentration at one of its surfaces equals C_0 and at the other equal 0 at t = 0, then the total amount of the substance entered the membrane during time t(m_t), is given by the equation:

$$\frac{m_\infty}{m_t} = 1 - \frac{8}{\pi^2}\exp\left[-\frac{D\pi^2 t}{l^2}\right] \qquad (6)$$

where m_∞ is the amount of substance, entered the membrane at t ∞, that is in the equilibrium, D is the coefficient of substance diffusion in the membrane, and l is the membrane thickness.

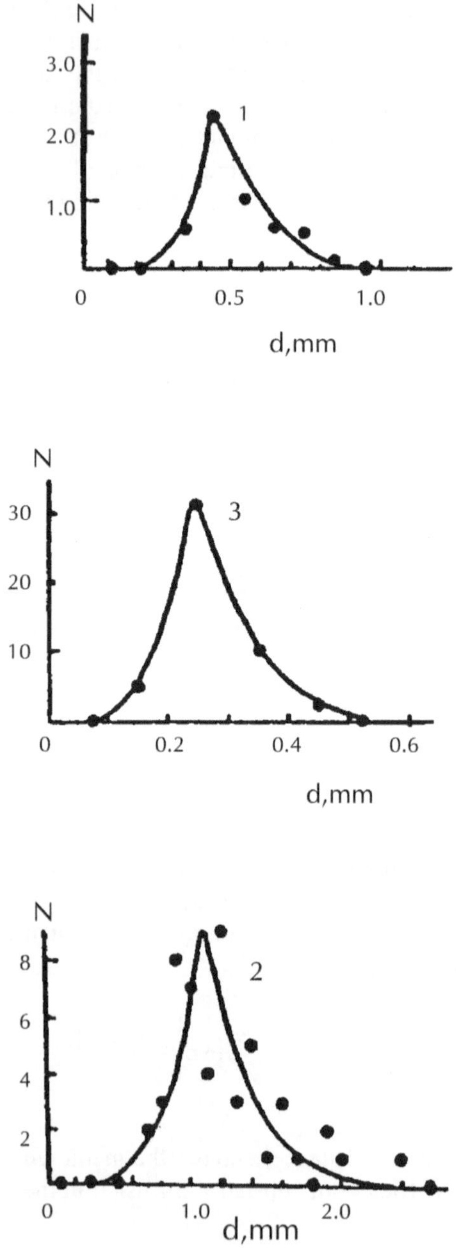

FIGURE 6 *(Continued)*

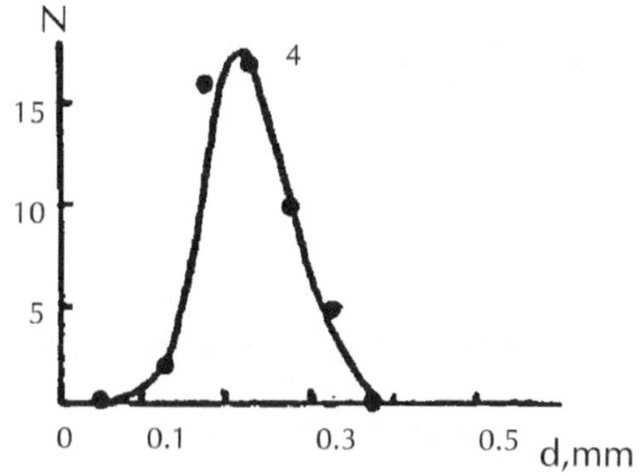

FIGURE 6 Curves of distribution of pores by sizes for different dressings–(1) Farmexplant, (2) Syncrite, (3) Collagen dressing Braun, and (4) Syspurderm.

The following correlation is satisfactory for the initial part of the kinetic curve of sorption:

$$\frac{m_t}{m_\infty} = 2\left[\frac{Dt}{\pi l^2}\right]^{1/2} \tag{7}$$

The Equation (6) was obtained for homogeneous material assuming that D does not depend on concentration of the substance in membrane. From Equation (6) it is possible to calculate D value and time of membrane saturation by the substance up to

a definite limit. For example, time of saturation up to $\frac{m_t}{m_\infty} = 0.85$ is calculated from the correlation:

$$\tau_{0.85} = \frac{l^2}{6D} \tag{8}$$

and $\frac{m_t}{m_\infty} = 0.97$ is reached during the time.

$$\tau_{0.97} = \frac{l^2}{3D} \tag{9}$$

Both these correlations may be applied for practical calculations.

As it was mentioned, the calculation of diffusion coefficient in heterogeneous systems is very difficult. According to ideas accepted in the present time, the penetration of liquid into porous body is ruled by the laws of capillarity. These ideas are successfully applied for interpreting the penetration of water into paper, leather, fabrics, and so on [46]. Capillary pressure, which is the driving force of liquid to rise, is determined from the Jurene equation [47]:

$$P_k = \frac{2\gamma_i \cos\theta}{r} \tag{10}$$

where γ_i is the surface tension of liquid, θ is the wetting angle, and r is the capillary radius.

The equation, taking into account real structure of porous body, was obtained by Deriagin [47].

EXPERIMENTAL

The kinetics of sorption of water and blood plasma was investigated using the device for the maximal sorption of water. Figure 7 shows typical kinetic curves of sorption of water and plasma by various dressings. All curves are satisfactorily described by

Equation (7). Table 7 shows the values of $\frac{D_{eff}}{l^2}$ and the value of $\tau_{0.85}$, calculated according to Equation (8). The following conclusions could be made from the Table 7 data:

- Burn dressings differ significantly in their rates of sorption of liquid media.
- The rate of sorption is determined by the pores size and the material hydrophilicity.

14.6 DETERMINATION OF VAPOR PENETRABILITY OF BURN DRESSINGS

With multilayered dressings, the external layer of is more dense than lower one which regulates the mass transfer of water from the wound into the surrounding. The process of mass transfer of water through the material layer is often called aquapenetrability or vapor penetrability. The penetrability and diffusion of water in polymers were the subject of numerous investigations, the results of which are generalized in the list of reviews and monographs [43, 49] and are presented in Table 8. The mass transfers of water molecules in polymers possess a list of features. In hydrophilic matrices the interaction between water molecules and the material matrix is weak (low solubility). Nevertheless, the interactions of water molecules with each other stipulate a specific transfer mechanism.

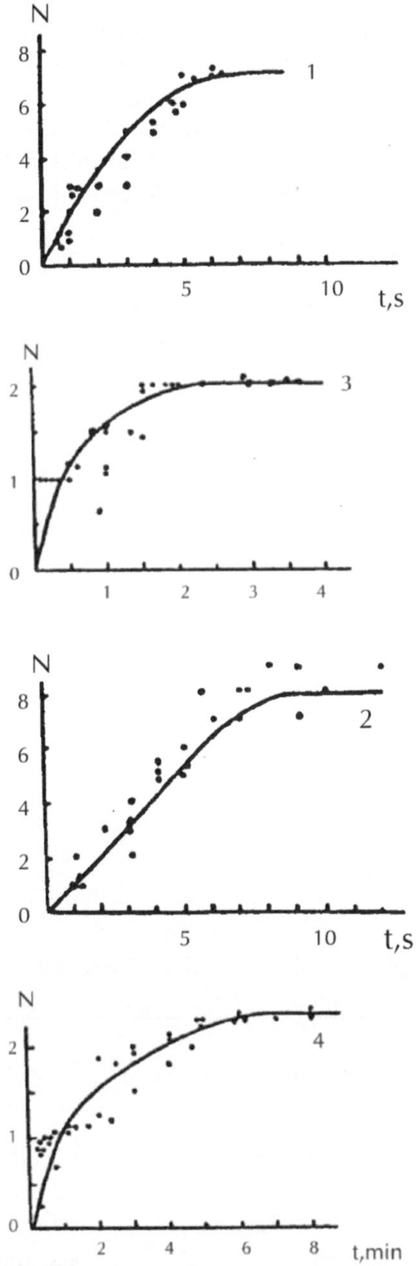

FIGURE 7 Curves of solution of water and blood plasma by different burn dressings—(1) water–by needle perforated material,(2) plasma–by needle perforated material,(3) water–by polyurethane dressing Syspurderm, and (4) plasma–by Syspurderm dressing.

In hydrophilic materials the interaction between water molecules and hydrophilic groups of the material matrix stipulates high solubility of water in the matrix and increased aquapenetrability. Thus, high aquapenetrability may be the property of hydrophobic as well as of hydrophilic materials, however the causes will be different. For example, in hydrophilic polydimethylorgano-siloxane the high mobility of water molecules is stipulated by high mobility of the chain units in this polymer. That is why despite low solubilities of water in polydimethylorganosiloxane the coefficient of aqua penetrability is significant.

TABLE 7 Values of $\frac{D_{\text{eff}}}{l^2}$ and $\tau_{0.85}$ for different burn dressings in water and plasma at 37°C

Dressing name (material)	$\frac{D_{eff}}{l^2}\cdot 10^3$, s		$\tau_{0.85}\cdot 10^{-3}$, s	
	Water	**Plasma**	**Water**	**Plasma**
Helitrex (collagen)	2.5 ± 0.3	1.3 ± 0.2	3.9 ± 0.2	7.0 ± 0.3
Helitrex (collagen sponge)	6.6 ± 0.6	3.3 ± 0.4	1.5 ± 0.1	3.0 ± 0.15
Corretium-2 (collagen)	4.7 ± 0.4	2.5 ± 0.3	4.1 ± 0.2	8.0 ± 0.4
Combutek (collagen)	7.0 ± 0.7	3.5 ± 0.4	1.4 ± 0.1	3.0 ± 0.15
Syspurderm (foamy polyurethane)	2.3 ± 0.3	1.1 ± 0.1	4.3 ± 0.2	8.6 ± 0.4
Syncrite (foamy polyurethane)	(9.8 ± 0.8)*	(5.0 ± 0.5)*	90 ± 10	200 ± 10
Switin (cellulose)	(4.5 ± 0.4)*	(2.2 ± 0.2)*	210 ± 20	420 ± 35
Needle-perforated (cellulose)	35 ± 3	17 ± 1.7	0.3 ± 0.02	0.6 ± 0.05

* Must be multiplied by 10 Syspurderm $^{-2}$.

TABLE 8 Penetrability and diffusion of water vapors in polymers [49]

Polymer	T, K	$\frac{p}{p_0}$	$P\cdot 10^{15}$, mol·m/m²·s·Pa	$D\cdot 10^{12}$, m²/s
Cellulose	298	1.0	8500	—
Regenerated cellulose	298	0.2	5700	0.1

TABLE 8 *(Continued)*

Cellulose acetate	303	0.5–1.0	2000	1.7
Cellulose diacetate	298	1.0	15.7	—
Cellulose triacetate	298	1.0	5.5	—
Ethylcellulose	298	0.84	7950	18
Polydimethylorganosiloxane	308	0.2	14400	7000
Polyethylene (= 0.922)	298	0–0.1	30	23
Polyethyleneterephthalate	298	0–0.1	58.6	0.39
Polypropylene	298	0–0.1	17	24
Polyvinyl chloride	303	—	—	2.3
Polycaproamide	298	0.5	134	0.097

On the opposite, in regenerated cellulose the diffusion coefficient is low because only the dissolved water molecules which are not connected with the matrix of this polymer participate in the mass transfer. In this case high value of aquapenetrability is stipulated by increasing the dissolved water content in regenerative cellulose which increases the part of water molecules participating in the mass transfer. This in turn leads to the increase of both diffusion coefficient and penetrability coefficient.

THEORETICAL

The mass transfer of water through a porous body is practically equal to that of gases in a polymer, provided there is no interaction between water molecules and the matrix of polymeric material. Since the hydrophilic materials are commonly used for the production of dressings, which actively interact with water molecules, the diffusion should be considered simultaneously with absorption.

As a rule, the rate of the absorption process is significantly higher than the diffusion rate. Therefore, it can be assumed that the absorption equilibrium is immediately reached, and the concentration of water in the material C_{H_20} is obtained from the following equation:

$$\frac{\partial C_{H_2O}}{\partial t} = D_{H_2O}\frac{\partial^2 C_{H_2O}}{\partial x^2} - \frac{\partial C^a_{H_2O}}{\partial t} \qquad (11)$$

where D_{H_20} is the coefficient of water diffusion in the material; x is the diffusion coordinate; $C^a_{H_2O}$ is the concentration of the absorbed water.

The concentration of the absorbed water can be calculated for the particular cases. For example, if the concentration of functional groups, capable to link water molecules irreversibly, is limited and equals C_f, we can assume that the bonded water molecules no longer participate in the diffusion process, but form domains on which fast absorption occurs. For the case when the concentration of water on one of the surfaces

(x = 0) is constant and equals $C^0_{H_2O}$, the reaction zone reaches the second surface of the membrane, which is l thick, during the time t [50].

Thus, during the time t there will be no water flow through the surface X = 1 on the membrane exit and then stationary flow will be set immediately. The amount of water passed through the membrane will equal

$$m_{H_2O} = D_{H_2O} \frac{\Delta C_{H_2O}}{l} \cdot S \cdot t \tag{12}$$

where S is the square of the membrane, $\frac{\Delta C_{H_2O}}{l}$ is the concentration gradient. If the solubility of water in the material is ruled by the Henry law:

$$C_{H_2O} = \sigma P \tag{13}$$

where P is the pressure of water vapors over the material; then substituting (6.13) into (6.12) we obtain the following equation:

$$m_{H_2O} = D_{H_2O} \cdot \sigma_{H_2O} \frac{\Delta P}{l} St \tag{14}$$

Considering the diffusional coefficient D_{H_2O} being equal:

$$P_{H_2O} = D_{H_2O} \cdot \sigma_{H_2O} \tag{15}$$

we obtain

$$P_{H_2O} = \frac{m_{H_2O}}{\Delta P \cdot S \cdot t} \tag{16}$$

EXPERIMENTAL

The aquapenetrability of burn dressings was determined on the device, described in the chapter. The values of penetrability coefficients were cal culated from the Equation (16). Table 9 shows the values of coefficients of aquapenetrability P_{H_2O} for various burn dressings.

14.7 DETERMINATION OF THE AIR PENETRABILITY OF BURN DRESSING

As it was mentioned in the chapter, active sorption of the wound exudate occurs during several minutes after the burn wound is closed by dressing. Further on, there proceeds the evaporation of water from the external side of the dressing. This leads to the change of the state of exudate in the material mass. On the whole, this changes penetrability of the dressing with respect to air. In this case in order for anaerobic conditions not to be created in the wound, it is necessary to provide optimal air penetrability during the entire period of application.

The data on penetrability of dressings according to dry air are known in literature. Thus, for example, it is recommended [51] to determine air penetrability with the help of industrially produced VPTM-2 device. This device records automatically the amount of the air, passed through the dressing of the known square during time t at pressure oscillations of about 5 mm H_2O. However the application of such device does not allow us to investigate the air penetrability of dense materials such as foamy polyurethane compositions, and most importantly of dressings in wet state.

TABLE 9 Values of aquapenetrability coefficients of burn dressings at 37°C

Dressing name (material)	$P_{H_2O} \cdot 10^9$, mol·m/m²·s·Pa
Helitrex (collagen)	1.6 ± 0.1
Helitrex sponge (collagen)	11.0 ± 1.0
Brown dressing (collagen)	6.6 ± 0.6
Syspurderm (foamy polyurethane)	0.8 ± 0.2
Syncrite (foamy polyurethane)	1.2 ± 0.2
Epigard (foamy polyurethane)	4.3 ± 0.4
Farmexplant (foamy polyurethane)	3.3 ± 0.3
Biobrant (silicon-polyamide)	1.6 ± 0.16
Johnson-Johnson (cellulose)	2.0 ± 0.2
Perforated metallized dressing (cellulose)	9.0 ± 0.7
Face mask dressing (cellulose)	2.5 ± 0.2
Burn towel (cellulose)	5.4 ± 0.5
50% cotton + 50% viscose	8.0 ± 0.7

TABLE 9 *(Continued)*

70% cotton + 30% viscose	8.0 ± 0.7
100% viscose	7.0 ± 0.7

The construction and principle of action of the device, developed by the authors, allowing the determination of air penetrability of any material in any states and conditions, were described in chapter.

TABLE 10 Penetrability of polymer

Polymer	**Penetrability coefficient,(mol·m/m²·s·Pa) × 10¹⁵**		**Separation coefficient,**
	O_2	N_2	O_2–N_2
Polycaproamide	0.013	0.0033	3.8
Polyvinylchloride	0.022	0.008	2.8
Polyurethanic elastomer	0.032	0.10	3.2
Polyethylene (ρ = 0.922)	0.35	0.13	2.7
Polystyrene	3.13	0.73	2.9
Teflon	2.07	0.67	3.1
Ethylcellulose	3.2	0.93	3.4
Polydimethylsiloxane	168	83.0	2.0
Silicon rubber	200	87.0	2.3

14.7.1 PENETRABILITY OF VARIOUS MATERIALS ACCORDING TO OXYGEN AND NITROGEN

The coefficient of gas penetrability (as well as of vapor penetrability) is calculated according to the Equation (16). The literature data on penetrability or various polymers according to the oxygen and nitrogen are given in Table 10. As it is seen from the data presented in the Table 10, the penetrability of polymers may differ by four orders of magnitude. Special attention should be paid to high gas penetrability of polydimethylsiloxane and compositions on its basis, which is the result of the increased solubility of gases in them at high rate of diffusion (Table 11).

TABLE 11 Values of penetrability coefficients P (molm/m^2·s·Pa), diffusion D (m^2/s) and solubility σ (mol/m^3·Pa) of gases into polydimethylsiloxane at 20°C [52]

Gases	P·10^{15}	D·10^{10}	σ·10^6
O_2	83	23.3	36
N	164	30	55.6
CO_2	720	—	

14.7.2 PENETRABILITY OF POROUS MATERIALS, FILLED BY LIQUID MEDIUM

A short list of studies considering the investigations of gas penetrability of polymeric membranes, in con tact with a liquid is given in [53]. It was observed, that the sorption of liquid by a polymer leads to a decrease of gas penetrability coefficient in comparison with the liquid free polymer.

THEORETICAL

Let us consider the mass transfer of the air through porous body in two cases—one in which the free volume of all pores is filled by the air and the other with free volume filled by liquid medium. Porous body may be presented as consisting of two phases—the material forming the body's carcass and free space. We also assume that pores possess cubic form and are disposed in the volume of the body, not joining each other. Such model is sufficiently suitable for porous burn dressings.

Let us determine the total thickness of the body in the direction of the mass transfer, total thickness of free space, occupied by pores, and total thickness of the layer, occupied by the material. The total thickness of the body in the direction of the mass transfer is:

$$l_{\Sigma} = \frac{V}{S} \tag{17}$$

where V and S are the volume and surface, respectively. Total thickness of the free space occupied by pores is:

$$Q_{\text{pores}} = l_{\Sigma} Q^{1/3} = \frac{V}{S} \cdot Q^{1/3} \tag{18}$$

where $Q_{\text{pores}} = \frac{V_{\text{pores}}}{V}$ is porosity.

The total thickness of the layer, occupied by the material is:

$$l_{\text{mat}} = l_{\Sigma} - l_{\text{pores}} = \frac{V}{S}\left(1 - Q^{1/3}\right) \tag{19}$$

Thus the air passing through the porous body will overcome the resistance of two layers, each possessing its own penetrability coefficient with respect to air.

TABLE 12 Values of the coefficients of penetrability, diffusion and solubility of the oxygen in air, water, plasma and blood at 37°C (dimensions as in Table 11)

Medium	P	D	σ
Air	$2.5{\cdot}\times 10^{-9}$	$2.7{\cdot}\times 10^{-5}$	$9.4{\cdot}\times 10^{-5}$
Water	$7.4{\cdot}\times 10^{-14}$	$3.0{\cdot}\times 10^{-9}$*	$2.5{\cdot}\times 10^{-5}$*
Plasma	—	$2.0{\cdot}\times 10^{-9}$*	—
Blood	$1.4{\cdot}\times 10^{-14}$	$1.4{\cdot}\times 10^{-9}$*	$1.0{\cdot}\times 10^{-5}$*

* The values were taken from [53].

Total penetrability coefficient P_{Σ} or the porous body equals:

$$\frac{1}{P_{\Sigma}} = \frac{l_{\text{pores}}}{l_{\Sigma} P_{\text{pores}}} + \frac{l_{\text{mat}}}{l_{\Sigma} P_{\text{mat}}} \tag{20}$$

where P_{pores} and P_{mat} are penetrability coefficients of the medium, presenting in pores, and the material, forming the body's carcass, respectively.

Let determine, the ratio of penetrability coefficients of the porous body according to air, when its pores are filled by liquid and air:

$$\frac{P_{\Sigma(l)}}{P_{\Sigma(\text{air})}} = \frac{\frac{P_{\text{mat}}}{P_{\text{air}}} + 1}{\xi \frac{P_{\text{mat}}}{P_{l}}} \tag{21}$$

where

$$\xi = \frac{Q^{1/3}}{1 - Q^{1/3}} \tag{22}$$

Values of P_{mat} are shown in Table 12.

Values of P_{air} and P_l can be estimated from the coefficients of diffusion and solubility of oxygen in air, water, plasma and blood at 37°C.

Values of penetrability coefficients of the oxygen in various media may be calculated according to the following expression:

$$P = D \cdot \sigma \tag{23}$$

For any material $P_{air} >> P_{mat}$, so we obtain more simple expression:

$$\frac{P_{\Sigma(air)}}{P_{\Sigma}(l)} \approx \xi \frac{P_{mat}}{P_l} + 1 \tag{24}$$

As for the majority of the dressings $\xi >> 1$, and P_{mat} and P_l are of the same degree, the decrease of air penetrability of the dressing at pores filling by liquid must be then significant.

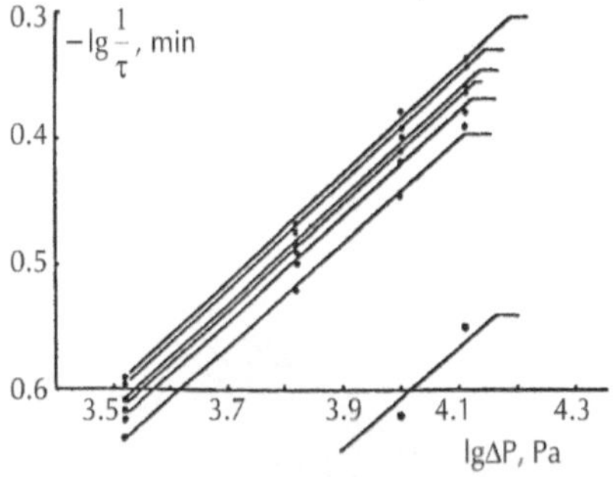

FIGURE 8 The dependence of l/lg(τ) on pressure in the system for dry air—(1) needle perforated material, (2) collagen sponge, (3) Syspurderm, (4) Syncrite, (5) cellulose dressing Svulyn, (6) Farmexplant, and (7) Epigard.

EXPERIMENTAL

Air penetrability of burn dressings was determined using the de vice developed for this purpose. Two types of experiments were performed: determination of air penetrability for dry air and determination of air penetrability of dressings, preliminarily saturated by water (in conditions of maximal sorption of water), for humid air. It has been shown by special experiment that air humidity (from 40 to 100%) does not practically influence the rate of penetration.

The experiments were performed according to the following scheme. At first we determined the time of filling by air of polyethylene sack of 45 volume in conditions, when the sample was not in the cell. This time (the constant of the device) depended on pressure in the system (p):

$$\lg\frac{1}{t} = -2.00 + 0.44\lg p \tag{25}$$

The time of polyethylene sack filling at p = 100 Pa was selected as the standard. At T= (21 ± l)°C, and $t_0 = 16.0 \pm 0.1$ min.

Subsequently, the time of polyethylene sack filling with the sample was placed into the cell was similarly determined. It was observed (Figure 8) that the dependence of t_x on p is described with the same slope as in (25) for all investigated dressings in conditions of the dry air penetration:

$$\lg\frac{1}{t_x} = -A_x + 0.44\lg p \tag{26}$$

where A_x is the constant depending on structure and properties of the dressing material.

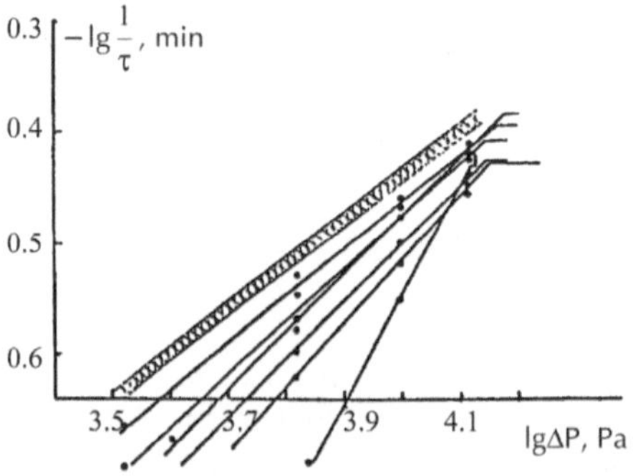

FIGURE 9 The dependence of 1/lg(τ) on pressure in the system for humid air—(1) needle perforated material, (2) Farmexplant, (3) Combutek, (4) Syspurderm, (5) compositional dressing Biobrant (silicon-polyamide), and (6) Epigard.

At bubbling humid air through the dressing saturated by water, the slope increased significantly. That is why it is necessary to perform several experiments for each dressing at different pressures in order to extrapolate t_x to the pressure of 100 Pa with the required accuracy (Figure 9). The increase the slope of $\lg\left(\frac{1}{t_x}\right) - \lg p$ at air bubbling through the dressing saturated by water was attributed to the change of the material structure of the dressing resulting from the change of forms and sizes of macropores. This is often accompanied by a decrease of the total volume of the dressing. The coefficient of air penetrability of the dressing (P_x) was calculated according to the equation:

$$P_x = \frac{ml_x}{S \cdot t \cdot (t_x - t_0)} \tag{27}$$

where m is the polyethylene sack bulk equal 2 moles of air at (21 ± 1)°C, S the surface square contacting with the bubbling air, equal $1.8 \cdot 10^{-3}$ m², and p = 100 Pa.

Thus,

$$P_x = 11 \frac{l_x}{(t_x - t_0)} \tag{28}$$

The value of air penetrability coefficients for the dressings dry and saturated by water are shown in Table 13. From the data shown in Table 13 we see that on saturation with water a significant decrease of air penetrability takes place for all dressings, except for the "Biobrant". Penetrability coefficient for dry dressings can be calculated according to the following equation:

$$\frac{1}{P_{\Sigma(\mathrm{air})}} = \frac{Q^{1/3}}{P_{\mathrm{air}}} + \frac{(1+Q^{1/3})}{P_{\mathrm{mat}}} \tag{29}$$

P_{air} is obtained from the Equation (23) using D = $2.7 \cdot 10^{-5}$ m²/s and 45 mol/m³ (solubility at the atmosphere pressure). The value of P_{air} is $1.2 \cdot 10^{-3}$ mol·m/m²·s·Pa. The values of $P_{\Sigma(air)}$ were taken from Table 13. Values of P_{mat} were calculated from Equation (24).

TABLE 13 Coefficients of air penetrability for dry and water-saturated burn dressings at temperature of (21 ± 1)°C.

Dressing name (material)	Coefficient of air penetrability,molm/m²·s·Pa	
	Dry	**Swollen**
Helitrex (collagen)	2.7·× 10^{-5}	0
Combutek (collagen)	1.1·× 10^{-3}	0
Epigard (foamy polyurethane)	1.3·× 10^{-4}	1.3·× 10^{-5}
Syspurderm (foamy polyurethane)	1.3·× 10^{-4}	1.0·× 10^{-6}
Syncrite (foamy polyurethane)	1.1·× 10^{-3}	4.0·× 10^{-5}
Farmexplant (foamy polyurethane)	4.5·× 10^{-5}	0
Biobrant (polyamide + silicon)	1.8·× 10^{-4}	7.0·× 10^{-5}

TABLE 13 *(Continued)*

Johnson-Johnson (cellulose)	$1.6\cdot\times 10^{-4}$	$3.0\cdot\times 10^{-6}$
Needle-perforated material (cellulose)	$1.1\cdot\times 10^{-3}$	–

The values of $P_{\Sigma(H_2O)}$ can be obtained from the following equation:

$$\frac{1}{P_{\Sigma(H_2O)}} = \frac{Q^{1/3}}{P_{H_2O}} + \frac{\left(1+Q^{1/3}\right)}{P_{\text{mat}}} \tag{30}$$

According to the calculations of $P_{\Sigma(H_2O)}$, its values fall close to 10^{-8} mole·m/m²·s·Pa for the majority of dressings. It is this result which reveals the extremely low air penetrability for the listed dressings. For some dressings the value of $P_{\Sigma(H_2O)}$ is significantly higher than 10^{-8} mole·m/m²· s·Pa. This can be explained by two effects:

1. The presence of the air flow along the surface of pores (surface flow) [50];
2. The pressure of channels in the materials that is free of water.

To test these suppositions additional investigations are required.

14.8 DETERMINATION OF ADHESION OF BURN DRESSINGS

Adhesion properties play a key role in the dressing performance. On one hand the lower layer of the dressing must be easily wetted, providing good adhesion of the dressing to the wound. On the other hand, the surface energy on the dressing–wound interface must be minimal to provide the smallest trauma on its removal from the wound.

THEORETICAL

The adhesive strength characterizes the ability of an adhesive structure to preserve its integrity. The adhesive strength as well as the strength of homogeneous solids is of kinetic nature. That is why the rate of tension increase and temperature affect the adhesive strength and the scale factor (that is sample dimensions) are also of great importance. Different theories of adhesion of polymers were previously suggested [55].

MECHANICAL THEORY (MACBAIN)

According to which the main role is devoted to mechanical filling of defects and pores of the surface (dressing) by the adhesive (blood).

ADSORPTIONAL THEORY (MAC-LOREN)

Considering adhesion as a result of the performance of molecular interaction forces between contacting phases. According to this theory low adhesion, for example, may be reached between a substrate (dressing) with nonpolar groups and polar adhesive (blood).

ELECTRICAL THEORY (DERIAGIN)

It is based on the idea that the main factor controlling the strength of adhesive compounds rests in the double electrical layer which is formed on the adhesive-substrate interface.

DIFFUSIONAL THEORY (VOJYTZKY)

It considers the adhesion as a result of interweaving of the polymers chains.

MOLECULAR-KINETIC THEORY (LAVRENTIEV)

It assumes that a continuous process of restoration and breakage of bonds proceeds in the zone of adhesive-substrate contact. Thus, the adhesive strength is defined by the difference between activation energies of the breakage and formation of bonds, and also depends on the correlation between the amount of segments participating in the formation of bonds and averge amount of molecular bonds per unit of the contact area.

In recent years, the thermodynamic concept received the most attention. There, the main role is devoted to the correlation of surface energies of adhesive and substrate. Thermodynamic work of adhesion of a liquid to a solid (W_a) is described by the Dupret–Jung equation:

$$W_a = \gamma_l \left(1 - \cos \theta\right) \tag{31}$$

where γ_l is the surface tension of liquid and θ is the wetting angle.

Substituting Jung's equation in Equation (31);

$$\gamma_{s-l} = \gamma_s - \gamma_{s-l} \cos \theta \tag{32}$$

we obtain the correlation:

$$W_a = \gamma_s + \gamma_l - \gamma_{s-l} \tag{33}$$

where γ_s and γ_{s-l} are the surface tension of solid and of the solid-liquid interface, respectively.

It follows from Equation (33) that the higher is W_a, the larger are the values of γ_s, and γ_l while γ_{s-l} are smaller. However, according to Equation (33) the increase of γ_s must, on one hand, lead to the growth of W_a, and on the other hand, to an increase of γ_{s-l}. That is why the increase of the surface tension of the substrate is accompanied by the action of two effects. The necessary condition of the adhesive strength is $\gamma_l >> \gamma_s$. Values of γ_l and W_{s-H_2O} for different materials are shown in Table 14.

TABLE 14 Values of the surface tension and thermodynamic work of adhesion of various materials [28]

Material	γ_s, mN/m	W_{s-H_2O} mN/m
Polytetrafluoroethylene	18.5	83
Silicon rubber	21.0	78
Polyethylene	31.0	99
Polystyrene	33.0	105
Polymethylmethacrylate	39.0	103
Polyvinyl chloride	39.0	101
Polyethyleneterephthalate	43.0	104
Polycaproamide	46.0	107
Glass	170.0	222

TABLE 15 Adhesive strength (a) and the angle of wetting by water (θ) of various burn dressings

Dressing name (material)	A, mN/m	θ°
Corretium (collagen)	220 ± 20	75 ± 2
Syspurderm (foamy polyurethane)	210 ± 20	—
Epigard (foamy polyurethane)	350 ± 50	125 ± 3
Farmexplant (foamy polyurethane)	200 ± 20	130 ± 2
Bern-pack (cellulose)	170 ± 20	—
Biobrant (silicon-polyamide)	70 ± 10	—

TABLE 15 *(Continued)*

Johnson-Johnson (cellulose)	20	—
Blood stopping non-adhesive dressing (cellulose)	20	—
Dressing with metallized lower layer (cellulose)	170 ± 50	—

The value of γ_l for blood is 55.0 [56].

EXPERIMENTAL

Adhesive strength of burn dressings was determined according to the method, described in the chapter. Table 15 shows adhesive strength of various burn dressings and the angle of wetting by water also.

14.9 THE MODEL OF A BURN DRESSING ACTION

Three main processes proceed at the application of a dressing to the wound:

1. Sorption of the wound exudate by dressing.
2. Water evaporation from the dressing surface.
3. Mass transfer of gases through the dressing in conditions of the proceeding of sorption and evaporation processes.

Processes 1 and 3 were analyzed in detail in the chapter. It was found that sorption of the liquid media (water, plasma) proceeds fast and reaches the limiting value (maximal sorption ability) after several minutes for the most number of dressings, that is during the time sufficiently lower than that of the dressing action (2–3 days).

The mass transfer of gases (oxygen and nitrogen) through the dressing, is 2–4 orders of magnitude slower with wet samples, than with the dry ones in similar conditions. Next, we consider the water evaporation from the dressing surface.

14.9.1 EVAPORATION OF WATER FROM THE DRESSING SURFACE

The dressing is saturated with water in air at 20°C temperature and 50% humidity. The temperature of the dressing surface is 32°C. This condition is chosen to take into account the temperature gradient in the matrix of the dressing. Let us determine the amount of water, evaporating from the surface of the dressing during a given time period under stationery air, atmospheric pressure, and the dressing surface is completely saturated by water. The partial pressure of air at 20°C and 50% relative humidity equals:

$$P_{H_2O} = 1.26\times10^{-3}\ \text{kg/cm}^2,$$

$$P_{\text{air}} = 1.02 \text{ kg/cm}^2 .$$

For air at 32°C in the saturation state:

$$P_{H_2O} = 4.85 \times 10^{-2} \text{ kg/cm}^2 ,$$

$$P_{\text{air}} = 0.98 \text{ kg/cm}^2 .$$

The values for density, viscosity, heat conductivity and heat capacity of the air at the average temperature of 26°C equal:

$$\rho = 1.185 \text{ kg/cm}^3 ;$$

$$\mu = 1.861 \times 10^{-6} \text{ g/m} \cdot \text{s} ;$$

$$\lambda = 6.1 \times 10^{-6} \text{ kcal/m} \cdot \text{s} \cdot \text{grad} ;$$

$$\lambda = 6.1 \times 10^{-6} \text{ kcal/m} \cdot \text{s} \cdot \text{grad} \cdot$$

After mathematical transformations using the method, described in [57], we may obtain the following equation for the mass transfer of water in dressing:

$$W = a_m \frac{P}{R \cdot T \cdot \rho_{av}} (p_1 - p_2) , \tag{34}$$

where a_m is the coefficient of heat conductivity; p_{av} is the average value of the mixture density over the surface and near the surface of the dressings; p_1 and p_2 is the partial pressure p_{H_2O} at 37°C and 20°C, respectively; R is the universal gas constant; P is normal pressure.

Substituting numerical values for the dressing of 1 × 1 m size, we obtain:

$$W = 1.2 \times 10^{-1} \text{ g/m}^2\text{s} .$$

If the dressing surface is not completely occupied by water, we should apply the following correlation:

$$w = \frac{C_{\text{surf}(H_2O)}}{C^0_{\text{surf}(H_2O)}} \times 1.2 \times 10^{-1} \text{ g/m}^2\text{s}, \tag{35}$$

where $C_{\text{surf}(H_2O)}$ and $C^0_{\text{surf}(H_2O)}$ are the surface concentration of water on the external side of the dressing and free water surface, respectively.

14.9.2 SORPTION OF FLUIDS BY BURN DRESSING FROM BULK CONTAINING A DEFINITE AMOUNT OF FLUID

Let us consider the case, where the burn dressing is applied to a wound containing a definite amount of liquid. Assume that the dressing of a membrane of a given size (thickness and surface area S) is in contact with the solution of the restricted bulk V, which contains a $C_{0(s-s)}$ concentration of diffusive substance. As the dressing is saturated by this substance, the concentration of the latter in the bulk will decrease.

The solution of the diffusional equation has the following form [57]:

$$\frac{m}{m_\infty} = 1 - \frac{2a(1-a)}{1+a+a^2q^2} \exp\left[\frac{4Dq^2t}{l^2}\right], \tag{36}$$

where q is the positive solution of the characteristic equation

$$tgq = -aq; \; a = \frac{V}{\sigma Sl},$$

where σ is the coeffflcient of distribution of the substance between the membrane and the solution.

When a sufficient part of the substance in solution is sorbed by the membrane, the value of a is small and a more simple expression can be used:

$$m_t \approx m^\infty \left[1 - \frac{a}{\left(4\pi Dt/l^2\right)^{1/2}}\right]. \tag{37}$$

From Equation (36) and Equation (37) we obtain two important correlations. Sorptional ability of the dressing, that is the part of the substance sorbed from the solution in the equilibrium conditions, equals:

$$\frac{m^\infty}{m^0} = \frac{1}{1+a}. \tag{38}$$

Thus, for the efficient action of the dressing it is necessary that the concentration C to be as high as possible in relation to the products of metabolism and toxins. Relating to water $C_{H_2O}{\sim}1$, it is desirable the dressing volume (l·S) to be close to that of the exudate of wounds V.

The time of reaching 0.85 degree of maximum sorption of liquid media by the dressing equals:

$$\tau_{0.85} = 12\frac{a^2 \cdot l^2}{\pi D} = 12\frac{V^2}{\pi D \sigma^2 S^2} \tag{39}$$

It depends on many parameters, each being able to affect in order the time of completion of the sorption process.

14.9.3 MASS TRANSFER OF WATER FROM WOUND INFO THE SURROUNDING

Generally the change of the water amount under the dressing in the wound (m_{H_2O}) is determined from the correlation derived from Equation (35) and Equation (37):

$$m_{(H_2O)} = V \cdot C^0_{H_2O} - m_{H_2O(\text{dressing})}\left[1 - \frac{a}{\left(4\pi Dt / l^2\right)^{1/2}}\right] - \frac{C_{\text{surf}(H_2O)}}{C^0_{\text{surf}(H_2O)}} \times 1.2 \cdot 10^{-1} \cdot S \cdot t. \tag{40}$$

Let consider, the application of the correlation Equation (40) for the following case. The wound characteristics are:

$$S = 10^{-2}\ \text{m}^{-2},$$

$$C^0_{H_2O} = 10^6\ \text{g/m}^3 \cdot m^0_{H_2O} = 50\ \text{g}\ V\ =\ 5 \times 10^{-5} m^3,$$

$$S = 10^{-2}\ \text{m}^{-2} =\ 10^{-3}\,\text{m},\ \sigma_{H_2O} = 1$$

$$\frac{C_{\text{surf}(H_2O)}}{C^0_{\text{surf}(H_2O)}} = 0.5\,.$$

For these conditions

$$\tau_{0.85} = 12\frac{25\cdot 10^{-10}}{\pi\cdot 10^{-9}\cdot 10^{-4}} = 9.5\cdot 10^{2}\ \text{s}\ (\text{or} \sim 15\ \text{min}),$$

$$\frac{m^{\infty}}{m^{0}} = \frac{1}{1+\dfrac{5\cdot 10^{-5}}{10^{-2}\cdot 10^{-3}}} = 0.17\ \text{or}\ 8.5\ \text{g}$$

During, the same time the following amount of water will evaporate from the dressing surface:

$$m_{\text{evap}(H_2O)} < 0.5\cdot 1.2\cdot 10^{-1}\cdot 950\cdot 10^{-2} = 0.6\ \text{g},$$

that is the rate of evaporation is signillcantly (14 times) lower than that of water sorption by the dressing. All the amount of water from the wound (wound exudate) will evaporate during the time:

$$t = \frac{50}{0.5\cdot 1.2\cdot 10^{-1}\cdot 10^{-2}} = 8.3\times 10^{4}\ \text{s}\ \ \text{or} \sim 23\ \text{hours}.$$

14.10 CRITERIA OF EFFICIENCY OF THE FIRST AID BURN DRESSINGS

14.10.1 THE REQUIREMENTS TO THE FIRST AID DRESSINGS

The first aid burn dressing must meet the following criteria:

- Sorption of the wound exudate, containing products of metabolism and toxic substances, during the period of the dressing action (24-48 hours).
- Wound isolation from infection of the external medium.
- Optimum air and water transfer between the wound and the surrounding.
- Easy removal from the wound, causing no damage to the wound surface.

14.10.2 CHARACTERISTICS OF BURN DRESSINGS

Below is the list of characteristics of the burn wounds based on approximate estimations, discussed above. Note that no quantitative data are reported in the literature.

14.10.3 RATIONAL CRITERIA FOR THE EFFICIENCY OF THE FIRST AID BURN DRESSINGS

SORPTIONAL ABILITY OF DRESSINGS

The burn wound (II-III degree) releases on the average 5×10^3 g/m^2 of the exudate. As it is seen from the Table 3, the water amount is about 90%. The sorption of different components of the exudate proceeds with different rates. In this case, the free volume of the dressing material will be first filled with water. The diffusion of proteins and cells takes place in space occupied by water.

Modem burn dressings possess porosity of 0.9 (Table 4) and almost the entire free volume can be filled with water (Figure 5). Maximum sorption ability for such dressings equals

$$C_{H_2O} \approx \frac{\rho_{H_2O}}{\rho},$$

and the amount of the liquid sorbed per square unit is:

$$m_\rho \approx C^{\infty}_{H_2O} \cdot \rho \cdot l = \frac{\rho_{H_2O}}{\rho} \cdot \rho \cdot l \approx 10^6 \cdot l \text{ g/m}^2$$

because $\rho_{H_2O} = 10^6$ g/m^3.

As the burn dressing of the first aid must sorb 5×10^3 g/m^3, it follows that

$$5 \cdot 10^3 \approx 10^6 \cdot l. \tag{41}$$

Therefore the thickness of tile first aid bum dressing equals

$$l \approx \frac{5 \cdot 10^3}{10^6} \approx 5 \times 10^{-3} \text{ m } (0.5 \text{ cm})$$

Thus, the first criterion of the efficiency of the first aid burn dressing can be formulated as follows:

The burn dressing of the first aid must use its entire free volume for sorption. This volume must be 0.9 or more of the total volume of dressing. Dressing thickness must be 0.5 cm or more.

It should be mentioned that the majority of foreign first aid dressings fulfill this criterion.

TABLE 16 Characteristics of burn wounds [3]

Burn degree	Image of damage	Physiological process	Burn depth, mm
I	Redness and edema (medium edema)	Aseptic inflammatory process	0
II	Sac formation	Aseptic inflammatory process	0
III	Damage of skin cover, exudating wound surface	Skin necrosis, tissue necrosis	1–2
IV	Exudating wound surface	Full necrosis of tissues, carbonization of tissues	2–5

TABLE 17 Water losses by means of evaporation from burnt surfaces of different types

Surface type	Evaporation, ml/cm²·hour
Natural skin	1–2
I-st burn degree	1–2.5
II-nd burn degree with sac intacts	2.8
II-nd degree with no damage of fermentative layer	37
II - IV degree of burn	20–31

AIR PENETRABILITY OF DRESSINGS

The air penetrability of dry air of the most of dressings is ranged within 10^{-4}–10^{-5} mole·m/m^2·s·Pa (Table 13). The air penetrability of dressings filled with water is much lower and decreases to values between 10^{-6}–10^{-5} mole·m/m^2·s·Pa, that is 0.2–2 dm^3/m^2·s.[1]

Thus, the second criterion of the efficiency of the first aid bum dressings can be formulated as follows:

The first aid burn dressing must possess air penetrability of 10^{-5} mole·m/m^2·s·Pa or higher after the sorption of water. For example, Biobrant burn dressing fulfills this criterion.

1 This is for the dressing 5–10^{-3} m (0.5 cm) thick at pressure of 50 Pa (5 mm H_2O) according to GOST 12088-77 (former USSR Standards).

ADHESION OF DRESSING TO WOUND

Adhesion strength of dressings with respect to coagulated blood (Table 16) varies in a wide range, but it has the minimum value of ~20 N/m. This value should be accepted as the optimal one, because it corresponds to the minimal pain and damage on removal from the surface of natural skin. Thus, the third criterion of efficiency of the first aid burn dressings can be formulated as follows:

The first aid burn dressing must possess adhesive strength to the wound of 20 N/m or less after the end of its action.

The following burn dressings fulfill this criterion, for example: Biobrant, blood-stopping remedy, Johnson-Johnson.

ISOLATION OF WOUND FROM INFECTION FROM EXTERNAL MEDIUM

It is known, that microorganisms, causing the wound infection, do not penetrate through the filters, possessing average pores size ~ 0.5 μm. The fourth criterion of efficiency of the first aid burn dressings is then as follows:

The first air burn dressings must possess no open pores with average diameter larger than 5×10^{-7} m (0.5 μm).

Moreover, it is implied, that tile first aid bum dressings possess sufficient me chanical strength and elasticity both in dry and humid conditions [58-60].

14.11 CONCLUSION

The experimental methods to estimate the main physicochemical properties were worked out.

Based on theoretical and experimental data we found that maximal sorptional ability of burn dressing equals the free volume of the dressing material, calculated from the value of the material density.

We found that the study sorption ability water can be used as a model liquid instead of the blood plasma medium.

Kinetic parameters were determined from the sorption curves. These parameters showed that the first aid burn dressings markedly differ in the value of the rate or liquid media sorption at stages close to the sorption limits.

We established that the air penetrability parameter in wet state decreases abruptly by 2–3 orders of magnitude for the majority of tested dressings. This is due to the filling of porous space by liquid medium.

We recommended that the air penetrability parameter be determined in wet state which represents the common condition of action for the first air burn dressing.

The value of adhesive strength after the end of its action on the wound must not exceed 20 N/m.

From the data obtained in this study we formulated the following criteria to estimate the efficiency of the fist aid bum dressing: maximum sorptional ability for water must be at least 10 g/g; optimal thickness of dressings, fulfilling this value of sorptional capacity, must be ~5×10^{-3} m, the adhesive strength must not exceed 20 N/min, and the average diameter of open (connected) pores must not exceed 5×10^{-7} m.

KEYWORDS

- **Burn dressings**
- **Adhesion**
- **Physical-chemical criteria**
- **Methods of investigation**
- **Sorption**
- **Mechanical properties**

REFERENCES

1. Fel'dshtein, M. I., Yakubovich, V. S., Raskina, L. V., and Daurova, T. T. *Polymer Coatings for Wound and Burn Treatment*, (Russia), Moscow, Institute of Information Publ., p. 299 (1981).
2. Park, G. B. *Biomaterials*, Med. Dev., Art. Org., **6**, 1 (1978).
3. Rudkovsky, V., Nezelovsky, V., Zitkevich, V., and Zinkevich, N. Theory and Practice of Burn Treatment, (Russia), Moscow, *Meditsina*, p. 200 (1988).
4. Robin, A. and Stenzel. K. H. In: *Biomaterials*, N.-Y., Plenum Press, pp. 157–184 (1969).
5. Robin, A., Riggio, R. R., and Nachman, R. L. *Trans. Soc. Art. Intern. Organs*, **14**, 1669 (1968).
6. Grillo, H. C. and Gross, I. *Surg. Res.*, **2**, 69 (1962).
7. Oluwasanmi, J. and Chapil, M. *J. Trauma*, **16**, 348 (1976).
8. Zaikov, G. E. *Intern. J. Polym. Mater.*, **24**, 1 (1994).
9. Abbendhaus, J. I., McMahon, R. A., Rosenkranz, J. G. et al., *Surg. Forum*, **16**, 477 (1965).
10. US Patent No. 35666871, 1971.
11. FRG Patent. No. 1642112, 1973.
12. US Patent No. 380078222, 1974.
13. *Biomedical Business International*, **3**, 115 (1981).
14. US Patent No. 3566871, 1971.
15. US Patent No. 3521631, 1970.
16. US Patent No. 3648692, 1972.
17. UK Patent No. 1309768. 1973.
18. Wilks, G. L. and Samuels, L. L. J. *Biomed. Mater. Res.*, **7**, 541 (1973).
19. Agureev, I.A. *Voenno-meditsinskii Zh.*, (Russia), **6**, 74 (1963).
20. French Patent No. 2156068, 1973.
21. Gorkisch, K., Vaubel, E., and Hopf, K. *Comparative Clincial Studies on the Synthetic Wound Dressings*, Proc. II Intern. Congress on Plastics in Medicine, Netherlands, June, 1973.
22. Iordanskii, A. L., Zaikov, G. E., and Rudakova, T. E. *Interaction Between Polymers and Chemical and Biochemical Media*, Transport, Kinetics, Mechanism, Zeist (Utrecht), VSP Science Press, p. 288 (1993).

23. USSR Certificate No. 245281, 1969, *Bulletin of Certificates*, No. 19.
24. Chardack, W. M., Martin, M. M., Jewett, T. C., et al., *Plast. Reconstruc. Surg.*, **30**, 554 (1962).
25. Hall, C. W., Liotta, D., Chidoni, J.J., et al., *J. Biomed. Mater. Res.*, **6**, 571 (1972).
26. Guldarian, J. J., Jelenko, C., Calloway, D., et al., *J. Trauma*, **13**, 32 (1973).
27. US Patent No. 3875937, 1975.
28. Polymers for Medicine, (Russia), Ed. S. Madou, Moscow, *Meditsina*, p. 350 (1981).
29. USSR Certincate No. 267010, 1970, *Bullentin of Certificates*, No. 12.
30. US Patent No. 358967, 1971.
31. US Patent No. 3678933, 1972.
32. US Patent No. 3654929, 1972.
33. Kinkel, H. and Holzman, S. *Chirurgie*, **36**, 535 (1965).
34. USSR Certificate No. 6658148, 1979, *Bulletin of Certificates*, No. 15.
35. Kuzin, M. I., Sologub, V. K., Yudenich V. V., el al., *Khirurgiya*, (Russia), **8**, 86 (1979).
36. Yannas, I. V. and Burke, J. F. *J. Biomed. Mater. Res.*, **14**, 65 (1980).
37. Jacobson, S. and Rothenaw, U. *J. Plast. Reconstr. Surg.*, **10**, 65 (1976).
38. Spruit, D. and Malten, K. E. *Dermat.*, **132**, 115 (1966).
39. USSR Certificate No. 685292, 1979, *Bulletin of Certificates*, No. 34.
40. *Textbook on Polymer Materials*, (Russia), Moscow, Khilniya, p. 255 (1980).
41. Zaikov, G. E., Iordanskii, A. L., and Markin, V.S. *Diffusion of Electrolytes in Polymers*, Utrecht, VSP Science Press, p. 328 (1988).
42. Moiseev, Yu. V. and Zaikov, G. E. *Chemical Resistance of Polymers in Reactive Media*, New York, Plenum Press, p. 586.
43. Barrie, I. A. *Diffusion in Polymers*, London - New York, p. 452 (1968).
44. Papkov, S. I. and Fainberg, E. Z. Interaction of Cellulose and Cellulose Materials with Water, (Russia), Moscow, *Khimiya*, p. 231 (1976).
45. Barrer, R. M. *Diffusion in Gases*, London - New York, Academic Press, p. 432 (1968).
46. Voyutskii, S. S. Physico-Chemical Principles of Fiber Materials Sorption by Polymers Dispersions (Russia), Leningrad, *Khimiya*, p. 336 (1969).
47. Fridrikhsberg, D. A. Course of Colloid Chemistry, (Russia), Leningrad, *Khimiya*, p. 351 (1974).
48. Al'tshuller, M. I. and. Deryagin, B.V. in: Investigations in the Field of Surface Force, (Russia), Moscow, *Nauka*, p. 235 (1967).
49. Mikhailov, M. M. *Moisture Permeability of Organic Dielectrics*, (Russia), Moscow - Leningrad, Gosenergoizdat, p. 162 (1960).
50. Nikolaev, N. I. Diffusion in Membranes, (Russia), Moscow, *Khimiya*, p. 232 (1980).
51. *Textbook on Textile Materials*, (Russia), Moscow, Legkaya Industriya, p. 342 (1974).
52. Raigorodskii, I. M. and Savin, V. A. *Plasticheskie Massy*, (Russia), **1**, 65 (1976).
53. Manin, V. N. and Gromov, A. N. Physico-Chemical Resistance of Polymer Materials During Exploitation, Moscow, (Russia), *Khimiya*, p. 247 (1980).
54. Laifut, E. Transfer Phenomena in Living Systems, (Russia), Moscow, *Mir*, p. 520 (1977).
55. Basin, V. E. Adhesion Durability (Russia), Moscow, *Khimiya*, p. 208 (1981).
56. Sherstneva, L., Efremov, A. E., and Ogorodova, T. ISAO/IFAC Symposium Control Aspects of Artificial Organs. *Transactions*, **4**, 845 (1982).
57. Batuner, L. M. and Pozin, M. E. *Mathematical Methods in Chemical Technology*, (Russia), Leningrad, Khimiya, p. 822 (1971).
58. Zaikov, G. E. and Krylova, L. P. *Biotechnology, Biodegradation, Water and Foodstuffs*, New York, Nova Science Publishers, p. 334 (2009).

59. *Selected Investigation in the Field of Starch Science,* Ed. by V.P. Yuryev, P. Thomasik, A. Blennow, L.A. Wasserman, and G.E. Zaikov, New York, Nova Science Publishers, p. 299 (2009).
60. Zaikov, G. E. and Lomova, T. N. *Processes with Participation of Biological and Related Compounds,* Boston-Leiden, Brill Academic Publishers, p. 448 (2009).

CHAPTER 15

SEMICRYSTALLINE POLYMERS AS NATURAL HYBRID NANOCOMPOSITES

G. M. MAGOMEDOV, K. S. DIBIROVA, G. V. KOZLOV, and G. E. ZAIKOV

CONTENTS

15.1 INTRODUCTION

It has been shown that at semicrystalline polymers with devitrification amorphous phase consideration as natural hybrid nanocomposites their abnormally high reinforcement degree is realized at the expense of crystallites partial recrystallization (mechanical disordering), that means crystalline phase participation in these polymers elastic properties formation. It is obvious, that the proposed mechanism is inapplicable for the description of polymer nanocomposites with inorganic nanofillers.

As it is known [1], semicrystalline polymers, similar to widely applicable polyethylene and polypropylene, at temperatures of the order of room ones have devitrification amorphous phase. This means that such phase elasticity modulus is small and makes up the value of the order of 10 MPa [2]. At the same time elasticity modulus of the indicated polymers can reach values of ~1.0–1.4 GPa and is a comparable one with the corresponding parameter for amorphous glassy polymers. In case of the latters it has been shown [3, 4], that they can be considered as natural nanocomposites, in which local order domains (nanoclusters) serve as nanofiller and as loosely packed matrix of polymer within the framework of the cluster model of polymers amorphous state structure [5] is considered as matrix. In this case elasticity modulus of glassy loosely packed matrix makes up the value of order of 0.8 GPa and a corresponding parameter for polymer (for example, polycarbonate or polyarylate) ~1.6 GPa. In other words, the reinforcement degree of loosely packed matrix by nanoclusters for amorphous glassy polymers is equal to ~2, whereas for the indicated semicrystalline polymers it can exceed two orders. By analogy with amorphous [3, 4] and cross-linked [6] polymers semicrystalline polymers can be considered as natural hybrid nanocomposites, in which rubber-like matrix is reinforced by two kinds of nanofiller—nanoclusters (analog of disperse nanofiller with particles size of the order of ~1 nm [5]) and crystallites (analog of organoclay with platelets size of the order of ~30–50 nm [7]). The clarification of abnormally high reinforcement degree mechanism allows giving an answer to the question, would this mechanism be applicable to polymer nanocomposites, filled with inorganic nanofiller (for example, organoclay). Therefore, the purpose of the present chapter is the study of reinforcement mechanism of rubber-like matrix of high density polyethylene (HDPE) at its consideration as natural hybrid nanocomposite.

15.2 EXPERIMENTAL

The gas-phase HDPE of industrial production of mark HDPE-276, GOST 16338-85 with average weight molecular mass 1.4×10^5 and crystalline degree 0.723, measured by the sample density, was used.

The testing specimens were prepared by method of casting under pressure on a casting machine Test Samples Molding Apparate RR/TS MP of firm Ray-Ran (Taiwan) at material cylinder temperature 473K, compression mold temperature 333K and pressure of blockage 8 MPa.

The impact tests have been performed by using a pendulum impact machine on samples without a notch according to GOST 4746-80, type II, within the testing temperatures range T = 213–333K. Pendulum impact machine was equipped with a piezoelectric load sensor, that allows to determine elasticity modulus E and yield stress σ_Y in impact tests according to the techniques [8] and [9], respectively.

Uniaxial tension mechanical tests have been performed on the samples in the shape of two-sided spade with sizes according to GOST 11262-80. The tests have been conducted on a universal testing apparatus Gotech Testing Machine CT-TCS 2000, production of German Federal Republic, within the testing temperatures range $T = 293–363$K and strain rate of 2×10^{-3} sec^{-1}.

15.3 DISCUSSION AND RESULTS

In Figure 1 the temperature dependences of elasticity modulus E for the studied HDPE have been adduced. As one can see, at comparable testing temperatures E value in case of quasi static tests is about twice smaller, than in impact ones. Let us note that this distinction is not due to tests type. As it has been noted for HDPE with the same crystalline degree at $T = 293$K E value can reach 1252 MPa [10]. Let us consider the physical grounds of this discrepancy. The value of fractal dimension d_f of polymer structure, which is its main characteristic, can be determined by several methods application. The first from them uses the following equation [11]:

$$d_f = (d-1)(1+\nu), \tag{1}$$

Where, d is dimension of Euclidean space, in which a fractal is considered (it is obvious, that in case d=3), ν is Poisson's ratio, estimated according to the mechanical tests results with the aid of the equation [12]:

$$\frac{\sigma_Y}{E} = \frac{1-2\nu}{6(1+\nu)}. \tag{2}$$

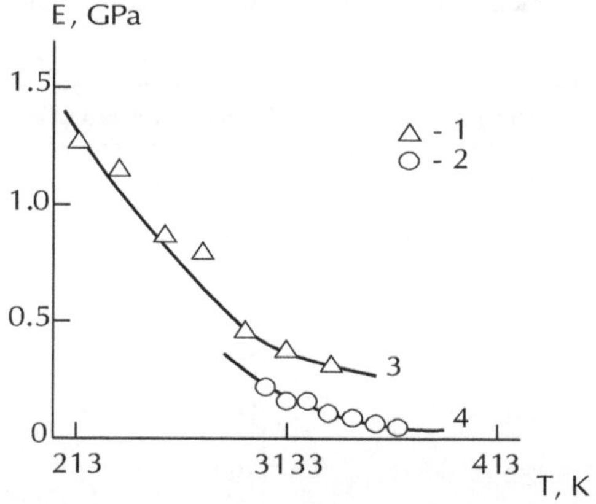

FIGURE 1 The dependences of elasticity modulus E on testing temperature T for HDPE, obtained in impact (1, 3) and quasi static (2, 4) tests. 1, 2—the experimental data and 3, 4—calculation according to the Equation (12).

The second method assumes the value d_f calculation according to the equation [5]:

$$d_f = d - 6.44 \times 10^{-10} \left(\frac{\varphi_{cl}}{C_\infty S} \right)^{1/2}, \tag{3}$$

Where, φ_{cl} is relative fraction of local order domains (nanoclusters), C_∞ is characteristic ratio, and S is cross-sectional area of macromolecule.

For HDPE $C_\infty = 7$ [13], $S = 14.4$ Å^2 [14], and φ_{cl} value can be calculated according to the following percolation relationship [5]:

$$\varphi_{cl} = 0.03(1-K)(T_m - T)^{0.55} \tag{4}$$

Where, K is crystalline degree and T_m and T are melting and testing temperatures, respectively. For HDPE $T_m \approx 400$K [15].

And at last, for semicrystalline polymers d_f value can be evaluated as follows [16]:

$$d_f = 2 + K \tag{5}$$

In Table 1 the comparison of d_f values, determined by the three indicated methods has been adduced (K change with temperature was estimated according to the data of paper [7]). As one can see, if in case of impact tests the calculation according to all three indicated methods gives coordinated results, then for quasi static tests estimation according to the Equations (1) and (2) gives clearly understated d_f values, especially with appreciation of possible variation of this dimension for nonporous solids ($2 \le d_f \le 2.95$ [11]).

TABLE 1 The values of fractal dimension d_f of HDPE structure, calculated by different met ods

Tests type	T, K	d_f, the Equation (1)	d_f, the Equation (3)	d_f, the Equation (5)
Quasi static	293	2.302	2.800	2.723
	303	2.296	2.796	2.723
	313	2.272	2.802	2.713
	323	2.353	2.801	2.693
	333	2.248	2.799	2.673
	343	2.182	2.799	2.663
	353	2.170	2.800	2.643
	363	2.078	2.808	2.633

TABLE 1 *(Continued)*

Impact	213	2.764	2.734	2.723
	233	2.762	2.741	2.723
	253	2.700	2.727	2.723
	273	2.750	2.756	2.723
	293	2.680	2.729	2.723
	313	2.624	2.743	2.713
	333	2.646	2.766	2.763

At present it has been assumed [1], that in case of semicrystalline polymers with devitrification amorphous phase deformation in elasticity region, that is at E value determination, the indicated amorphous phase is only deformed, that defines smaller values of both E and d_f. This conclusion is confirmed by disparity between d_f values, calculated on the basis of mechanical characteristics (the Equations (1) and (2)) and crystalline degree (the Equation (5)). And on the contrary, a good correspondence of d_f values, obtained by the three indicated methods (Table 1), assumes crystalline phase participants at HDPE deformation in elasticity region in case of impact tests (Figure 1).

In Figure 2 the temperature dependence of yield stress σ_Y has been adduced for HDPE in case of both types of tests.

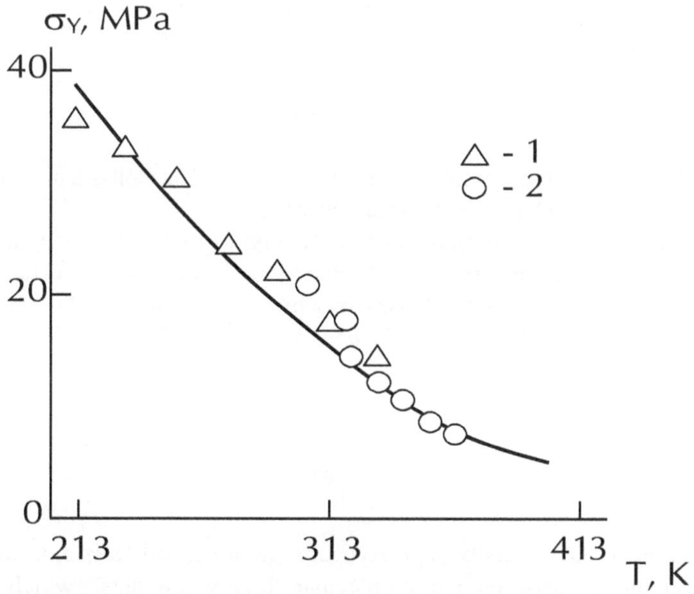

FIGURE 2 The dependence of yield stress σ_Y on testing temperature T for HDPE, obtained in impact (1) and quasi static (2) tests.

As one can see, the data for quasi static and impact tests are described by the same curve. As it has been shown in work [5], σ_Y values are defined by contribution of both nanoclusters that is amorphous phase and crystallites. Combined consideration of the plots of Figure 1 and 2 demonstrates, that the distinction of d_f values, determined according to the Equations (1), (2), and (5), is only due to the indicated structural distinction of HDPE deformation in an elasticity region.

The quantitative evaluation of crystalline regions contribution in HDPE elasticity can be performed within the framework of yield fractal conception [17], according to which the value of Poisson's ration in yield point ν_Y can be estimated as follows:

$$\nu_Y = \nu\chi + 0.5(1-\chi), \tag{6}$$

Where, ν is Poisson's ratio in elastic strains region, determining according to the Equation (2) and χ is a relative fraction of elastically deformed polymer.

For amorphous glassy polymers it has been shown that χ value is equal to a relative fraction of loosely packed matrix $\varphi_{l.m.}$. In case of semicrystalline polymers in deformation process partial recrystallization (mechanical disordering) of a crystallites part can be realized, the relative fraction of which χ_{cr} is determined by the Equation (7) [5]:

$$\chi_{cr} = \chi - (1-K). \tag{7}$$

If to consider HDPE as natural hybrid nanocomposite, then amorphous phase (the indicated nanocomposite matrix) elasticity modulus E_{am} can be determined within the framework of high elasticity conception, using the known equation [2]:

$$G_{am} = kNT , \tag{8}$$

Where, G_{am} is a shear modulus of amorphous phase, k is Boltzmann constant, and N is a number of active chains of polymer network.

As it is known [5], in amorphous phase two types of macromolecular entanglements are present—traditional macromolecular "binary hookings" and entanglements, formed by nanoclusters, networks density of which is equal to ν_e and ν_{cl}, respectively. The ν_e value is determined within the framework of rubber high elasticity conception [2]:

$$\nu_e = \frac{\rho_p N_A}{M_e}, \tag{9}$$

Where, ρ_p is polymer density, N_A is Avogadro number, and M_e is molecular weight of polymer chain part between macromolecular "binary hookings", which is equal to 1390 [18] and ρ_p value can be accepted equal to 960 kg/m³ [15].

In its turn, ν_{cl} value can be determined according to the following equation [5]:

$$\nu_{cl} = \frac{\varphi_{cl}}{C_\infty l_0 S}, \tag{10}$$

Where, l_0 is length of the main chain skeletal bond, for HDPE equal to 1.54 Å [13]. E_{am} and G_{am} are connected by a simple fractal formula [11]:

$$E_{am} = d_f G_{am} \cdot \tag{11}$$

Calculation according to the Equations (6) and (7) has shown that in case of HDPE quasi static tests χ_{cr} value is close to zero and in case of impact tests $\chi_{cr} = 0.400–0.146$ within the range of $T = 231–333$K. Now reinforcement degree of HDPE, considered as hybrid nanocomposite, can be expressed as the ratio E/E_{am}. In Figure 3 the dependence $E/E_{am}\left(\chi_{cr}^2\right)$ has been adduced (such form of dependence was chosen for its linearization). As one can see, this linear dependence demonstrates E/E_{am} (or E) increase at χ_{cr} growth and is described analytically by the following relationship:

$$\frac{E}{E_{am}} = 590\chi_{cr}^2, \text{MPa}. \tag{12}$$

The comparison of experimental E and calculated according to the Equation (12) E^T elasticity modulus values for the studied HDPE has been adduced in Figure 1. As one can see, the good correspondence between theory and experiment is obtained (the average discrepancy between E and E^T does not exceed 6%, that is comparable with an error of elasticity modulus experimental determination).

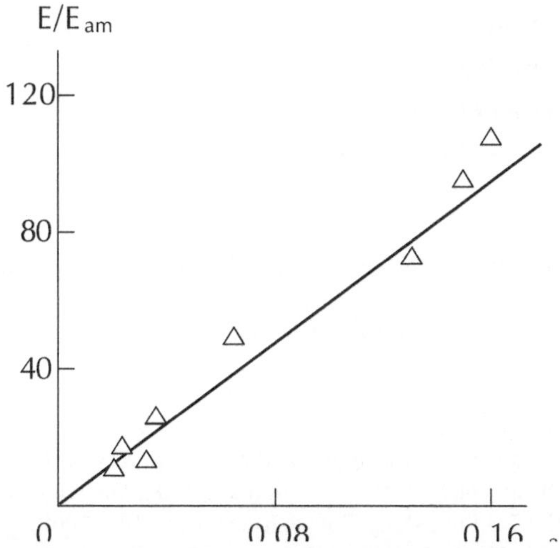

FIGURE 3 The dependence of reinforcement degree E/E_{am} on relative fraction of crystalline phase χ_{cr}, subjecting to partial recrystallization for HDPE in impact tests.

Thus, the performed estimations demonstrated, that high values of reinforcement degree E/E_{am} for semicrystalline polymers, considered as hybrid nanocomposites (in case of studied HDPE E/E_{am} value is varied within the limits of 10–110) were due to recrystallization process (mechanical disordering of crystallites) in elastic deformation process and, as consequence, to contribution of crystalline regions in polymers elastic properties formation. It is obvious, that this mechanism does not work in case of inorganic nanofiller (for example, organoclay). Besides, a nanofiller (crystallites) is formed spontaneously in a polymer crystallization process, that automatically cancels the problem of its dispersion at large K of the order of 70 mass %, whereas to obtain exfoliated organoclay at contents larger than 3 mass % is difficult [19]. Taking into consideration the indicated factors it becomes clear, why nanocomposites polymer/organoclay maximum reinforcement degree does not exceed 4 [20].

15.4 CONCLUSION

The performed analysis has shown that at the consideration of semicrystalline polymers with devitrification amorphous phase as natural hybrid nanocomposites their abnormally high reinforcement degree is realized at the expense of crystallites partial recrystallization (mechanical disordering), that means crystalline phase participation in the formation of these polymers elastic properties. It is obvious, that the proposed mechanism is inapplicable for the description of reinforcement of polymer nanocomposites with inorganic nanofiller.

KEYWORDS

- **Fractal dimension**
- **Natural nanocomposite**
- **Recrystallization**
- **Reinforcement degree**
- **Semicrystalline polymer**

REFERENCES

1. Narisawa, I. *Strength of Polymeric Materials*. Chemistry Publishing House (in rus.), Moscow, p. 400 (1987).
2. Bartenev, G. M. and Frenkel, S. Ya. *Physics of Polymers*. Chemistry Publishing House (in rus.), Moscow, p. 432 (1990).
3. Kozlov, G. V. *Recent Patents on Chemical Engineering*, **4**(1), 53–77 (2011).
4. Kozlov, G. V. and Mikitaev, A. K. *Polymers as Natural Nanocomposites: Unrealized Potential*. Lambert Academic Publishing, Saarbrücken, p. 323 (2010).
5. Kozlov, G. V., Ovcharenko, E. N., and Mikitaev, A. K. *Structure of Polymer Amorphous State* (in rus.). RKhTU Publishing House, Moscow, p. 392 (2009).
6. Magomedov, G. M., Kozlov, G. V., and Zaikov, G. E. *Structure and Properties of Cross-Linked Polymers*. A Smithers Group Company, Shawbury, p. 492 (2011).
7. Tanabe, Y., Strobl, G. R., and Fisher, E. W. *Polymer*, **27**(8), 1147–1153 (1986).

8. Kozlov, G. V., Shetov, R. A., and Mikitaev, A. K. *Russian Polymer Science* (in Rus.), **29**(5), 1109–1110 (1987).
9. Kozlov, G. V., Shetov, R. A., and Mikitaev, A. K. *Russian Polymer Science* (in rus.), **29**(9), 2012–2013 (1987).
10. Pegoretti, A., Dorigato, A., and Penati, A. *EXPRESS Polymer Lett.*, **1**(3), 123–131 (2007).
11. Balankin, A. S. *Synergetics of Deformable Body*. Ministry of Defense SSSR Publishing House, Moscow, p. 404 (1991).
12. Kozlov, G. V. and Sanditov, D. S. *Anharmonic Effects and Physical-Mechanical Properties of Polymers*. Nauka (Science) Publishing House (in rus.), Novosibirsk, p. 264 (1994).
13. Aharoni, S. M. *Macromolecules*, **16**(9), 1722–1728 (1983).
14. Aharoni, S. M. *Macromolecules*, **18**(12), 2624–2630 (1985).
15. Kalinchev, E. L. and Sakovtseva, M. B. *Properties and Processing of Thermoplastics* (in rus.). Chemistry Publishing House, Leningrad, p. 288 (1983).
16. Aloev, V. Z. and Kozlov, G. V. *Physics of Orientation Phenomena in Polymeric Materials* (in rus.). Polygraphservice and T, Nal'chik, p. 288 (2002).
17. Balankin, A. S. and Bugrimov, A. L. *Russian Polymer Science* (in rus.), **34**(5), 129–132 (1992).
18. Graessley, W. W. and Edwards, S. F. *Polymer*, **22**(10), 1329–1334 (1981).
19. Miktaev, A. K., Kozlov, G. V. and Zaikov, G. E. *Polymer Nanocomposites: Variety of Structural Forms and Applications*. Nova Science Publishers, Inc., New York, p. 319 (2008).
20. Liang, Z. M., Yin, J., Wu, J. H., Qiu, Z. X., and He, F. F. *Europ. Polymer J.*, **40**(2), 307–314 (2004).

CHAPTER 16

DEGRADATION MECHANISM OF LEATHER AND FUR

ELENA L. PEKHTASHEVA and GENNADY E. ZAIKOV

CONTENTS

16.1 INTRODUCTION

The chapter is including the information about biodamages and protection of leather and fur. The contributors did special emphasis on rawhide microflora, putrefied hide microflora, the essence of prior operations and their effect on bioresistance of leather, interesting facts, fur skin structure and properties changes by microorganisms and the methods of leather preservation against microorganism impact.

Solving of problems of polymers degradation and stabilization are very important task for pure and applied chemistry [1-32]. If we can increase the life spend of materials in two times it will be the same as increase the production of these materials in two times. It is special very important to predict (exactly) time of exploitation and time of storing of materials. Being the natural protein material, leather assumes commercial properties via a multistage treatment by various reagents and represents a culture medium where bacteria and mold fungi can develop [15, 16, 21, 27-32].

Animals' hide has a complex structure and various ways of microorganism permeation into it. Hide of an animal while alive carries lots and lots of microbes appeared from the environment (water, air, soil), because it has direct contacts with it (dipping, rolling on the ground, dust deposition). If the animal care is insufficient, its hide may carry a vast number of microbes (up to 1–2 billion/cm^2). Microbes found on the hide after flaying, had already appeared on the living animal and partly after flaying due to further contamination. After killing and flaying, the hide is secondarily contaminated by microbes sourced from dirty floor and baskets, animal dung and dirt. Specific microbial flora of killing chambers and raw leather warehouses consisting of microbes, greatly propagating on both molding production waste and stored raw materials and tools, is of great importance. Air in these facilities is usually full of this specific microflora, mold spores, salt microbiosis, and so on. [32].

Rawhide contamination by microorganisms largely depends on the purity of keeping of appropriate compartments and the hide handling. If the rawhide is poorly handled, the number of microbes on it significantly increases that may cause subsequent significant damage to the raw material. The external side of just flayed hides contains a lot of microbes, whereas the external side is sterile. However, already two hours after the rawhide completely loses commercial properties due to microorganism effect. It is therefore immediately preserved by solutions with introduced biocides. However, since before treatment by preservative agents, the hide is attacked by microbes, which can occur on it from both the epidermis side and the living hypoderm. While inspected, their external layers demonstrate various kinds of microorganisms, the upper layer of epidermis being most favorable for their habitation and consisting of separated flat keratinized cells losing connection with one another. The animal papillary dermis represents a dense connective tissue consisting of the main substance, intercellular fibrous formations and cellular elements; it is loose, unstable and non-resistant to microorganisms' impact [1].

The reticular layer primarily consists of complexly and densely tangled collagen fiber yarns; therefore, if the hide is just flayed, its reticular layer contains no microbes. However, the internal part of the layer adjacent to subcutaneous fat is loose and more permeable for microbes. Four structural levels of collagen, the dermis protein, are distinguished:

- Primary–polypeptide chain
- Secondary–spiral (the α-form)
- Tertiary–triple spiral (protofilament)
- Quaternary–supermolecular structure associated with the regulated packing of protofilaments (fibril).

Collagen organization level next to fibrils is represented by fibers. In the dermal layer of the skin, collagen fibers are submerged into glycosaminoglycan (GAG) structures, which act as an interfibril "cement" [27-32].

16.2 RAWHIDE MICROFLORA

The so-called putrefactive microorganisms, including cocci and rods, sporous and asporous, aerobes and anaerobes, are abundant in the rawhide. The general feature of these microorganisms is their ability to degrade proteins. Without getting into specifics of every species of microbes, several groups of them most commonly observed in the raw material should be highlighted. The most of them are rod-like, both sporous and asporous. The group which includes *Proteus r*epresents asporous mobile rods; this group has clearly proteolytic ability and degrades proteins to final products. A group composed by *E. Coli* bacteria, generally occurring from dung and being short rod shaped, represents the intestinal flora, both mobile and immovable. These group representatives induce peptone decay to amino acids with the indole formation. The sporogenous group, including *Bac. subtilis, Bac. mesentericus, Bac. mycoides*, and *Bac. megatherium*, mostly represent mobile rods generating high stability spores. These microbes also feature pronounced proteolytic ability and breakdown proteins to final products.

To a lesser extent, the group of cocci, including micrococci and pilchard generally producing pigments (yellow, ochreous, brown, red and white), is observed. Many of them produce enzymes affecting partly degraded protein. A group of actinomycetes has optimum development at pH 7.0–7.5. They are also able to degrade protein. This species is frequently observed in soils, wherefrom they, apparently, occurs in the hide. Sometimes bacteria of the fluorescent group occur on the hide. These are asporous gram-negative rods. Many species dissolve gelatin and decompose fats. They are mostly psychrophilic bacteria. These kinds of microbes most often occur in water, among which *Bact. Fluorescens*, and so on all the above groups of microbes are aerobes. Yeasts observed on the raw material are so-called wild, wide spread in the nature, namely, white, black and red yeasts [21].

Molds representatives often occur on the rawhide. Many of them have pronounced proteolytic ability. Fungi of *Mucor, Rhizopus, Aspergillus, Penicillium* and *Oidium* genus are observed on the hide. As mentioned above, microbes commonly occur on the surface of the raw, freshly flayed hide, both at the side reticular layer and epidermis. Microtomy on the rawhide shows the absence of microbes in the tissue, both in near surface and deep in layers. Single cocci may only be infrequently observed in hair bags.

16.3 PUTREFIED HIDE MICROFLORA

Unpreserved hide is easily putrefied. High temperature and humidity, stocking of uncooled hides and their contamination lead to fast propagation of putrefactive microbes in the hide [32].

Aerobic putrefaction starts from the surface and gradually penetrates deep into the layers. Three putrefaction stages are distinguished. The first stage is characterized by fast microbial propagation on the hide surface causing no visible reflex. The second stage has visible hide changes: sliming, color change, odor and doze. This period coincides with commencing permeation of microflora into the dermis thickness. The third stage is characterized by intensification of visible displays in line with hair and epidermis weakening and deep microbial permeation and spreading by the hide layers (Figure 1).

Hide putrefaction demonstrates gradual change of the species composition of the microflora. *Coccus* species of bacteria, which significant amount is observed in the rawhide, are gradually substituted by highly propagating rod forms, namely, *Proteus vulgaris, Bac. subtilis, Bac. mesentericus*, clubbing, and so on Microtomy of unpreserved putrefactive rawhides demonstrates particularly rod forms penetrated deep in the hide layers [19].

16.4 THE ESSENCE OF PRIOR OPERATIONS AND THEIR EFFECT ON BIORESISTANCE OF LEATHER

Rawhide preservation. Immediately after flaying, the raw skin is affected by microorganisms, which result in damages and reduces quality of rawhides and the yield of leather. Primary signs of the hide decomposition are surface sliming and flesh side color change. Then characteristic putrefactive odor occurs, hair root bonding with their bags is weakened, hairslip occurs followed by exfoliation. Finally, pigmentation occurs and mechanical strength reduces down to massive fracture [17].

All that necessitates rawhide preservation within two hours after flaying, otherwise leather will lose its commercial properties. The preservation goal is to create unfavorable conditions for bacterium and enzyme action. This may be reached by moisture elimination and impacting protein substances of the hide by reagents. At the preservation stage, aerobic bacteria of *Bacillus, Pseudomonas, Proteus* and *Achromobacter* genus possessing proteolytic enzymes manifest the highest activity. These bacteria are capable of damaging the hair side, its globular proteins, lipids, and carbohydrates. Some of them are capable of causing collagen decomposition [16].

To prevent putrefactive processes, rawhides are preserved by three methods: flint-dried, dry salting and wet salting. Flint-dried and dry salting cure is based on suppressing vital activity of bacteria and activity of proteolytic enzymes by reducing raw material humidity to 18–20% due to treatment by dry sodium chloride and sodium silicofluoride. Meanwhile, optimal condition of the flint-dried cure is a definite temperature mode +(20 to 35)°C, because the process at lower temperature may cause bacterial damages and hide decay due to slow water removal. Special requirements are also set to relative air humidity in the compartment which is to be 45–60%, and good circulation of air is required [15].

Wet salting cure is performed by sodium chloride salting from the internal side of the hide (fleshing side) or treatment by saturated sodium chloride aqueous solution brining, with further hides add salting in stock piles. Wet salting generally removes free water from the hides. Meanwhile, greater part of asporous bacteria dies, and development and propagation of other microorganisms and action of enzymes is terminated or suppressed. At dry salting, the preservative effect of sodium chloride is based on hide dehydration, and at wet salting – on breaking intracellular processes due to sodium chloride diffusion into cells. However, sodium chloride does not provide full protection against microorganisms and may even be a substrate for halophilic (salt-loving) bacteria and salt-tolerant bacteria (*Bacillus subtilis*), which possess the proteolytic capability. To protect against them at brining, sodium metabisulfide is added as bactericide [14].

In addition to the above listed preservation methods, rawhide freezing can be used as a temporary measure. At low temperature activity of bacteria and enzymes is terminated; however, procurement organizations must defrost the frozen material and cure it by wet salting. Irradiation is considered to be an effective method of rawhide protection against microorganism action. After rawhide irradiation in 1 kJ/kg dose, it can be stored during 7 days without noticeable signs of bacterial damage. Irradiation by 3 kJ/kg dose extends the storage period to 12 days. In this case, rawhides need no additional chemical curing. A combination of wet salting and irradiation of the rawhides, in turn, almost completely eliminates microflora, which activity in the rawhides is terminated for 6 months [32].

16.4.1 BASIC OPERATIONS OF RAWHIDE PROCESSING TO LEATHER

The rawhide processing to leather is a multistage process. At various stages of this process quite favorable conditions for microorganism growth and development on the leather may be created [13]. The danger of leather bacterial damage occurs already at the initial stage (soaking) aimed at preservatives removal from the rawhide and making it as close to just flayed state as possible. Soaking is water treatment at 30°C, mostly with electrolyte adding. Salt concentration in the leather thereby abruptly reduces that promotes occurrence and development of bacteria, which become active in water, especially at higher temperature. In this case, bacterial damage starts from the grain side, and biologically unstable components at this stage are globular proteins. Among 10 species of bacteria detected during soaking, more than a half of bacteria have proteolytic enzymes [12].

At this stage, sodium silicofluoride is used, which is active in neutral and weak acid media. At the next stage of liming, to remove interfiber protein substances and to loosen fibrous structure of the dermis, hides are treated by caustic lime solution. The aim of liming is to weaken hair and epidermis bonding to dermis that provides their further free mechanical removal. Moreover, dermis free from interfibrous proteins becomes more permeable, it volume is formed, and tannin diffusion into the dermis is accelerated. As regards bioresistance, the liming process is characterized by the fact that asporous bacteria die in the lime bath, and sporous ones stop growing and propagating [11].

Then hides without hairs (pelt) are subject to preliminary tanning operations, which are deliming and drenching. The first operation is performed at +20 to +30°C, most often using ammonium sulfate or lactic acid. Drenching in turn represents a short-term pelt processing by enzyme compounds in water at increased temperature +37 to +38°C. As a result, leather becomes soft and elastic, with interfiber proteins and collagen decomposition products eliminated, and becomes breathable. However, at this stage favorable conditions for bacterium propagation are formed. With respect to drenching fluid composition, bacteria of *Sarcina, Staphilococcus, Pseudomonas, Bacillus,* and so on genus have been extracted. Therefore, to avoid biodamages duration of operations should be controlled and preset treatment time should not be exceeded. Thus at all stages of rawhide processing to leather conditions may be formed that promote microorganism growth and development on the leather [7].

CURIOUS FACTS

When using fire, the ancient man has discovered preservative properties of smoke. For tanning hides Indians applied a mixture of chopped liver, brain and fat. After currying the leather was smoked. Clothes from such leather were not hardened under the effect of water and microbes and its odor repelled mosquitoes. Another tanning method also existed. Leather was wrapped around the leg, then wrapped by foot wraps and carried until it gained best properties; either leather was soaked in urine or dug in sheep manure.

16.5 THE TANNAGE PROCESS

Tannage concludes in injection of tanning substances into the dermis structure and their interaction with functional groups of protein molecular chains that results in formation of additional stable cross bonds. Tannage is one of the most important processes for leather manufacture. This stage in the leather industry radically changes dermis properties, transforming it to tanned leather. Finally, the dermis property change defines leather behavior during dressing, manufacture and operation of articles from it. As compared with the pelt structure, the tanned leather structure features increased fiber separation assumed during preparation and fixed by tanning. The fiber separation defines a number of basic physical and mechanical properties of leather:

- Tensile strength
- Compression strength
- Toughness
- Hardness
- Elasticity, and so on [5].

Tannage starts with tannic compound penetration into the collagen structure via capillaries and diffusion of these compounds from capillaries to direct response centers. Primarily, adsorptive interaction happens between tanning compounds and collagen, followed by stronger chemical bonding. Tannage causes strengthening of collagen spatial structure due to formation of cross bonds by tanning substances between molecular chains and the protein structure. It is known that collagen spatial cross-linked structure between molecular chains of the protein structure. Finally, tannage leads to formation of new, more stable cross bonds, which strength depends on the tanning compound origin [4].

Tannage is accompanied by increasing dermis resistance to hydrothermal impact that is increasing seal temperature. Tannage increases dermis resistance to swelling in water that causes significant impact on performance properties of leather. It also increases dermis resistance to proteolytic enzymes, that is leather biostability increases. Moreover, the interaction of tanning substances with functional groups of dermis protein leads to increased elastic properties of collagen and, consequently, reduced deformation rate of moist dermis, reduction of leather shrinkage, area and thickness when dried. Tanning compounds are intrinsically divided into two main groups: organic and inorganic. Inorganic ones are compounds of various metals: Chromium, zirconium, titanium, iron, aluminum, silicon, phosphorus, and so on, among which chromium compounds are of the prime importance. Organic compounds are natural tanning substances tannides, synthetic tanning matters, aldehydes, water-soluble amino resins, and blubber oils [1].

16.5.1 INORGANIC TANNING MATTERS

A large number of inorganic compounds possessing tanning properties are known. However, mostly chromium compounds are used as tanning matters, because they allow obtaining of high quality leather good both in manufacture and operation and storage. Primarily, tanning chromium compounds were mostly applied to leather manufacture for e shoe upper. At present, chromium compounds are used in manufacture of almost all kinds of leather both solely or combined with vegetable tanning matters, with syntans and amino resins. Tannage concludes is treatment of pelt by tanning chromium compound solution. Drums and worm apparatuses are used for tannage. When rotated in apparatuses, the semi-product is subject to quite intensive mechanical impact, because wooden wobblers or boards rise and drop the semi-product that along with continuous fluid mixing and temperature increase accelerates tannage. Tannage starts with diffusion of compounds into dermis structure. Primarily, tanning matters penetrate into capillaries, wherefrom chromium compounds diffuse to direct response centers and link to functional groups of protein. Meanwhile, tanning agent penetration into the dermis depends on a number of factors:

- Firstly, on sufficient separation of structural components, loosening of collagen in preliminary operations. The more loosened fibers are, the higher diffusion rate of chromium compounds into dermis is.
- Secondly, on tanning particle size: the smaller they are, the higher diffusion rate is.

One more important factor is pelt pH to be 4.5–5.5 to provide normal tanning matter diffusion into the dermis. Chromium tanning compound diffusion is controlled by pelt section color change varied from white to blue or green. Resulting from these procedures, chromium tanned leather assumes exclusive properties. It is highly resistant to acids and alkali and has high seal temperature + 110…+ 120°C, high tensile strength (11 MPa), high softness and elasticity, is resistant to high temperatures (compares with vegetable tannage leather). Moreover, chromium tannage dried leather cannot be soaked again either in hot or in cold water. Although water penetrates easily into the chromium dermis, its structural elements remain waterproof [25].

These negative properties are explained by the nature of chromium complexes, type of bonds between them, functional groups of collagen and the strength of cross bonds between molecular chains forming bridges. Assumed features in turn explain sufficiently high bioresistance of chromium tanned leather. It is proven that tanned leathers are mainly impacted by microscopic fungi, because development of bacteria is hindered. However, it is found that chromium tanned leather possess the highest resistance to molds; because it is deeply impregnated by oils, wax and fats, and its fibers assume hydrophobic properties [21].

Moreover, chromium salts are weak antiseptics that also play some role. However, despite relatively high resistance to biodamages of chromium tanned leather, danger of microorganism development is not completely eliminated. Biodamaging agents extracted from tanning solutions and from the semi product surface may be bacteria *Bacillus mesentericus* and some fungi: *Aspergillus niger, Penicillium chrysogenum, Penicillium cyclopium*. At this stage, sodium pentachlorophenolate and chloramines B may be used as biocides [32].

Beside usage of chromium as the self tanning agent, it is also used for more intensive bonding of aluminum compounds with collagen. Aluminum tanning is one of the oldest tannage methods. Basic aluminum sulfate, chloride and nitrate are used as tanning agents. However, this method is applied restrictedly, because these aluminum compounds are unstably bound to dermis proteins. This bond breaks under the water effect and leather becomes detanned. Although aluminum tannage leather differs from other types of tannage due to higher softness and white color, it is unsubstantial and due to aluminum compound washing out becomes wet rapidly and warps after drying [16]. These factors are significant for bioresistance of aluminum tanned leather; washing out of aluminum compounds makes semi product and final product accessible for microorganism penetration into dermis and their active propagation in it. However, application of aluminum compounds for retanning of vegetable tanned innersole leather, for instance, significantly increases its ability to resist mold development [14].

16.5.2 ORGANIC TANNING AGENTS

Organic tanning agents are simple and complex by structure. Simple tanning agents are mostly aliphatic compounds: aldehydes and some kinds of blubber oils (tanning oils). Complex tanning agents are aromatic derivatives and some heterochain polymers:

- Vegetable tanning agents (tannides)
- Synthetic tanning agents (syntans)
- Synthetic polymers (mostly amino resins)

Tannides are substances contained in various parts of many plants, are extracted by water and capable of transforming dermis into leather while interacting. With regard to species of plants tannides are accumulated in their different parts:

- Cortex
- Wood

- Leaves
- Roots
- Fruits

In this case, tannide content may vary in a very broad range, from parts to tens of percents.

Leather formation during rawhide tanning by complex organic compounds is the result of permeation of these substances into the semi product and their bonding with broadly developed inner surface of particular structural elements of collagen via both thermodynamic adsorption and chemical interaction with amino groups and peptide groups of protein, with formation of cross bonds of electrovalent, hydrogen and, apparently, covalent type [13].

Semi product seal temperature increases during tannage by tannides. This is explained by the fact that collagen has a multistage structure, and tanning particles having large size are incapable of penetrating to the smallest structural elements of the protein. These particles can easily be washed off, accompanied by simplest phenols and acids, which reduce the seal temperature. Moreover, similar to collagen, tannides have many reaction groups. As located on the collagen structure surface, tannide particle reacts with several structural elements of collagen forming cross bonds between them that, in turn, increase the seal temperature (up to +68 to +90°C).

It is desirable to use tannage by tannides in cases, when volume, hardness and rigidity are to be imparted to the leather or to increase its size stability in case of humidity variations, and when leather with high friction coefficient is required. However, such leather has low tensile strength, because large tannide particles when permeate into fibers expand them and reduce the number of protein substance per specific cross-section. It should be noted that biostability and performance characteristics of such leather and articles from it abruptly decrease with increasing tannide concentration. In this case, tannides represent nutrition for microorganisms and manifest their impact in the form of hydrolysis of tanning agents, pigment spots, grain roughness [12].

However, tannides that represent phenols derivatives possess some bactericide and fungicide action. Fungicide effect of vegetable tanned leather on *Trichophyton* genus fungi is shown. Another kind of complex tannides (syntans) is of two types:

1. Tannide substitutes
2. Auxiliary ones

Syntans of the first type are produced from raw phenols, and alongside with tanning properties they provide some leather protection against biodamages. Syntans of the second type are produced from hydrocarbon petroleum and gas refinery products and have no biocide properties. In the leather industry, the instances of severe damages of the semi product by mold fungi after vegetable tannage using syntans prepared from hydrocarbon material. Thus neither organic, nor inorganic tanning compounds have the absolute biocide which, along with providing biostability, would promote formation of sufficient physical, mechanical and chemical properties of ready leather [21].

16.6 FUR SKIN STRUCTURE AND PROPERTIES CHANGES BY MICROORGANISMS

Microbiological stability of fur skins at different stages of manufacture by the standard technology (tannage, greasing) was assessed on the example of mink skins (rawhide and semi products). Mink skins as rawhides, chrome non-oil and ready (that is greasy) semi-product were studied.Mink skin samples matured under conditions favorable for microorganism development showed clear sensory determined signs of degradation leather tissue samples became fragile and hairslip of the fur was observed.

Data on the influence of microorganisms, spontaneous microflora, on such physical properties of materials, as thickness, density and porosity (Table 1).

TABLE 1 The change of mink leather thickness, density and porosity under the effect of spontaneous microflora ($T = 30–32°C$, $\varphi = 100\%$) ($n = 10$, $P \leq 0.93$)

Sample	Exposure time days	Thick-ness, mm	Density, kg/m³		Porosity, %
			real	seeming	
Rawhide	init.	0.42	1336	701	47.5
	7	0.41	1340	719	46.3
	28	0.40	1345	747	44.5
Tanned semi product (non-greased)	init.	0.4	1342	647	51.8
	7	0.37	1357	681	49.8
	28	0.33	1363	719	47.2
Prepared semi product (greased)	init.	0.58	1353	632	53.3
	7	0.55	1355	755	44.3
	28	0.45	1366	825	39.6

As follows from the data obtained, both rawhide and non-greasy and ready semi product demonstrate density increase resulted from the spontaneous microflora effect.

Such change of the leather tissue density is apparently associated with the fact that molecular weight of collagen decreases and denser packing of structural elements (frequently, more ordered) becomes easier. This is testified by observed increase of real density of the leather tissue and abrupt reduction of its porosity with increasing duration of microbiological impact on both leather tissue of the rawhide and tanned semi product. The system packing observed leads to leather tissue thickness reduction [10].

Data on the structure density increase resulting from microbiological impact are confirmed by data obtained by EPR spectroscopy by radical probe correlation time [1]. The radical probe mobility decreases with the increase of microbiological impact duration. This may testify packing of the leather tissue structure as a result of microflora impact on the material.

Analysis of the chemical composition of mink's leather tissue before and after impact of microorganisms testifies reduction of the quantity of fatty matter with simultaneous relative increase of collagen proteins and mineral substances quantity (specifically observed in the raw material) (Table 2).

TABLE 2 Chemical composition of mink fur's leather tissue before and after spontaneous microflora impact (T = 30–32°C and φ = 100%)

Material	Impact duration, days	Humidity, %	Substance content, in % of abs. dry substance			
			fatty	mineral	collagen protein	non-collagen protein
Raw material	Init.	12.3	10.15	2.28	76.39	11.18
	7	12.8	7.25	2.53	83.06	7.16
	14	14.3	7.14	2.69	84.28	5.29
Tanned semi product (ungreased)	Init.	9.7	2.99	5.88	90.08	1.15
	7	10.1	2.78	6.11	90.10	1.01
	14	10.8	2.58	6.49	89.98	0.95
	28	11.3	2.03	6.67	90.37	0.93
Ready semi product (greased)	Init.	8.2	13.63	6.54	78.83	1.00
	7	9.8	9.99	6.88	82.14	0.99
	14	10.5	7.62	7.06	84.37	0.95
	28	11.1	6.61	7.06	85.42	0.91

One may suggest that affected by the microflora, non-collagen proteins degrade and are "washed off" from the structure. In this connection, relative content of collagen proteins in the leather tissue increases.

Thus, observed reduction of organic matter content (fats, non-collagen proteins) is obviously associated with the change of leather tissue structure. The destructive effect on the structure of studied materials may also cause alkalinity of the medium, which, as known, increases at natural proteins degradation [32]. Table 3 shows data of pH change of water extract of initial and test samples. In all cases, system alkalinity increase was observed.

TABLE 3 Water extract pH of mink's leather tissue at different stages of their manufacture before and after spontaneous microflora impact (T = 30–32°C and φ = 100%) (n = 10, P≤ 0.93)

Mink's sample	Water extract pH Microflora impact duration, days			
	0	**7**	**14**	**28**
Raw material	5.76	6.40	6.30	Not determined
Tanned semi product (ungreased)	3.84	4.27	5.24	6.40
Ready semi product (greased)	3.91	4.83	5.92	7.23

The evidence of the fact that observed changes in the material properties are caused by the change of supermolecular structure is electron microscopy study results.

Figures 1, 2 show electron microscopic images of the leather tissue sample surfaces (raw material and ready semi product) before and after spontaneous microflora impact.

Images clearly show the fibrous structure of initial samples. Fibrous yarns (within 20–40 μm in diameter), separate filaments forming yarns (up to 20 μm in diameter) and collagen fibrils (0.1–20 μm) are distinguished in collagen. Dermis represents an irregular three-dimensional entanglement of fibers and their yarns that are clearly seen in the images of initial samples of both mink's raw skins and ready semi product. Collagen yarns with clear contours are seen, fibers in yarns being higher structured for the semi product [21].

Samples affected by microorganisms show destruction of fibril formations and their transformation to laminated structures. Hence, it is noted that the leather tissue porosity was slightly reduced as a result of microorganism impact.

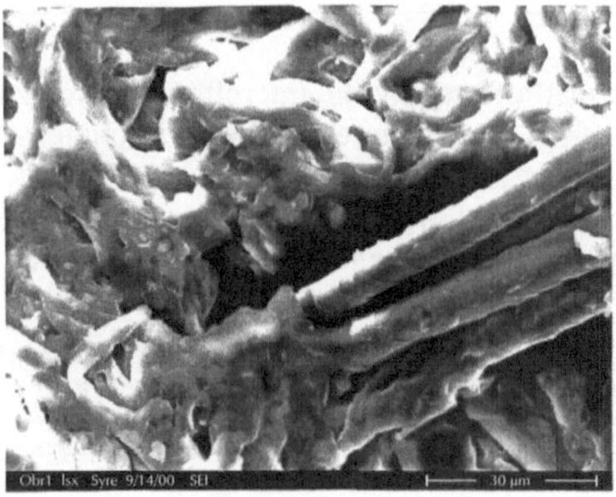

FIGURE 1 *(Continued)*

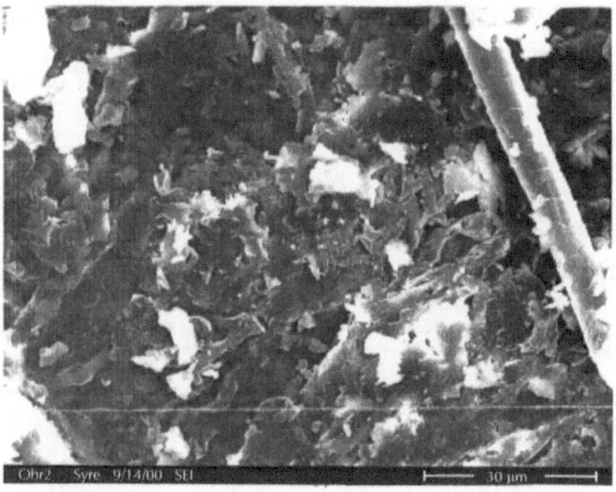

FIGURE 1 Microphotographs of the leather tissue of mink's rawhides (x1000): a) initial sample; b) 14 days after spontaneous microflora impact.

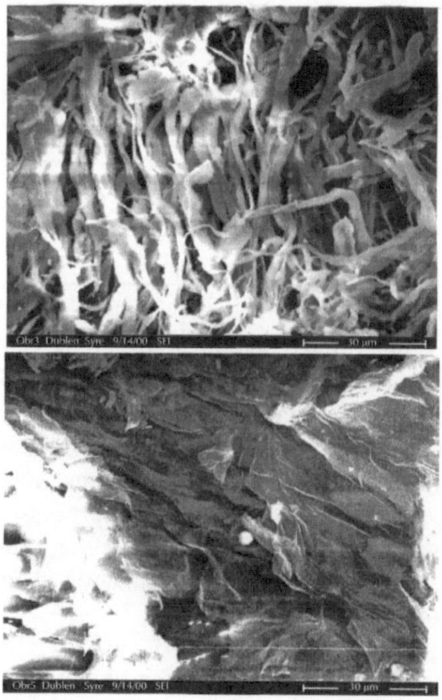

FIGURE 2 Microphotographs of mink's fur leather tissue, both tanned and greased (x1000): a) initial sample; b) 28 days after spontaneous microflora impact

Physicomechanical properties, first of all, strength and deformation characteristics of the leather tissue, as well as hair connection strength with the leather tissue, are the most important fur properties.

Table 4 shows results of our study of physicomechanical property change of ready mink semi product (with greasing) impacted by microorganism's spontaneous microflora, micromycetes *Aspergillus niger* and bacteria *Bac. subilis.*

TABLE 4 The change of physicomechanical properties of ready mink semi product before and after impact of microorganisms (at $T = 30–35°C$ and $\varphi = 100\%$)

Kind of impacting microflora	Exposure time, days	Breaking strain of leather tissue, σ_p, MPa	Change to init., %	Relative elongation at rupture of leather tissue, ε, %	Change to init., %	Hair bonding strength with leather tissue, σ_p, $1 \cdot 10^{-4}$ N	Change to init., %
Spontaneous microflora	7	11.5	–10.9	50.4	–12.0	64.4	–11.2
	14	11.0	–14.7	45.3	–20.9	53.3	–26.5
	28	9.8	–24.0	37.7	–34.2	34.1	–52.9
Asp. niger	7	9.6	–25.6	48.3	–15.7	57.5	–20.7
	14	4.7	–63.6	43.5	–24.1	45.4	–37.4
	28	2.4	–81.4	35.6	–37.9	29.3	–59.6
Bac. subilis	7	7.5	–41.9	47.1	–17.8	53.5	–26.2
	14	2.6	–79.8	41.5	–27.6	40.3	–44.4
	28	1.9	–85.3	34.3	–40.1	26.7	–63.2

The investigation has determined that the highest reduction of the breaking strain, leather tissue elongation at rupture and hair bonding strength with the leather tissue is observed under the impact of bacteria *Bac. subtilis.*

Thus it is observed that the properties of both rawhide and ready semi-product leather tissue change under the impact of microorganisms. Meanwhile, the true density increases, and porosity decreases due to degradation processes, which simplify packing of structural elements into more ordered formations as a result of steric hindrances reduction. The tensile strength, leather tissue elongation at rupture and hair bonding strength with the leather tissue are also reduced [6, 7].

16.7 THE METHODS OF LEATHER PRESERVATION AGAINST MICROORGANISM IMPACT

16.7.1 RAWHIDE AND CURED RAW LEATHER PROTECTION

To increase biostability of leather and articles from it, it is recommended to protect leather against microorganisms at all stages of its treatment, starting with the rawhide.

Due to termination of oxygen delivery and metabolism, tissue degradation in just flayed hides is accelerated. The medium response reaches its optimal value for protease action. First breakdown of proteins forming the hide base is initiated, and then carbohydrates, fats and other organic compounds degrade. As a result, chemical composition and the structure of tissues change. As impacted by microorganisms and enzymes, rawhides go bad rapidly at temperature above 18°C, and tissue putrefaction starts [4].

At further development of putrefaction process, epidermis is destroyed and delaminated, and "damaged grain", the absence of grain in some areas, occurs. Putrefactive microbes damage subcutaneous fat. Then occurring in the reticular dermis, they rapidly spread in the interfascicular space, and then degrade collagen and elastin fibers. Resulting these processes, dermis delaminates that, in turn, leads to complete destruction of leather [2].

To preserve high quality of the raw material at this stage, to make it stable to putrefactive microbe impact, the hide shall be thoroughly cured, that is all contamination, slices of fat and meat, shall be removed, and proper curing performed. As mentioned above, rawhides are cured by three methods:

- Flint-dried
- Dry salting
- Wet salting

The main substance used for curing is sodium chloride. Common salt microflora is represented by microbes developed in salt solutions – brines, and microbes occurring in salt during its production and transportation.

Under natural conditions, this salt possesses microbes of the halophobe group, as well as salt tolerant species. Moreover, salt contains representatives of sporous microflora, yeasts, mold fungi spores, micrococci, variously dyed bacteria of *Flavobacterium* genus. When occurring on hides with salt during curing, microorganisms induce various defects; therefore, antiseptic agents are formed alongside with sodium chlorides. They possess bactericide, fungicide, bacteriostatic and fungistatic properties. Antiseptic agents used for curing must be toxic for microorganisms, well soluble in water and in sodium chloride solution, cause no negative effect on the hide quality and leather semi products. In this connection, paradichlorobenzene and sodium silicofluoride are most widespread [1].

When affected by paradichlorobenzene, some microbes developing in the wet salt rawhide die, and development of others is terminated. Hence, gaps between hides are filled with vapors of this substance. These vapors are heavy, slowly removable and hindering microorganism propagation for a long time. Some part of antiseptic agent is dissolved in fat and penetrates inside the dermis; therefore, its typical odor is preserved for a long time. Sodium silicofluoride is quite effective. It possesses high bactericide properties and causes no negative effect on dermis. At rawhides brining with simultaneous treatment by sodium silicofluoride, a long, over a year, storage of hides without additional salting is provided. However, this antiseptic agent is poisonous and care should be taken, when operating it.

It is found that antiseptic agents also give high effect, when combined with each other. Moreover, goods results are given by application of sodium hypochloride, boric acid, sodium borate, zinc chloride, sodium fluoride, benzene and phenol chlorine derivatives, antibiotics and other antiseptic agents to rawhides curing.

However, beside chemical protection means for the raw material against microorganisms, of importance is meeting conditions and technology of curing. If the raw material was tainted when cured, then despite the absence of tissue destruction, the microflora of such raw material is richer. The presence of a great layer of reticular dermis, musculature and especially fat inclusions hinders diffusion processes, decelerates curing, has negative impact on the raw material quality and promotes development of microbes [30]. Moreover, the rawhide of flint-dried and dry salt curing requires ideal conditions during transportation and storage, because high humidity forms favorable conditions for bacterium and mold development [21].

16.7.2 PROTECTION OF LEATHER AND LEATHER ARTICLES AGAINST BIODAMAGES

The problem of biological damaging of natural materials, especially upper leather for shoes used in increased humidity conditions, is of great importance. Alongside with direct action associated with leather structure damage, microorganisms also manifest indirect adverse effect on the leather articles. Microscopic fungi promote leather hygroscopic property increase. As a result, relative humidity inside the shoes increases. This causes untimely wear of joints and development of pathogenic microorganisms inside the shoes [17].

In the world practice, rawhide and ready leather is widely protected by the following compounds: phenylmercury, bromoacetophenone, n-chloro-m-cresol, alkyl naphthalene-sulphodiacid, sodium borate, zinc oxide, 2-oxydiphenyl, salicylanilide, and some other compounds. However, a wide application of some biocides is restricted by specific requirements for leather protection: biocides must be soluble in fats, thermostable at stuffing temperature and compatible with other components used for leather treatment [15].

It is also found that most of the above biocides do not manifest a long-term antimicrobial action, because the antiseptic agent injected at the stuffing stage sweats out with fat during the use, and frequently, the fungicide simply evaporates. In this connection, the optimal protection may be provided by biocides introduced into finish coating composition and compounds capable of chemically bonding with collagen. In this regard, β-naphthol and β-oxy-naphthaldehyde injected into leather when proofing, have approved themselves. Leather treatment after tannage by catamine AB from the class of quaternary ammonium compounds is worthy. Linking to collagen by coordination or salt and adsorptive bonds, catamine AB prevents development of microscopic fungi on the ready leather surface.

Among microorganisms, mold fungi from *Aspergillus* and *Penicillium* genus are the most active and widespread destructors of the real leather. To eliminate mold development much can be done by temperature regulation, leather humidity decrease below 12% with the help of hygroscopic agents (example silicagel). However, such regulation is not always practical. The most effective means to eliminate mold development on the leather is the use of chemical agents [15, 17, 21, 30].

It is found that natural fungus resistance of the leather is increased by staffing materials based on organochlorine products; leather with fungicide properties can be obtained by using sulfochlorinated paraffins during stuffing. Both semi products and the ready leather are effectively protected by benzoguanamine formaldehyde resins

(BGAF). These compounds containing 40% of the basic substance are water-soluble that allows their application at all stages of leather manufacture process.

The ability of BGAF-resins to suppress mold micellium growth was also determined, the resin with higher content of sulphosalicylic acid having higher fungicide activity [32].

16.8 DISCUSSION AND RESULTS

Hydrated GAG cause very high effect on the structure of collagen fibers:

They protect protein fibers of fibrils against fusion and increase their mobility, thus providing fiber integrity.

It is known, for instance, that fairly preserved samples of medieval vellum consist of densely packed collagen fibrils submerged into the interfibril amorphous substance. However, some cases have been described, when the vellum transformed to bonded protein due to damages caused by microorganisms [1, 32]. After storing bovine dermis in water during 9 months, the intense differentiation of fibrils and their isolation into fibrils is observed. This testifies predominant decay of interfiber and interfibril substance, that is basically GAG [17, 32]. At the initial stages, microbiological degradation is similar to hydrolysis forming particles, which consist of amino acid groups. These particles, along with separate amino acids, are rapidly subject to further transformation. Microbiological protein degradation products commonly contain ammonium, fatty acids, amino acids, aldehydes, and amines [30-32].

For the microorganism development, subcutaneous fat is particularly favorable medium, where they propagate and from which penetrate into dermis. When appearing in the reticular layer, in the interfilament space, microbes may induce the putrefactive process deep in the hide. Meanwhile, blood remained in vessels due to poor bleeding and intercellular substance of the dermis and the reticular layer are proper nutrition for microorganisms. Regarding environmental conditions and sponginess of the reticular layer microbes can form numerous colonies on its surface or penetrate into upper layers, propagating and destroying them. They can move easily and rapidly by interfiber and interfilament spaces destroying the dermis substance and various intercellular components. For example, moving by the fiber, rod-shaped microbes penetrate into collagen filaments and then spread in the surrounding tissues, whereas cocci penetrate into hair bags. Mold mycelium spreads either along collagen fibers or, occurring in the interfilament space, in all directions forming dense entanglement [30-32].

By chemical composition, leather tissue represents a medium favorable for fast microorganism propagation. The rawhide contains both inorganic and organic substances. Inorganic substances in the leather tissue are water (50–70%) and mineral substances (0.35–0.5%). Organic substances in it are lipids (fats and fat like compounds), carbohydrates, non-protein nitrogen-containing components and proteins forming the histological structure base of the leather tissue. The most important components of this structure are fibrous proteins, such as collagen, keratin, elastin and reticulin. In addition, cutaneous covering contains globular (albumins, globulins) and conjugate proteins. Albumins and globulins, which sufficient amounts are contained in the rawhide, degrade most easily. Fats are affected by specific lipoclastic microbes [32].

High amounts of proteins present in the leather tissue are one of the factors which cause its extreme sensitivity to destructive impact of putrefactive microbes. This is also promoted by the medium response (rawhides have pH 5.9–6.2). The skin coating contains vitamins, enzymes catalyzing chemical reactions and affecting development of biochemical processes in tissues, and substances which increase (activators) and suppress (inactivators) enzyme activity in the hide. If one of activators or inactivators is absent, biochemical processes in the hide change. For example, rawhide aging increases protease activity which induces protein breakdown and promotes development and propagation of microorganisms.

Increasing fat content in the leather tissue entails relative decrease of water content which increases leather resistance to the action of various microorganisms. Depending on chemical composition of the leather tissue, that is whether amounts of proteins, fats, and so on in it are high or low, its resistance to microbial activity is different. It has been found that the kind of microorganisms inhabiting in the rawhide depends on animal feeding. In the absence of vitamin B_2 and biotin, for example, dermatitis and loss of hair by the skin coating are observed that promotes permeating of microbes into the hide [1, 32].

KEYWORDS

- **Biodamages**
- **Bioresistance**
- **Fur**
- **Methods of preservation**
- **Microorganisms**
- **Rawhide microflora**

REFERENCES

1. Emanuel, N. M. and Buchachenko, A. L. Chemical physics of degradation and stabilization of polymers. *VSP International Science Publ.*, Utrecht, p. 354 (1982).
2. Bochkov, A. F. and Zaikov, G. E. *Chemistry of the glycosidic bonds. Formation and cleavage*. Oxford, Pergamon Press, p. 210 (1979).
3. Razumovskii, S. D. and Zaikov, G. E. Ozone and its reactions with organic compounds, Amsterdam, Elsevier, p. 404 (1984).
4. Emanuel, N. M., Zaikov, G. E., and Maizus, Z. K. *Oxidation of organic compounds. Medium effects in radical reactions*. Oxford, Pergamon Press, p. 628 (1984).
5. Afanasiev,V. A. and Zaikov, G. E. *In the realm of catalysis*., Mir Publishers, Moscow, p. 220 (1979).
6. Moiseev, Yu. V. and Zaikov, G. E. *Chemical resistance ofpolymers in reactive media*. New York, Plenum Press, p. 586 (1987).
7. Zaikov, G. E., Iordanskii, A. L., and Markin, V.S. *Diffusion of electrolytes in polymers*, Utrecht, VNU Science Press, p. 328 (1988).

8. Minsker, K. S., Kolesov, S. V., and Zaikov, G. E. *Degradation and stabilization of polymers on the base of vinylchloride*, Oxford, Pergamon Press, p. 526 (1988).
9. Aseeva, R. M. and Zaikov, G. E. *Combustion of polymer materials*, Munchen, Karl Hanser Verlag, p. 389 (1986).
10. Popov, A. A. Rapoport, N. A., and Zaikov, G. E. *Oxidation of stressed polymers*, New York, Gordon & Breach, p. 336 (1991).
11. Bochkov, A. F., Zaikov, G. E., and Afanasiev, V. A. *Carbohydrates*, Zeist-Utrecht, VSP Science Press, VB, p. 154 (1991).
12. Afanasiev, V. A. and Zaikov, G. E. *Physical methods in chemistry*, New York, Nova Science Publ., p. 180 (1992).
13. Todorov, I. N., Zaikov, G. E., and Degterev, I. A. *Bioactive compounds: biotransformation and biological action*, New York, Nova Science Publ., p. 292 (1993).
14. Roubajlo, V. L., Maslov, S. A., and Zaikov, G. E. "*Liquid phase oxidation of unsaturated compounds*, New York, Nova Science Publ., p. 294 (1993).
15. Iordanskii, A. L., Rudakova, T. E., and Zaikov, G. E. "Interaction of polymers with bioactive and corrosive media", Utrecht, VSP International Publ., p. 298 (1994).
16. *Degradation and stabilization of polymers. Theory and practice*, G. E. Zaikov (Ed.), Nova Science Publ., New York, p. 238 (1995).
17. Polishchuk, A. Ya. and Zaikov, G. E. *Multicomponent transport in polymer systems*, Gordon & Breach, New York, p. 231 (1996).
18. Minsker, K. S. and Berlin, A. A. *Fast reaction processes*, Gordon & Breach, New York, p. 364 (1996).
19. Davydov, E. Ya., Vorotnikov, A. P., Pariyskii, G. B., and Zaikov, G. E. *Kinetic pecularities of solid phase reactions*, John Willey & Sons, Chichester (UK), p. 150 (1998).
20. Aneli, J. N., Khananashvili, L. M., and Zaikov, G. E. *Structuring and conductivity of polymer composites*, Nova Science Publ., New York, p. 326 (1998).
21. Gumargalieva, K. Z. and Zaikov, G. E. *Biodegradation and biodeterioration of polymers. Kinetical aspects*, Nova Science Publ., New York, p. 210 (1998).
22. Rakovsky, S. K. and Zaikov, G. E. *Kinetics and mechanism of ozone reactions with organic and polymeric compounds in liquid phase*, Nova Science Publ., New York, p. 345 (1998).
23. Lomakin, S. M. and Zaikov, G. E. *Ecological aspects of polymer flame retardancy*, VSP International Publ., Utrecht, p. 158 (1999).
24. Minsker, K. S. and Zaikov, G. E. *Chemistry of chlorine-containing polymers: synthesis, degradation, stabilization*, Nova Science Publ., New York, p. 198 (2000).
25. Jimenez, A. and Zaikov, G. E. *Polymer analysis and degradation*, Nova Science Publ., New York, p. 287 (2000).
26. Kozlov, G. V. and Zaikov, G. E. *Fractal analysis of polymers*, Nova Science Publ., New York, p. 244 (2001).
27. Zaikov, G. E., Buchachenko, A. L., and Ivanov, V. B. *Aging of polymers, polymer blends and polymer composites*, Nova Science Publ., New York, **1**, 258 (2002).
28. Zaikov, G. E., Buchachenko, A. L., and Ivanov, V. B. *Aging of polymers, polymer blends and polymer composites*, Nova Science Publ., New York, **2**, 253 (2002).
29. Zaikov, G. E., Buchachenko, A. L., and Ivanov, V. B. *Polymer aging at the cutting* edge, Nova Science Publ., New York, p. 176 (2002).
30. Semenov, S. A., Gumargalieva, K. Z., and Zaikov, G. E. *Biodegradation and durability of materials under the effect of microorganisms*, VSP International Science Publ., Utrecht, p. 199 (2003).

31. Rakovsky, S. K. and Zaikov, G. E. *Interaction of ozone with chemical compounds*. New frontiers, Rapra Technology, London (2009).
32. Pekhtasheva, E. L. Biodamages and protections of non-food materials (in Russian). Moscow Masterstvo Publishing House, p. 224 (2002).

CHAPTER 17

DEVELOPMENT OF THERMOPLASTIC VULCANIZATES BASED ON ISOTACTIC POLYPROPYLENE AND ETHYLENE-PROPYLENE-DIENE ELASTOMER

E. V. PRUT and T. I. MEDINTSEVA

CONTENTS

17.1 INTRODUCTION

Reactive polymer blending is very important contributor in the development of new polymer materials. The market for thermoplastic vulcanizates (TPV) created by reactive blending (dynamic vulcanization) is probably the fastest growing sector in the polymer market. The peculiarities of the process of dynamic vulcanization of polyolefins blended with elastomers are analyzed. The advantage of TPV over conventional rubber is the ease (and therefore low cost) of processing, the wide variety of properties available, and the possibility of recycling and reuse. The influence of components ration and the nature of vulcanizing agents on the morphology, mechanical and rheological behavior of thermoplastic vulcanizates are considered.

Polymer chemistry and physics discovers new properties of macromolecules with important practical results. Basic advances reveal key relationships in the development of new materials that exhibit previously inconceivable combinations of properties. A principal difference between fundamental science and technologies is intent.

Polymers influence the way we live, work, and play. There are currently more than 50 families of polymers, many of whose uses still go unrecognized because of their diversity. Polymers are true man-made materials that are the ultimate tribute to man's creativity and ingenuity. It is now possible to create different polymers with almost any quality desired in the end product some similar to existing conventional materials but with greater economic value, some representing significant property improvements over existing materials, and some that can only be described as unique prime materials with characteristics unlike any previously known.

Macromolecular science covers a fascinating field of research, focused on the creation, the understanding, and the tailoring of materials formed out of very high molecular weight molecules. Such compounds are needed for a broad variety of important applications. Due to their high molar masses, macromolecules show particular properties not observed for any other class of materials.

Nowadays, polymer compositions grow to take on a very significant role in the major application areas for polymers. This is connected with the significant expansion of the polymer applications with comparatively limited assortment of monomers. Therefore, a lot of applications in the packing, electronics/electrical, transportation, and construction industries have been instrumental in allowing polymeric materials to expand against other, more traditional materials.

The combination of polymers using different methods provides one with a means of varying their properties. These combinations of polymers allow one to obtain polymer blends. However, changes in polymer properties cannot be achieved by simply blending them with one another, as this is limited by reciprocal solubility and the compatibility of materials. The blending of polymers is a complicated physical and chemical process that occurs under the mechanical and temperature action. The problem can be solved by modifying and blending polymers, during which time chemical reactions occur that lead to changes in their composition and structure. This in turn allows one to combine a wide range of diverse properties in a single material, example, heat stability and elasticity, hydrophobic and hydrophilic natures, and crystalline and amorphous structures.

The processes in which chemical reactions occur during modification and blending are called reactive blending or reactive extrusion [1, 2]. The technique of the reactive blending/extrusion of polymers presents us with not only new possibilities for polymer products that are already available, but also enables us to form polymer blends at lower costs, which was previously considered to be commercially unfavorable with the use of other methods.

Isotactic polypropylene (IPP) is a versatile polymer widely used in many consumer and engineering applications. Its area of application has been extended by blending with other polymers. Polymer compositions based on IPP have been very actively studied in the last two decades [3, 4]. As second components in most of these blends, ethylene propylene rubber (EPR), ethylene propylene diene terpolymer (EPDM), butyl and polyisobutylene (PIB) elastomers, and so on are generally used. At low contents of elastomer, impact resistant polypropylene (PP) is obtained, whereas thermoplastic elastomers (TPEs) are produced at higher elastomer contents. Both classes of materials find wide application that requires detailed consideration of different factors, such as the influence of the chemical and molecular structures of the elastomer, the molecular weight distribution, and so on, of the properties of the finished products.

The TPEs constitute a commercially relevant and fundamentally interesting class of polymeric materials. The emergence of TPEs provided a new horizon to the field of polymer science and technology. They combine the properties of crosslinked elastomers, such as impact resistance and low temperature flexibility, with the characteristics of thermoplastic materials, example, the ease of processing. In general, TPEs are phase separated systems consisting of a hard phase, providing physical crosslinks, and a soft phase, contributing to the elastomeric properties. In many cases the phases are chemically linked by block or graft copolymerization. In other cases, a fine dispersion of the hard polymer within a matrix of the elastomer by blending also results in TPEs like behavior. Consequently, TPEs exhibit mechanical properties that are, in many ways, comparable to those of a vulcanized rubber, with the exception that the network and hence the properties of the TPEs are thermally reversible. This feature makes TPEs ideally suited for high-throughput thermoplastic processes, such as melt extrusion and injection molding. Mainly three classes of commercial TPEs can be distinguished: polystyrene-elastomer block copolymers, multiblock copolymers, and polymer-elastomer blends.

The polymer-elastomer blends consist of a semicrystalline polymer (IPP or a propylene copolymer) as the hard phase. For the soft phase often EPR or EPDM is used. In addition, there are also systems based on butyl, nitril, and natural rubber elastomers.

In a U.S. patent issued in 1962, Gessler [5] claimed a high tensile strength, semi rigid plastic prepared by the dynamic vulcanization of a blend of 5–50 parts of chlorinated butyl rubber and 95–50 parts of IPP with curing agents which did not contain peroxide. This was one of the first references to dynamic vulcanization as an essential element to producing a practical flexible plastic [6].

17.2 DYNAMIC VULCANIZATION

Dynamic vulcanization is an *in situ* process in which the vulcanization of an elastomer proceeds with upon its mixture in melt with a thermoplastic polymer. Among reactive blending processes in an extruder reactor, dynamic vulcanization is the most promising for industrial use. Dynamic vulcanization produces so called thermoplastic vulcanizates (TPVs).

The TPVs combine the mechanical properties of vulcanized elastomers at room temperature with a capacity for processing, which, for linear thermoplastic polymers, is characterized by a higher melting temperature, as well as the following: compression molding, blow molding, extrusion, calendering, and so on. This technique allows for the production of materials with excellent operating characteristics. Since they are quite competitive with conventional rubbers on the market, interest in these materials is currently increasing [2]. The TPVs have shown a consistently strong market growth (~ 12% per year) over the last two decades.

Fischer [7] introduced the first commercial polyolefin thermoplastic elastomer under the trademark TPR® in 1972. These were mechanically mixed blends of EPDM and IPP partially dynamically vulcanized [6]. The TPVs based on IPP/EPDM blends are most important from a commercial point of view. Further improvement of the thermoplastic process ability of these blends was reached by Coran, Das and Patel [8-14] by fully crosslinking of the rubber phase under dynamic shear. IPP/EPDM blends were further improved by Abdou-Sabet and Fath [15] in 1982 by the use of phenolic resins as curatives, to improve the rubber like properties and the processing characteristics.

Dynamic vulcanization has allowed us to develop TPVs based on flexible and rigid chain thermoplastics with different saturations and rubbers that are vulcanized by different systems, including peroxides, sulfur, divinylbenzene, phenol resins, bismaleimides, and so on. The combination of different elastomers and plastics resulted in the preparation of more than 75 types of TPVs [16].

A dynamic vulcanization technique allows the preparation of blends with unique morphology characterized by dispersion of the particles of the vulcanized elastomer in a continuous thermoplastic matrix.

To prepare a material with a homogeneous distribution of components, blending on a micro level is needed. This process is conducted in several stages. First, spatial irregularities in the distribution of components are equalized without changing the sizes of mixing particles at the same time as they are dispersed (agglomerates degradation to smaller particles). This physical process is superimposed on the cross-linking reaction of the elastomer phase. In this case, the viscosity of the medium increases and this hampers the homogenization of the blend. For the preparation of more homogeneous material with a micron distribution of the particles of the dispersed phase, a necessary condition is that the characteristic time of mixing must be shorter than the characteristic time of the reaction. The cross-linking process for elastomers during their mixture with a thermoplastic occurs at elevated temperatures (150–220°C); the availability of high-speed mixing equipment is necessary to carry out this task. To prepare this type of materials, one can treat them as though they are periodic and, thus, use the continuous mixing method. Thus, the mixing of IPP and elastomers for the purpose of preparing TPVs takes place in the temperature range of 180–210°C with a shear rate of 90–120 rev/min.

It has been found experimentally that, for chemically reacting systems, twin screw extruders are preferable to single screw extruders, due to their more efficient mixing and heat removal. In this case the rate of chemical reactions is defined by blending character, temperature gradients, diffusion of reagents. The complexity of the chemical and physical processes that occur in an extruder reactor hampers the choice of the molding parameters and equipment. As this takes place, the composition of products prepared from a polymer blend can be regulated by varying the viscosity during the reaction, as well as the molecular weights and temperatures of components and the introduction of corresponding additives [2].

Both physical and rheological properties depend on the blend ratio and nature of the curatives. Each and every curative has its own merits and demerits.

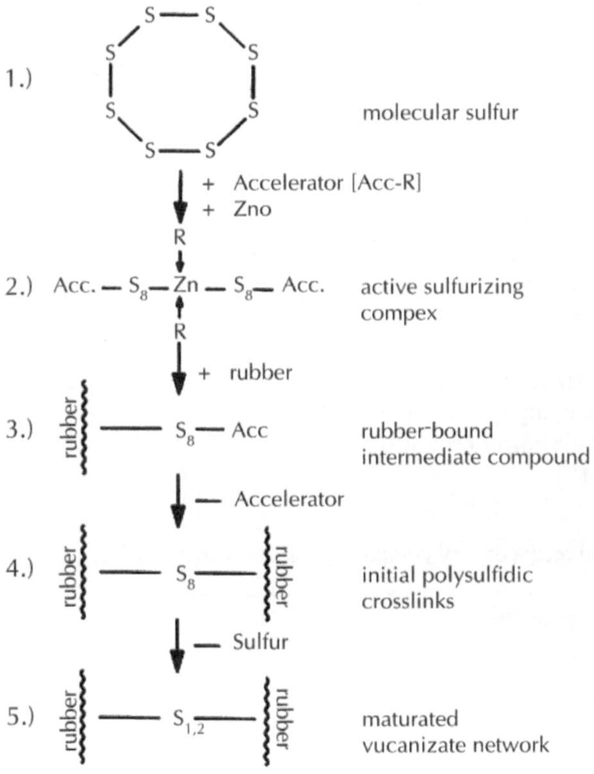

FIGURE 1 Scheme of crosslinking reaction with sulfur.

Initially, vulcanization was accomplished by using elemental sulfur at a concentration of 8 parts per 100 parts of rubber (phr). It required 5 hours at 140°C.The addition of zinc oxide reduced the time to 3 hours. The use of accelerators in concentrations as low as 0.5phr has since reduced the time to as short as 1 to 3 min. As a result, elastomer vulcanization by sulfur without accelerator is no longer of much commercial significance [17].

Accelerated sulfur vulcanization is the most widely used method. For many applications, it is the only rapid crosslinking technique that can, in a practical manner, give the delayed action required for processing, shaping, and forming before the formation of the intractable vulcanized network. Accelerated sulfur vulcanization of EPDM rubber and other elastomers is normally performed in the presence of activators (ZnO and stearic acid) and accelerators example, dibenzothiazole disulfide (MBTS), tetramethylthiuram disulfide (TMTD) and so on. It is proven, that the mechanism of sulfur vulcanization of EPDM is more or less similar to the mechanism, which is normally accepted for polydiene elastomers (Figure 1) [18, 19].

By means of accelerators and activators the sulfur cycle opening (1) occurs providing for an active sulfurizing complex (2). The self-destruction of this complex results in an active sulfur oligomer (3), which reacts with the rubber molecules crosslinking them (4). Due to the high vulcanization temperatures thermo-oxidative degradation of the rubber molecules as well as the sulfidic bridges can take place. Maturation (5) is the separation of sulfur from the polysulfidic bridges. Reshuffling the crosslinks may form intramolecular bridges (cyclic structures).

Sulfur and accelerator combinations have in principle been demonstrated to be applicable to dynamic vulcanization of IPP/EPDM blends by Coran and Patel [8]. A typical recipe of TPV composition is shown in Table 1 where, X, phr of polyolefin resin, and Y, phr of sulfur, were varied. An increase in the amount of sulfur from 0 to 2.0 phr in the above recipe resulted in a dramatic improvement of the elasticity of an IPP/EPDM = 40/60 blend.

Particle size in the dynamically vulcanized blends had a pronounced effect on mechanical properties, also demonstrated by Coran and Patel [12]. As the particle size was reduced, the elongation at break and tensile strength increased rapidly. The most disadvantage of the sulfur crosslinking system is the development of an undesirable, sulfurous odor [19].

TABLE 1 Typical recipe of TPV composition (sulfur cured) (in phr)

EPDM (Epsyn 70A)	100
Polyolefin resin (PP)	X
Zinc oxide	5
Stearic acid	1
Sulfur	Y
TMTD (Tetramethylthiuram disulfide)	Y/2
MBTS (2-benzothiazyl disulfide)	Y/4

The crosslinking systems most commonly used for the production of IPP/EPDM and TPVs are activated phenol formaldehyde resins, commonly known as resol resins.

Phenol formaldehyde resins are used to vulcanize diene elastomers in the absence of sulfur (Figure 2). The crosslinking reaction occurs through the free phenol groups in the presence of stannous or iron chloride as catalysts. Vulcanizates with good mechanical properties and enhanced resistance to moisture and heat are obtained.

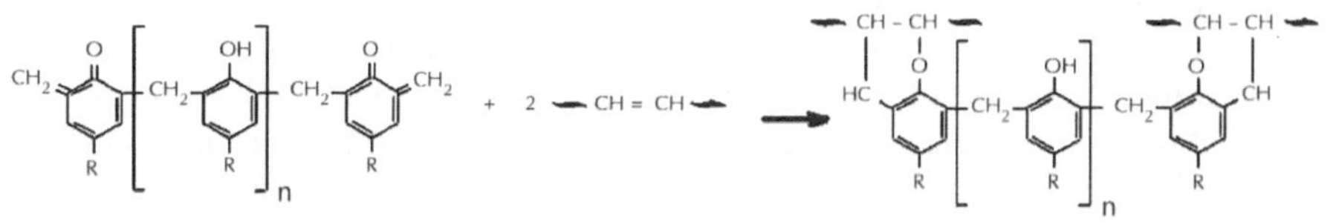

FIGURE 2 Crosslinking reaction of a diene elastomer with phenolic resin.

Resols are used extensively for the crosslinking of EPDM in the production of TPVs. The application of resols as crosslinking agents for TPVs is gaining importance due to the good high temperature properties of the corresponding products [20-23].

Abdou-Sabet and Fath [24] demonstrated that the rubber like properties of IPP/EPDM TPVs could be improved by using dimethylol octyl phenol curing resin. Improvements in compression set, oil resistance and processing characteristics of the material were achieved.

The degree of EPDM crosslinking during dynamic vulcanization of IPP/EPDM TPV was modified by varying its phenolic curative content. The rise in TPV viscosity, the drop in its swelling, and the change in its nuclear magnetic resonance, magic-angle spinning (NMR MAS) line shape, and the increase in its EPDM domain atomic force microscopy (AFM) force modulation hardness verified the increase in EPDM crosslink density with increasing curative content. A narrowing of the EPDM domain size distribution, with a decrease in the third moment of the distribution, was observed with increasing EPDM crosslink density. Correspondingly, the PP ligament number average thickness was raised slightly [25].

There are two major problems associated with this crosslinking system:

- Hygroscopicity of the TPV produced with this system; the tendency to absorb moisture, even at ambient temperature, that must be removed through lengthy, high-temperature drying procedures before processing, to eliminate product defects;
- Appearance of a dark brown color, which is difficult to mask and sometimes necessitates the use of two different pigment systems in order to achieve a desired color.

The above-mentioned disadvantages of the resol resins create a demand for other alternatives. Crosslinking of TPVs with peroxides is a potential alternative.

Crosslinking elastomers with peroxides have been known already for many years [19]. A wide variety of peroxides can be used to crosslink most types of elastomers. The crosslinking reaction of an EPM with peroxide is shown in Figure 3. The mechanism of crosslinking using peroxides is a homolytical one. At the beginning of the vulcanization process, the organic peroxide splits into two radicals. The free radicals formed as a consequence of the decomposition of the peroxide, abstract hydrogen atoms from the elastomer macromolecules, converting them into macroradicals. In the case of an ethylene propylene copolymer, the radical formed by the decomposition of the peroxide attacks the tertiary carbon atom. The resulting macroradicals react with each other forming carbon-carbon intermolecular bridges.

Simultaneously with these crosslinking processes, side reactions occur, which reduce vulcanization. Thus, peroxides can react with the components of the compound, that is, antioxidants, plasticizers, extenders, and so on, and can be deactivated. Other side reactions can take place on the radical centers formed on the elastomer backbone. These radicals can disproportionate leading to a saturated molecule and an unsaturated one.

The main factors which influence the peroxide crosslinking efficiency of EPM and EPDM are: Type and amount of termonomer, ethylene/propylene ratio, randomness of monomer distribution, polymer molecular weight and molecular weight distribution. For instance, the higher the ethylene/propylene ratio, the better is the crosslink density of EPM or EPDM.

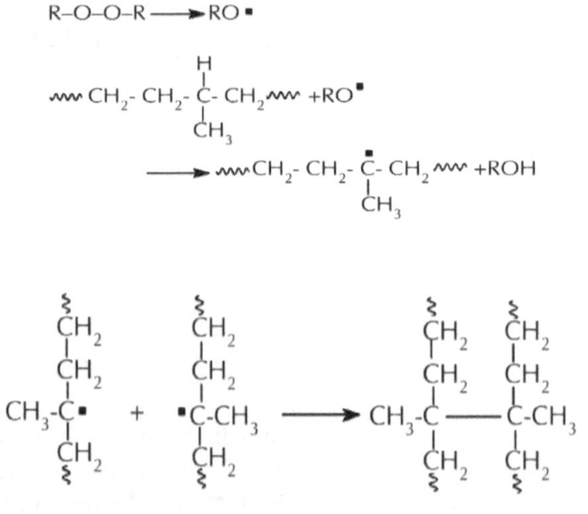

FIGURE 3 Crosslinking reactions with peroxide.

However, there exist some disadvantages of this type of crosslinking system as well. Those disadvantages of peroxide crosslinking are as follows:

1. Sensitive to oxygen under curing conditions;
2. Certain components of the rubber compounds and IPP under certain conditions consume peroxide free radicals and therefore reduce the crosslink efficiency.

Consequently, when peroxide is added to IPP/EPDM blend, two competing processes may take place simultaneously:

- EPDM-crosslinking
- IPP-degradation

The degradation of IPP leads to a loss of properties. The damaging effects of β-scission of propylene-rich sequences in EPDM can be partially avoided and consequently the crosslinking efficiency of a peroxide enhanced, by the addition of a suitable co-agent. Co-agents are basically multifunctional reactive compounds, in most cases possessing multiple allylic groups [26].

Saturated as well as unsaturated rubbers or copolymers can generally be crosslinked by free radicals, induced by silanes and irradiation.

An alternative crosslinking system for TPVs was developed based on the hydrosilation reaction of a silicon hydride (SiH) functional silane with a vinyl containing polymer in presence of platinum catalyst [27]. This crosslinking technology has been adapted to TPV by using unsaturated organic polymer (EPDM, styrene-butadiene rubber (SBR), and so on) with the silicon-hydride functional siloxane containing at least two Si-H bonds per molecule. With the help of a Pt catalyst, SiH functions will react with carbon-carbon double bond and will therefore bridge two polymer chains. Although this technology is limited to unsaturated rubbers, a major advantage of this technology is the fact there is no by-product formed, that is, no low molecular weight species generated during compounding. Hydrosilylation was most probably the first attempt in the use of silicon technology for TPV's crosslinking. However, it was shown that there is still room for further development using organosilanes [28].

During crosslinking with silanes a polyfunctional network structure is formed, in which the polymer chains are crosslinked via siloxan bridges (Si-O-Si). In case of EPDM two steps are necessary. First unsaturated organosilane molecules are grafted onto a polymer chain, which was first activated with peroxide. Secondly the crosslinking reaction via hydrolysis and condensation result in Si-O-Si bridges between the polymer chains.

A more recent method of crosslinking, which can be carried out without curatives, is radiation crosslinking. The crosslinking of polymers can be brought about by high energy radiation such as X-ray, proton, electron, and neutron beams. Upon irradiation, high local concentrations of free radicals are formed in the polymer molecules. The free radicals can combine to form crosslinks as is the case with peroxide crosslinking.

Electron induced reactive processing of EPDM/IPP blends presents such novel approach where the EPDM phase is vulcanized and grafted to IPP by high energy electrons during melt mixing with IPP. In this novel approach, the electron accelerator is directly coupled with the mixing device and the crosslinking of EPDM is done during melt mixing with IPP without any co-agent. Usually, this electron beam treatment is done under ambient temperature and utilizes the energy input of high energy electrons to achieve the desired material properties via radical induced chemical reactions. Numerous reports are available in the literature on the electron beam treatment of polymeric materials under stationary condition and at ambient temperature [29-32].

Naskar et al. [33] described the effect of absorbed dose and electron treatment time on the properties and morphology of IPP/EPDM TPVs. It showed that the absorbed dose of 100 kGy and electron treatment time of 15 s give the best set of mechanical properties together with the finer dispersion of EPDM particles in IPP. Thakur et al.

[34] showed that the absorbed dose and dose per rotation showed significant influence on the morphology and the properties of the TPVs. The microscopic pictures showed that there was a widespread distribution of the EPDM particle size in the TPVs prepared. The particle size varied from 1 μm to few hundreds of nanometers. This type of inhomogeneous morphology showed the best set of mechanical properties.

The blend ratio played an important role in the morphological development and final properties of the TPVs prepared. It was found that at 30/70 IPP/EPDM blend ratio, EPDM and IPP formed co-continuous phase morphology and hence, the properties were not increased significantly. On the other hand, at 50/50 blend ratio, EPDM was dispersed in the continuous PP phase and therefore, it showed excellent mechanical properties. At higher EPDM concentration, the viscosity ratio was too high which resulted incomplete phase inversion and hence, average mechanical properties of the TPVs prepared. This study reveals the positive influence of the electron induced reactive processing in the preparation of TPVs and paves a new technology for the development of various types of TPVs in an eco-friendly direction.

TABLE 2 Effect of curatives on physical properties IPP/EPDM=40/60 TPV

Property	**Curative**			
	None	**Sulfur**	**Dimethyl alkyl phenol**	**Peroxide**
Hardness, Shore D	36	43	44	39
Tensile Strength, MPa	4.96	24.3	25.6	15.9
Stress at 100% elongation, MPa	4.83	8.00	9.72	8.07
Ultimate elongation, %	190	530	350	450
Oil swell, %	disintegrated	194	109	225
Compression set, %	91	43	24	32
Odor	+++	–	++	++
Fabricability	++	–	+++	+++

The effect of the curatives on the physical properties is shown in Table 2. As can be seen, TPV prepared with the use phenolic system exhibit better characteristics (especially, oil swell and compression set).

Thus, dynamic vulcanization is a route to new thermoplastic vulcanizates which have properties as good or even , in some cases, better the an those of elastomeric block copolymers.

17.3 MORPHOLOGY OF TPVs

One of the main factors affecting the properties of heterophase systems, including those based on thermoplastics and elastomers, is their morphological features: the component ratio; the size, shape, and distribution character of dispersed-phase particles in the matrix; and the presence of the interfacial layer.

The structure and properties of IPP blends with unvulcanized EPDM have been studied in sufficient details [36-43]. However, the morphology of such materials is unstable and can undergo substantial changes during processing. It was shown that the interpenetrating-phase morphology is observed immediately after blending [44]. The structure of the molded samples dramatically change elastomer particles become significantly larger. It is likely that the enlargement results from the intense flow of less viscous IPP under the action of temperature and pressure, as well as from the coalescence of small elastomer particles into large domains in the absence of the three-dimensional network that is formed by vulcanization. Abdou-Sabet et al. [24] studied the morphology of a blend containing IPP and 80 wt% EPDM by means of scanning electron microscopy (SEM). It was found that IPP forms the disperse phase in the initial unvulcanized system. During dynamic vulcanization, phase inversion takes place as the crosslink density of the rubber increases: IPP becomes the continuous phase and EPDM becomes the dispersed phase, the crosslinked rubber particles with a size of the order of 0.8–2.0 μm being closely packed in the continuous thermoplastic matrix.

The most suitable method for studying the TPV structure is AFM which received wide acceptance in recent years and is currently one of the most popular methods among the variety of scanning probe microscopies [45]. The methodological aspects of AFM investigation of the structure of elastomer materials, including TPVs, have been considered in [46]. It was shown that AFM makes it possible to visualize regions with different crosslink densities of the elastomer phase and to study the effect of the distribution of various fillers and a plasticizer (paraffinic oil) on the TPV morphology.

The peculiarities of TPV morphology were investigated by means of AFM in some papers [44, 47, 48]. It was shown that the morphology of the unvulcanized specimen containing 25 wt% of oil-free EPDM is characterized by a homogeneous distribution of finely divided elastomer particles in IPP matrix (Figure 4 (a)). With the increasing of EPDM content, its domains grow larger, most likely due to their coalescence / aggregation as a process competitive to the dispersion one (Figure 4 (b) and Figure 4(c)). And finally, material morphology changes to the type of co-continuous phases (Figure 4 (c)). A further increase in the amount of EPDM (> 60 wt%) leads to phase inversion, that is, elastomer becomes a matrix (Figure 4 (d)).

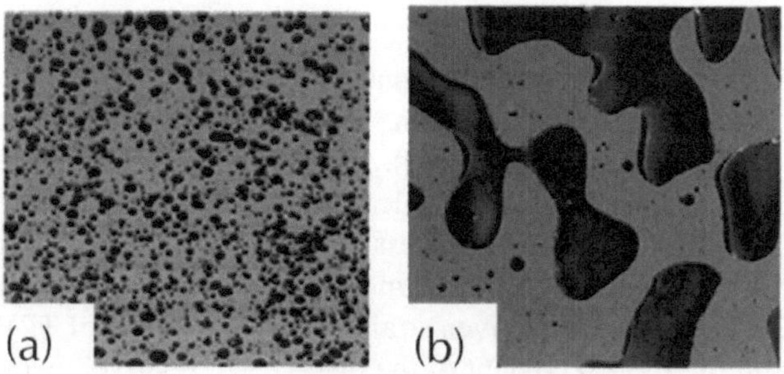

FIGURE 4 *(Continued)*

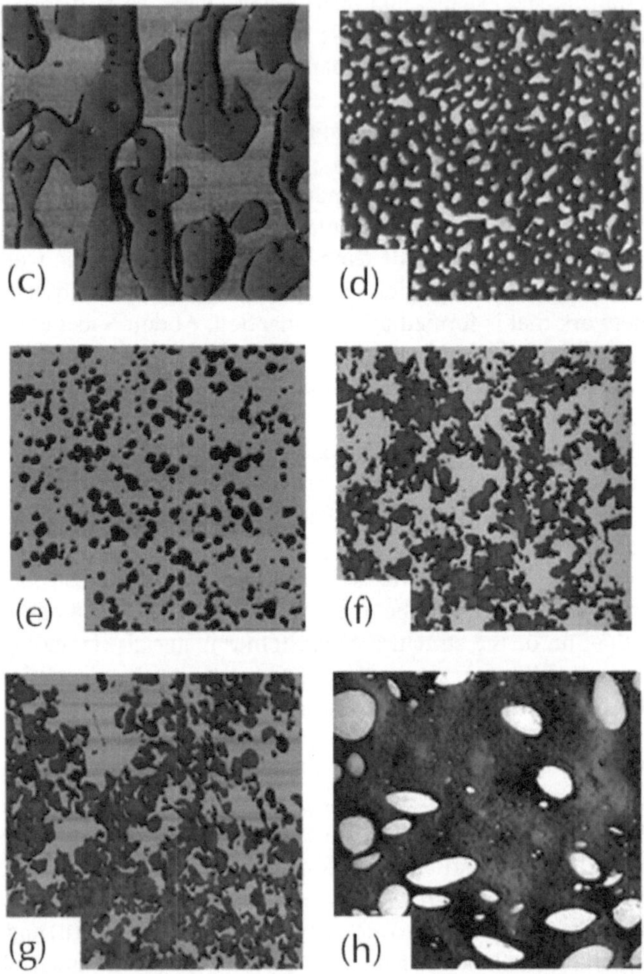

FIGURE 4 AFM images of molded samples of (a–d) unvulcanized blends and TPVs based on IPP and (a, e) 25, (b,f) 50, (c, g) 60 and (d, h) 75 wt% of oil-free EPDM. The size of images is (a–d) 50 × 50 and (e–h) 15 × 15 μm.

Dynamic vulcanization substantially changes the structure of IPP/oil-free EPDM blends (Figure 4 (e)–(g)). Rubber domains are smaller (less than 1 μm) than those in unvulcanized blends (Figure 4 (e)–(g)). It meets expectation because increasing of crosslinking degree during dynamic vulcanization makes agglomeration of rubber particles much more difficult. The shapeless character of crosslinked rubber particles probably reflects the irreversible breakdown of larger initial particles to sizes that correspond to stresses appearing at a given strain rate in the course of TPV preparation.

An increase in the EPDM content of the blend has a relatively weak effect on the size of elastomer particles. The jagged form of the envelope of these domains renders the area of their interaction with the matrix much more developed than that of unvulcanized blends. The crosslinked rubber particles form a three-dimensional framework, whose density increases with an increase in their concentration. This development is presumably due to the agglomeration of EPDM particles. The structure of TPVs that contain up to 75 wt% rubber is similar to that of filled composites. It is stable and does not undergo substantial changes even after several processing cycles, unlike the case of blends with unvulcanized EPDM [49]. Thus, dynamic vulcanization displaces the phase inversion boundary toward a higher elastomer content (75 wt%) relative to the blends with unvulcanized EPDM, in which the transition to the morphology of co-continuous phases and, then, to their inversion is observed at lower elastomer contents (~ 50–60 wt%).

The increase of EPDM content more than 60 wt% leads to the phase inversion as in the case of unvulcanized blends (Figure 4 (h)). However, the shape of IPP domains becomes more regular, a change that is presumably a consequence of IPP crystallization in the matrix of vulcanized EPDM, which has a higher hardness.

Components of crosslinking systems which used for TPVs creation can influence on thermoplastic structure. It is known that dynamic vulcanization by peroxides effectively crosslinks EPDM but at the same time causes a degradation of IPP, which inevitably affects mechanical properties of the final material, such as tensile strength and tear [50, 51]. Dynamic vulcanization by accelerated sulfur also has an effect on the IPP crystallization process (thermophysical properties and crystalline modification) and changes the morphology of thermoplastic [52, 53]. Thus, the introduction of a vulcanizing system composed of sulfur, zinc oxide, and organic compounds into IPP facilitates the partial formation of the β-modification of IPP and changes in its stress-strain properties.

To obtain TPVs with particular properties, various fillers and plasticizers are used. Low molecular mass oligomers (oils) are frequently added to high-viscosity elastomers. These additives exert a substantial effect on the structure and rheological behavior of TPVs [54-56] The size of the rubber particles and their distribution in TPVs based on oil-extended EPDM are about the same as for blends contained oil-free elastomer [44, 47, 48].

Upon mixing of IPP with oil-extended elastomer oil redistributes between the polymers [56-60]. It was found that oil distributed between the thermoplastic and elastomer by equal parts [44, 57, 58].

The high-resolution AFM studies display finer structural features of TPVs containing oil-free and oil-extended EPDMs (Figure 5). There are both zones corresponding to the IPP matrix (the brightest areas) and oil-free EPDM domains, inside which regions with different contrasts are distinctly visualized (Figure 5a). The fact of appearance of different contrasts in elastomer inclusions indicates the presence of mechanical heterogeneity, which is evidently due to different crosslink densities of the rubber phase. The lighter areas refer to regions with a higher crosslink density and vice versa.

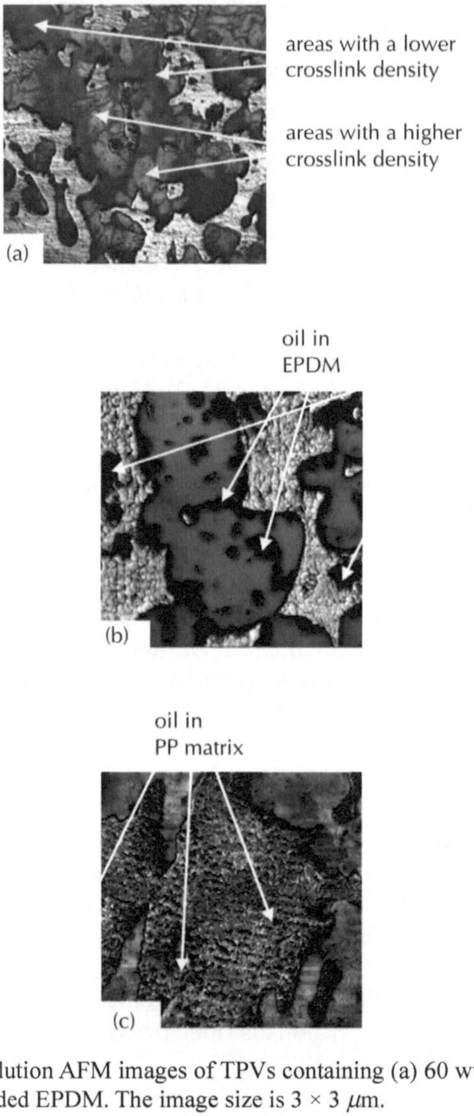

FIGURE 5 High-resolution AFM images of TPVs containing (a) 60 wt% oil-free EPDM and (b, c) 75 wt% oil-extended EPDM. The image size is 3 × 3 μm.

The AFM images of TPVs based on oil-extended EPDM show that both rubber domains and IPP matrix contain nanosized dark inclusions, which presumably correspond to oil (Figure 5 (b) and Figure 5 (c)). This finding suggests that the oil is redistributed between EPDM and PP during blending; that is, a portion of oil remains in the elastomer and the other portion diffuse into amorphous regions of IPP. Thermoplastic/oil mixtures are heterogeneous systems which are thermodynamically consistent in the melt [60]. Increase of oil content in the mixture leads to a change in the size of spherulites and lamellae size reduction, respectively, to reduce the melting temperature of IPP.

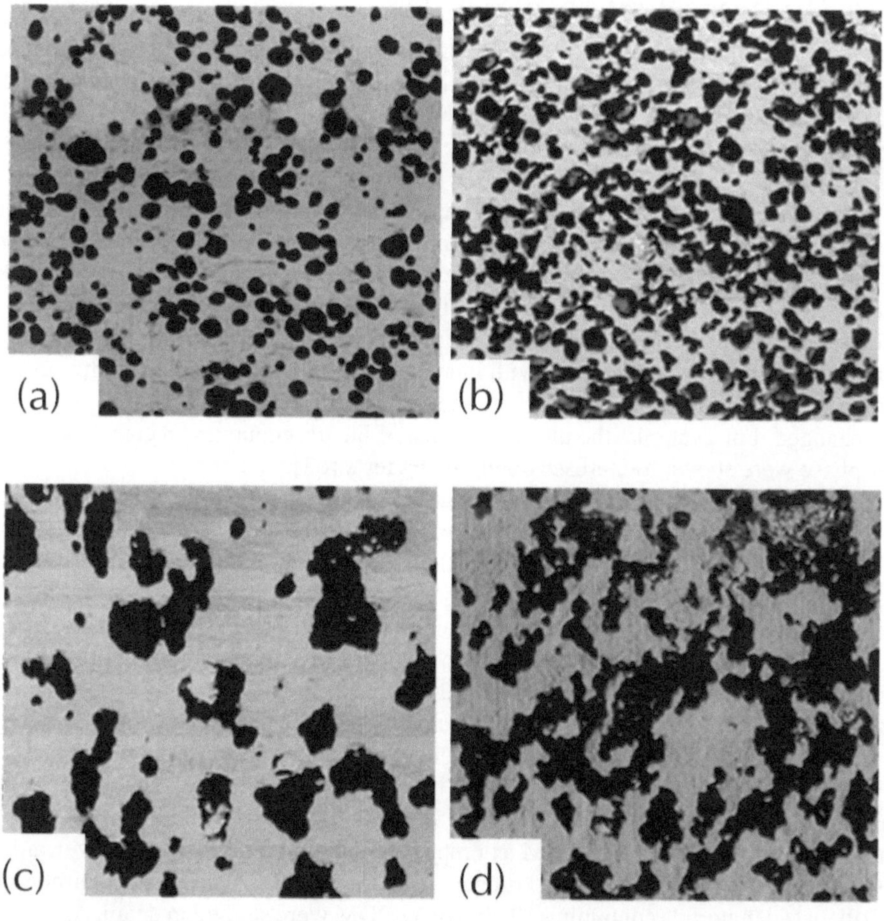

FIGURE 6 AFM phase images of TPVs dynamically cured by (a, b) sulfur system and (c, d) phenolic resin system. The content of oil-free EPDM is (a, c) 25 wt% and (b, d) 50 wt%. The size of images is 15 × 15 μm.

The nature of vulcanizing system influences on the size of rubber particles and their distribution in thermoplastic matrix. The morphology of TPVs dynamically vulcanized with a phenol-formaldehyde resin and sulfur systems was investigated in [61]. It was shown that the size distribution of EPDM particles in blends dynamically vulcanized with sulfur system is more narrow and homogeneous than in blends prepared by phenol-formaldehyde resin (Figure 6). An irregular shape of rubber domains with more coarse external contours in TPV obtained by phenolic resin system indicates that they have been dispersed in another way, experiencing different level of tearing stress than EPDM cured by sulfur system. So, viscoelastic characteristics of elastomers cured by various crosslinking agents are different.

17.4 MECHANICAL PROPERTIES OF TPVs

Numerous publications were devoted to studying the structure, deformation mechanism, and properties of high-impact materials based on IPP and unvulcanized EPDM [3, 4, 35, 37, 38]. As was shown, depending on the structure of a blend and conditions of tensile drawing, deformation proceeds either *via* crazing or formation of shear bands (shearing).

The mechanical properties of TPEs based on IPP and unvulcanized EPDM were studied as a function of their composition and phase morphology in the entire range of elastomer contents [40]. As was found, the morphology of the blend as controlled by the method of its preparation exerts no effect on the dependence of elastic modulus, yield stress, and strength on the concentration of rubber phase. The values of the above parameters are shown to decrease with increasing the elastomer content in the composition. As a result of vulcanization, the mechanical properties of the above materials are changed. For example, the ultimate strains of blends containing a crosslinked rubber phase were shown to increase upon impact tests [62].

The effect of the degree of crosslinking on the deformation of the PP/EPDM compositions was studied [43]. The mechanical characteristics of the blends containing a vulcanized EPDM appear to be higher than the corresponding parameters of TPEs based on an unvulcanized EPDM. At the same time, the effect of the crosslinking degree of the rubber phase on the deformation mechanism of blends has not been studied in all details. As is known, the deformation behavior of a composition is controlled by dimensions of disperse phase particles [63-65]. Therefore, to study the effect of vulcanization of a rubber phase on the mechanical properties of blends, the blends with a homogeneous distribution of finely dispersed rubber particles should be prepared.

As is known, the deformation of IPP based blends with a low content of an uncrosslinked EPDM proceeds via necking [38], whereas the deformation of TPVs containing 70 wt% crosslinked EPDM as prepared by the method of dynamic vulcanization is similar to that of vulcanized rubbers [66]. In [67, 68], the mechanical properties of IPP based materials containing 10–40 wt% EPDM were studied in detail. As a result of dynamic vulcanization, the deformational and strength properties of the as prepared blends were shown to be markedly improved.

The mechanical properties of IPP/EPDM blends with rubber content from 5 to 85 wt% were studied in [69]. It was shown that dynamic vulcanization initiates specific deformational processes in TPVs. The character of such processes is determined by the ratio between the components in the blends. At a low content of EPDM or IPP (less than 25 wt%), the mechanical characteristics are primarily controlled by matrix: IPP or EPDM, respectively. At a content of IPP 25–75 wt% deformation is initially provided by the PP matrix, whereas further deformation proceeds *via* an almost independent deformation of matrix (IPP) and disperse phase (crosslinked EPDM). In this case, the particles of the rubber phase are separated from the IPP matrix at the stage of neck formation; after the yield point, the deformation of PP and EPDM proceeds separately.

As a result of dynamic vulcanization, the boundary of phase inversion is shifted. Elastic modulus, strength, and elongation-at-break decrease with decreasing the content of polypropylene down to concentration corresponding to the phase inversion point. As the concentration of sulfur is increased, the elastic modulus of the blend decreases, whereas its strength increases [69, 70].

As a result of dynamic vulcanization, the modulus of elasticity varies in different manners for the systems containing the two types of rubber:

1. The modulus decreases for the oil-free blends
2. Increases for the oil-extended compositions

It is likely that the different character of change in the elastic modulus depending on the composition is due to the EPDM structure.

The modulus of elasticity and the tensile stress of the oil extended TPVs are substantially lower than those for the oil-free materials, probably due to the diffusion of a part of oil from the rubber to the thermoplastic. However, the values of the elongation at break for TPVs of both types are approximately of the same order (within the limits of experimental error) and do not depend on the presence of oil in the blend.

The obtained relationships between the mechanical characteristics and the composition were rationalized in terms of the models proposed for heterophase systems [44]. Of the Equations used to calculate the elastic modulus of a two-phase blend, the Kerner [71] and Uemura–Takayanagi [72] Equations derived for dispersion morphologies are most frequently applied. Applicable to calculation of the dependence of the elastic modulus on the composition of heterogeneous blends with the morphology of interpenetrating continuous phases is the Davies equation obtained on the basis of analysis of the existing models of macroscopically isotropic and homogeneous two-phase composite materials [73].

All these Equations for thermoplastic elastomer blends can be written as follows:

$$\text{Kerner equation, } E = E_{PP}\frac{1}{1+(\varphi_{el}/\varphi_{PP})[15(1-\mu)/(7-5\mu)]} \quad (1)$$

$$\text{Uemura–Takayanagi equation, } E = E_{PP}\frac{(7-5\mu)\,\varphi_{PP}}{(7-5\mu)+(8-10\mu)\,\varphi_{el}} \quad (2)$$

$$\text{Davies equation, } E^{1/5} = E_{PP}^{1/5}(1-\varphi_{PP}) + E_{el}^{1/5}\,\varphi_{el}, \quad (3)$$

where E is the elastic modulus of a blend; E_{PP}, E_{el}, φ_{PP}, and φ_{el} are the elastic moduli and the volume fractions of IPP and EPDM, respectively; and μ is the Poisson ratio of the matrix, equal to 0.35. The following experimentally measured elastic moduli of the polymers were used in the calculations—E_{PP} = 1400 MPa, E_{el} = 1.3 MPa (oil-free EPDM), and 0.4 MPa (oil-extended EPDM) for the unvulcanized elastomers, and E_{el}

= 2.0 MPa (oil-free EPDM) and 0.6 MPa (oil extended EPDM) for the vulcanized rubbers.

Later Coran and Patel proposed an equation for calculating the elastic modulus of TPEs as a function of the volume fraction of hard or soft filler, which is a superposition of the upper and lower bound moduli [74]:

$$E = \varphi_{\mathrm{I}}^{n}\,(n\varphi_{\mathrm{el}}+1)(\varphi_{PP}E_{\mathrm{PP}}+\varphi_{\mathrm{el}}E_{\mathrm{el}}-\frac{1}{(\varphi_{PP}/E_{pp})+(\varphi_{\mathrm{el}}/E_{el})})+\frac{1}{(\varphi_{PP}/E_{PP})+(\varphi_{\mathrm{el}}/E_{el})} \quad (4)$$

In this Equation, the parameter n takes into account the geometry of the system:

- The filler particle size
- Agglomeration of particles, and so on.

Higher values of n correspond to softer systems. However, it is impossible to determine on the basis of this value the type of structure existing in a given blend.

Figure 7 depicts the dependence of the experimental values of the modulus of elasticity and the values calculated by Equations. (1)–(3) on the composition IPP/oil-free EPDM. It is seen that the experimental moduli of elasticity in the region of $\varphi_{\mathrm{el}} \leq 0.5$, that is, in the region of dispersion morphology, satisfactorily agree with the values calculated according to the Kerner Equation for the unvulcanized materials and the Uemura–Takayanagi equation for the dynamically vulcanized blends. As the amount of unvulcanized oil-free EPDM in the blends increases, the structure alters and transition to the morphology of interpenetrating continuous phases takes place. In this case, the experimental values of the modulus of elasticity approach the values calculated by the Davies equation.

In the case of dynamically vulcanized TPVs, the density of the structural framework increases, and the framework may be considered a continuous phase, a development that is responsible for the closeness of the experimental elastic moduli of TPVs and the values calculated by the Davies Equation.

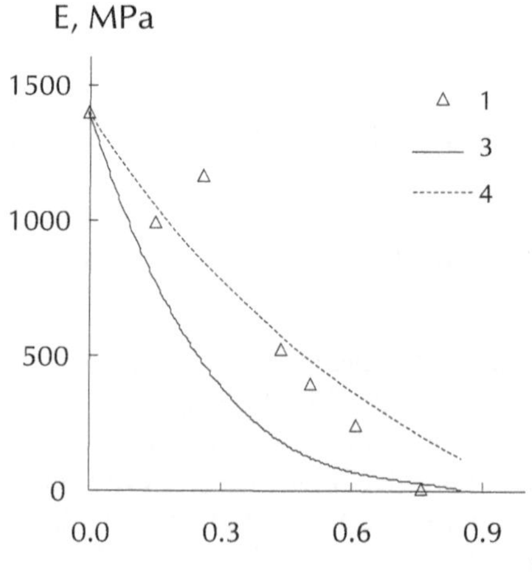

FIGURE 7 *(Continued)*

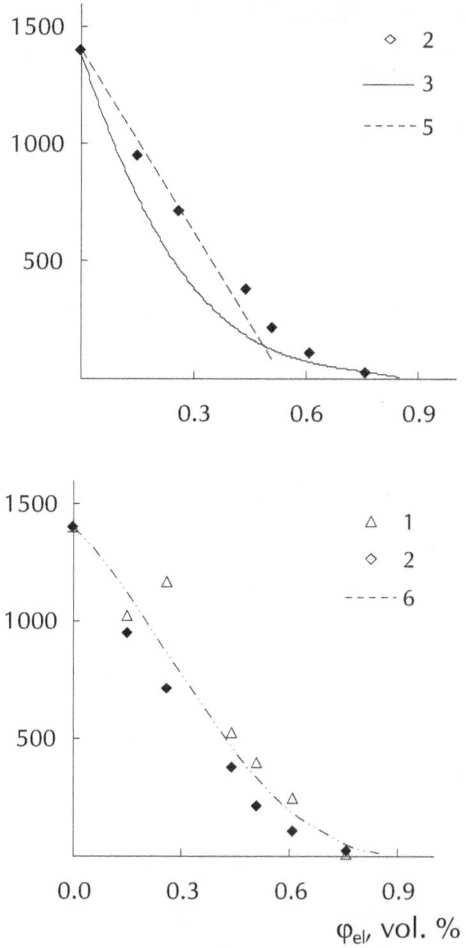

FIGURE 7 Elastic modulus E of IPP/oil-free EPDM blends as a function of elastomer content φ_{el}. The symbols represent the experimental data for (*1*) unvulcanized blends and (*2*) TPVs. Curves 3–6 are the results of calculation according to Equations (3), (1), (2), and (4), respectively.

In addition, Figure 7 shows the dependence of the experimental values of the elastic modulus and the values calculated by Equation (4) at $n = 2$ on the composition of the blends containing unvulcanized and vulcanized oil-free EPDM. There is good agreement between the data over the entire range of elastomer concentrations. The curves for the dependence of the elastic modulus on the composition for the blends of oil-extended EPDM lie below the theoretical curves, example, those calculated in terms of the Davies model (Figure 8). This disagreement is evidently due to the redistribution of oil during PP blending with EPDM. Therefore, the value corresponding to oil extended IPP, not the elastic modulus of pure IPP, should be used for the calculation of the elastic modulus of oil extended TPVs. The corresponding quantities for IPP oil blends were obtained in [60].

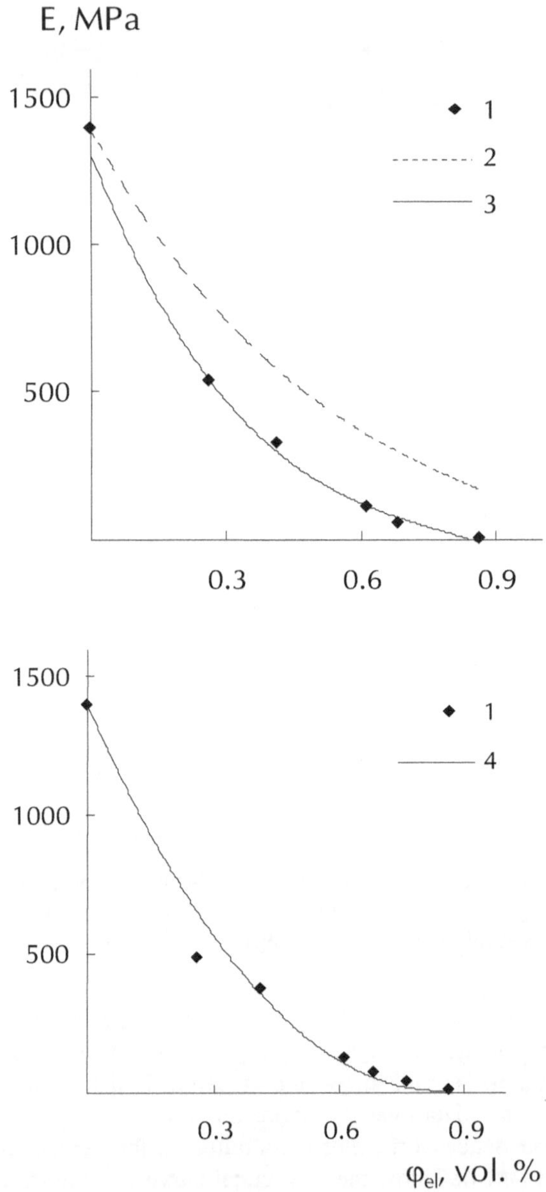

FIGURE 8 Elastic modulus E of TPVs based on oil-extended EPDM as a function of elastomer content φ_{el}. The symbols represent the experimental data (*1*) and the curves are the results of calculation by Equation (3) ((*2*) without and (*3*) with allowance for oil redistribution) and (*4*) by Equation (4).

A satisfactory fit of the experimental values to the values predicted in terms of the Davies model and the Coran–Patel Equation was obtained for the blends based on oil extended EPDM (Figure 8). Thus, both the Davies and Coran–Patel models can be used for qualitative assessment of the dependence of the elastic modulus on the composition for blends that contain oil extended EPDM. It is particularly important in this case that, knowing the elastic moduli of the individual polymers and their blend with one component ratio, it is possible to predict the elastic modulus of a blend with another ratio of the components.

The type and content of individual components have significant effect on the deformation behavior of TPV [75]. In particular, IPP molecular weight is an important characteristic which determines the mechanical properties of TPVs based on IPP and EPDM [76, 77]. It was found that Young's modulus of dynamically vulcanized IPP/EPDM blends decreases and their tensile strength increases with molecular weight of IPP (Figure 9). However, the modulus of the blends with high rubber content is independent on the IPP M_w. The dynamic vulcanization decreases the elongation at break of the blends with low EPDM content and increases this parameter for blends with high rubber content. The value of elongation at break depends on the IPP M_w. So, it is possible to produce materials with more high tensile properties by using of IPP with high molecular weight as a basic component.

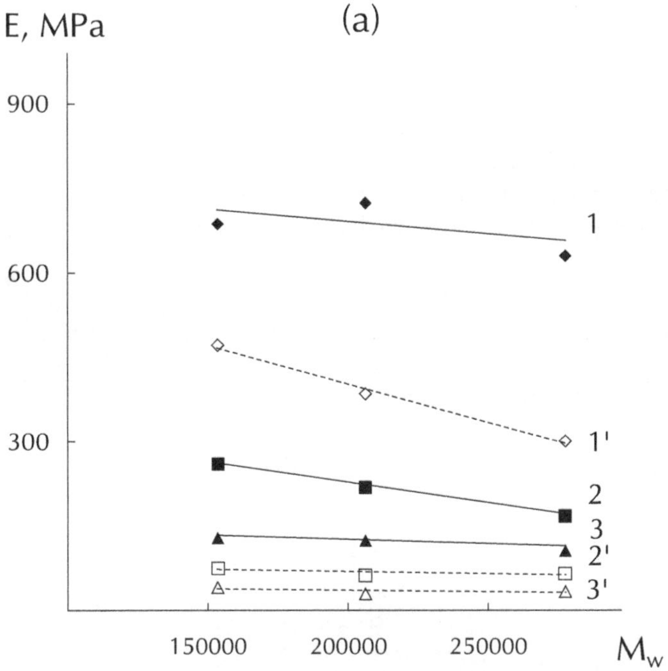

FIGURE 9 *(Continued)*

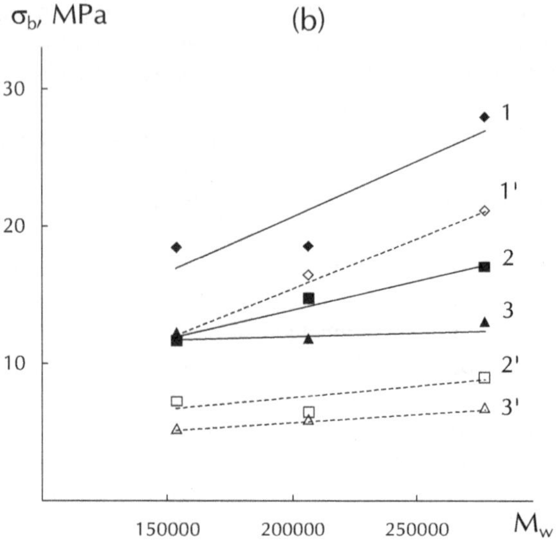

FIGURE 9 The dependences of (a) Young's modulus E and (b) ultimate tensile strength σ_b for oil-free blends (*1*, *2*, *3*) and oil extended TPVs (*1'*, *2'*, *3'*) as the function of IPP M_w. IPP/EPDM ratio: 3.00 (*1*, *1'*), 1.00 (*2*, *2'*) and 0.67 (*3*, *3'*).

The nature and concentration of vulcanizing agent determines mechanical parameters of TPVs. Dynamic vulcanization by sulfur system practically does not change elastic modulus and tensile strength of the blends based on oil-free elastomer, but reduces their elongation at break [61, 78]. The dynamic vulcanization by phenolic resin system leads to some decrease in tensile properties in comparison with sulfur vulcanization. TPVs prepared by phenolic resin system have lower values of mechanical parameters.

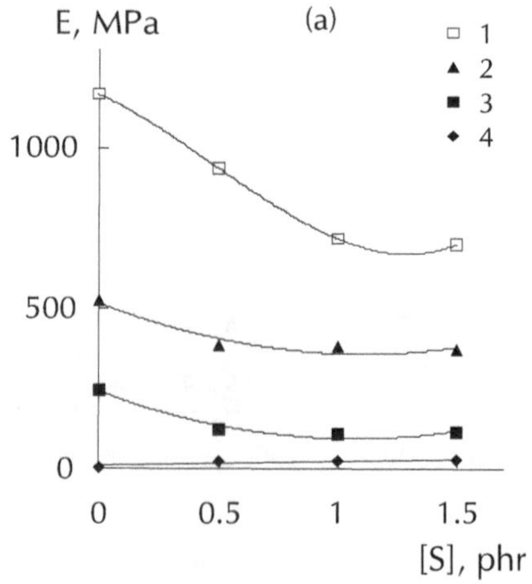

FIGURE 10 *(Continued)*

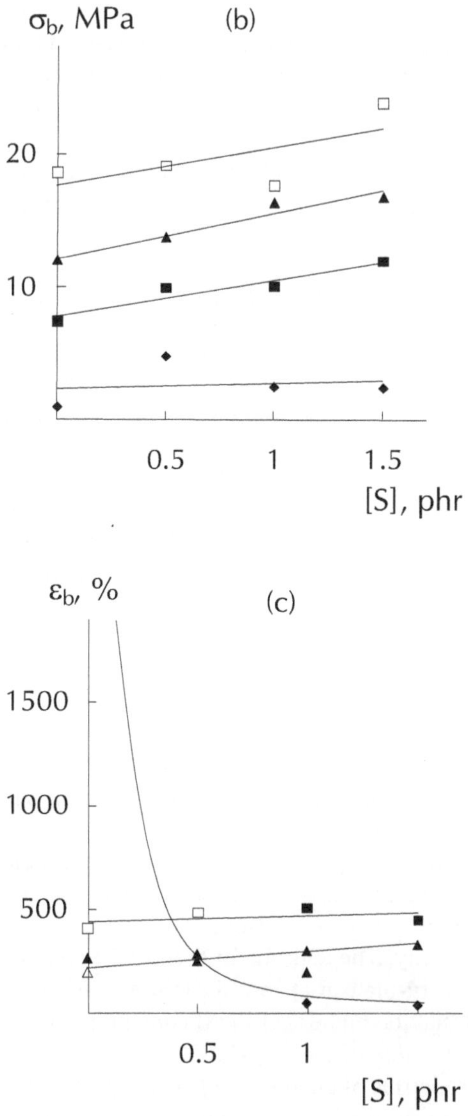

FIGURE 10 (a) Elastic modulus E, (b) ultimate tensile strength σ_b, and (c) elongation at break ε_b versus sulfur concentration [S] for dynamically vulcanized blends with 25 (*1*), 45 (*2*), 60 (*3*), and 75 wt% (*4*) of oil-free EPDM.

The relationship of mechanical parameters versus concentration of vulcanizing agent (sulfur) was studied in [69, 70, 79]. It was found that Young's modulus of TPVs depends on IPP/EPDM ratio—it is practically constant for the blends contained 40–60 wt% oil-free EPDM, decreases with a sulfur concentration for the samples with 25 wt% EPDM, and increases slightly for the blends with rubber content more than 75 wt% (Figure 10). In the latter case, the matrix is represented by the rubber phase.

Therefore, with growth of the sulfur concentration, the elastic modulus of the rubber increases and, correspondingly, the elastic modulus of cured blends increases too. IPP as a more rigid filler also increases this parameter. The elongation at break is virtually independent of the amount of the crosslinking agent; the only exception concerns the blends with high content of EPDM. For such compositions, this parameter decreases upon vulcanization.

17.5 RHEOLOGICAL BEHAVIOR OF TPVS

Rheological behavior is one of the key parameters to design new TPV grades with diverse flow characteristics and to facilitate better processability. Successful processing of polymeric materials requires a good understanding of their rheological behavior [80]. The melt rheological properties of TPEs and TPVs can provide fundamental insights about the processability of these materials.

Rheological behavior of toughening IPP and TPEs with low (5–25 wt%) and high (50–75 wt%) content of uncured rubber respectively was investigated in papers [47, 81-86]. It was shown that their rheological properties qualitatively follow a general rules, known for blends of other polymers, and depend on morphology of blends and properties of components.

The blends of IPP and small amount of EPR or EPDM (up to 25 wt%) are compatible in the melt [81, 82]. Their rheological properties close to IPP ones, and viscosity of blends slightly increases with elastomer content. Hence, these materials can be processed as individual IPP. Compatibility of the components in melt is limited, but sufficient for the formation of multiphase systems with micron particle size, determined only by of IPP/EPDM ratio [82].

The viscosity of uncured IPP/EPDM blends drops at sufficiently high shear stress values. As the shear stress decreases, the shear viscosity is essentially constant, and Newtonian behavior is observed [47, 83-86]. An increase of elastomer content leads to a rise of blends' viscosity. The addition of the EPDM content exhibits the decrease of the value of a viscosity-plateau at low shear stress, followed by a shear thinning behavior. It is evident that the higher is EPDM content in the blends, the greater is the deviation from the Newtonian behavior [47, 85, 86].

The rheological properties of uncured IPP/EPDM blends depend on the nature and molecular characteristics of elastomers. Their initial viscosity is proportional to molecular weight of elastomer. In addition, an increase in EPDM molecular weight leads to a growth of viscosity deviation of blends from the logarithmic additivity rule [85]. It was found that the viscosity of IPP/EPDM blends was higher than the viscosity of IPP mixed with acrylonitrile-butadiene copolymer (NBR), while the NBR viscosity was higher than EPDM ones [86].

Some papers addressed to the rheological behavior of TPVs based on IPP and EPDM [47, 48, 55, 86, 87]. It was shown that the type of initial components, plasticizer concentration, the amount and composition of crosslinking system essentially influence on the rheological properties of PP/EPDM blends.

One of the first investigation on the rheological properties of TPVs was performed by Goettler et al. [87]. Chung et al. studied the effect of dynamic vulcanization on the viscoelastic properties of PP/EPDM and PP/nitrile rubber systems with rubber concentration 20 wt% and 60 wt%, respectively, using plate-plate and capillary rheometers [86]. Han and White devoted a comparative study using different devices, namely sandwich, cone-plate, and capillary rheometers, to three materials:

1. IPP
2. A blend of IPP/EPR
3. A commercial available dynamically vulcanized IPP/EDM blend [55].

As was shown, TPVs may exhibit an increased viscosity. The rheological properties of TPVs largely resemble those of composites containing a highly disperse filler, since both polymer materials exhibit the yield stress. TPVs in molten conditions behave like to particle filled polymer melts. These materials exhibit not only non-Newtonian viscosity, but a yield value in shear flow which virtually does not depend on the crosslinking degree of the rubber. The rheological properties of IPP/EPDM blends with component ratios varying from 75/25 to 25/75 was investigated using a capillary rheometer [47]. Viscosity of TPVs increases dramatically with a decrease in shear stress (Figure 11). It is a characteristic of highly structured systems, in particular for polymer compositions containing highly disperse fillers for which the yield stress makes itself evident at sufficiently low shear stress values. The melt viscosity of dynamically cured blends at their flow through a capillary is higher than that of uncured blends. It is influenced by several factors. One of them is a decrease of cured rubber particles deformability in the melt flow due to increase of their rigidity. The viscosity of TPVs based on oil extended EPDM is several times lower than oil-free TPVs, apparently due to weakening of the interaction of dispersed phase at the introduction of a plasticizer in rubber.

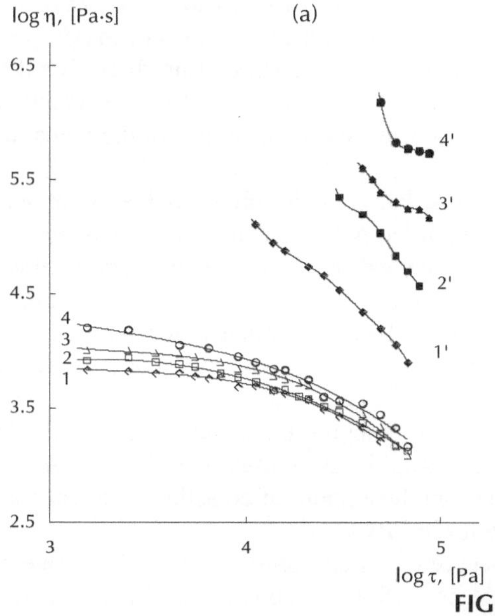

FIGURE 11 *(Continued)*

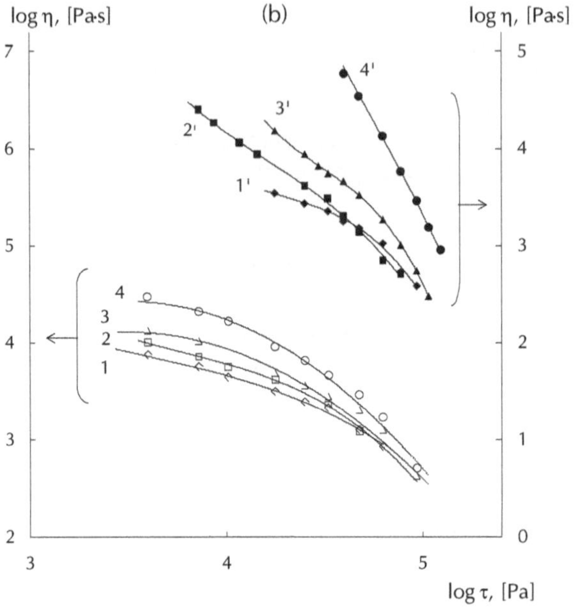

FIGURE 11 Logarithmic plots of viscosity η versus shear stress τ for IPP blended with (a) oil-free and (b) oil-extended EPDM; (*1–4*) unvulcanized blends, (*1'-4'*) dynamically vulcanized blends; IPP/EPDM = (*1, 1'*) 1.33, (*2, 2'*) 1.0, (*3, 3'*) 0.67, and (*4, 4'*) 0.33.

In general, the viscosity of TPVs increases with concentration of vulcanizing system, but it depends on EPDM type. Thus, depending on the blend composition and the amount of the crosslinking agent, the rheological properties of TPVs may be similar to the properties of blends of unvulcanized polymers or the properties of filled polymer composites containing highly disperse structuring filler. The distinction between the rheological behavior unvulcanized blends and TPVs is evidently related to differences in their structures and viscoelastic characteristics of the vulcanized and unvulcanized EPDM copolymers.

A principal difference between the melts of blends of unvulcanized polymers (polymer emulsions) and filled polymer composites (polymer suspensions) is that the flow of the first is accompanied by the irreversible deformation of particles of both phases, which facilitates the flow. In the second case, solid undeformable particles of the filler hamper the flow of the polymer matrix for purely hydrodynamic reasons. As for TPVs, the presence of crosslinks in the particles of the rubber phase precludes their irreversible deformation and impedes their flow compared to the unvulcanized blend. These difficulties should become more distinct when the rigidity of rubber particles increases and their deformability decreases with the increasing crosslink density. It results from an increase in the amount of crosslinking agent and in the concentration of rubber phase particles in TPVs.

A crosslinking degree of oil extended EPDM vulcanizates is much smaller than that of oil-free EPDM ones [88]. Possibly, this is due to a partial dissolution of vulcanization system components in the plasticizer contained in the rubber. It seems that this dissolution eventually promotes the reduction in the viscosity of TPVs based on oil-extended EPDM compared to oil-free TPVs. As is known, the deformability of crosslinked elastomers can be improved upon the introduction of plasticizers [89]. This effect should also contribute to the decrease in the viscosity of oil extended TPVs.

Along with the elasticity of crosslinked rubber particles, the viscosity of TPVs is also affected by a three-dimensional structural framework formed by these particles. This leads to the development of the yield stress. It is this framework of TPVs that increases the viscosity of the material. However, in the flow of TPVs, the breakdown of the framework under the action of high shear stresses should cause a more pronounced viscosity anomaly than that of TPEs.

In order to reduce viscosity and to improve processability of TPV, oils are added to the rubber component. As is known, the deformability of crosslinked elastomers can be improved upon the introduction of plasticizers [90]. This effect should also contribute to the decrease in the viscosity of TPVs based on oil-extended EPDM. Rheological properties of IPP blended with oil-extended EPDM are strongly determined by amount of oil and its distribution between EPDM and IPP phases [47]. It should be noted that a method of incorporation of a paraffinic oil into EPDM rubber strongly determined rheological behavior of TPVs. For example, a mechanical addition of oil into elastomer leads to a significant decrease of melt viscosity of unvulcanized blends, but in fact does not improve that of TPVs [79]. But TPVs contained a great content of EPDM oil extended during synthesis possess a better melt flowability than the corresponding TPVs based on oil-free EPDM.

The difference between the melt shear viscosities of unvulcanized blends and TPVs depends on IPP molecular weight. The melt viscosity of TPVs exceeds IPP melt viscosity at any content of oil-free EPDM but for oil extended TPVs only when oil extended EPDM content is high than 40 wt%. This parameter for both types of TPVs is sensitive to a change of IPP molecular weight.

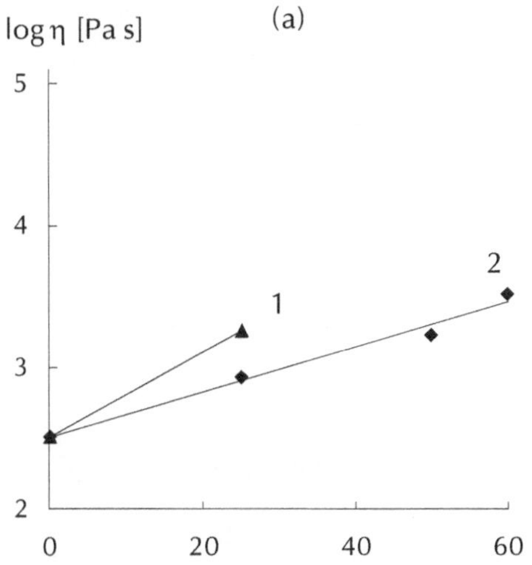

FIGURE 12 *(Continued)*

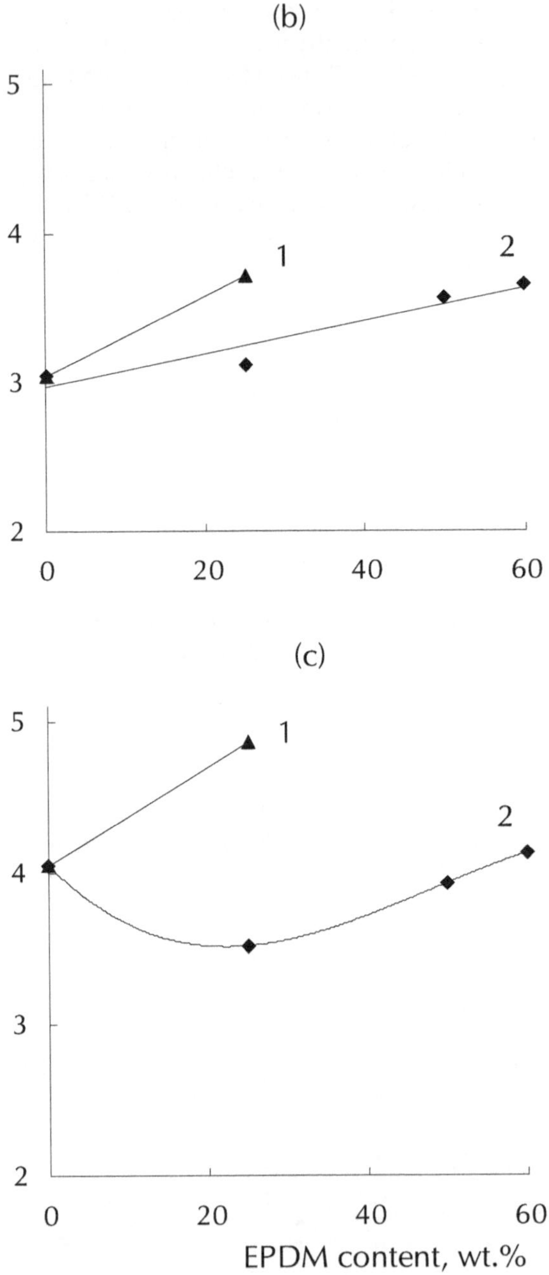

FIGURE 12 Logarithmic plots of melt shear viscosity η versus oil-free EPDM content for TPVs based on (a) PP–1, (b) PP–2, (c) PP–3 and dynamically cured by (*1*) sulfur and (*2*) phenolic resin systems. $\log \tau = 4.22$ [Pa].

The nature of crosslinking agents has a significant influence on melt rheology of dynamically vulcanized blends. TPVs vulcanized by sulfur system lose flowability as the oil-free EPDM content is higher than 25 wt%, but TPVs vulcanized by phenolic resin system flow up to 50 wt% of rubber at whole range of the studied shear stress [78].

It is known, that phenolic resin is less soluble in rubber than sulfur, and their vulcanization mechanisms are dissimilar [22, 23]. Formation of complex heterocyclic chroman compounds is the main principle of phenolic resin vulcanization of non-saturated rubbers [21, 91]. In this case the structure of intermodal chains is more complicated due to their chemical nature in comparison with sulfur vulcanization. Therefore, it's quite possible to assume that a rubber network formed by phenolic resin system has lesser density. Accordingly, the elasticity of cured rubber particles and their crosslinking density are lower than in case of sulfur vulcanization. Apparently, this is the main factor which determines size, shape of rubber particle, and their distribution during dynamic vulcanization.

Another factor which might effect on the melt viscosity of dynamically cured blends is a density of three-dimensional network of the material as whole [23, 47]. Blends containing 50 wt% oil-free EPDM or more have a sufficiently denser three-dimensional structural network formed by the cured rubber, which hinders their flowability at given shear stress range [61, 78].

However, for dynamically cured blends based on both types of EPDM, in case of identical component ratio, viscosity and mechanical parameters of blends cured by sulfur system are higher than those vulcanized by phenolic resin system, probably, due to higher crosslinking density of sulfur vulcanizates.

So, the usage of phenolic resin system as a crosslinking agent and oil extended EPDM allows to produce more soft materials with lower viscosity at higher rubber content in comparison with compositions based on oil-free elastomer.

17.6 TPVs BASED ON STEREOBLOCK ELASTOMERIC PP AND EPDM

The TPVs based on IPP and elastomers have been well investigated and have wide application owing to their properties and chemical stability. They are low-cost and have excellent low temperature impact strength, rigidity, flowability, low shrinkage, high ozone, water, and corrosion resistance. The Shore A hardness of these materials ranges from 65 to 90 units. However, the growing market has necessitated an expanded assortment along with creation of new polymer materials characterized by controlled properties, in particularly, reduced hardness.

Original blend polymer materials with reduced hardness prepared through the method of dynamic vulcanization were created in the Semenov Institute of Chemical Physics of Russian Academy of Sciences [92]. TPVs are based on a stereoblock elastomeric polypropylene (ELPP) synthesized in the presence of asymmetric ansa-metallocenes with different characteristics and a 50–70 wt% oil extended EPDM.

The mechanical parameters and melt flow index (MFI) values of various TPVs are given in Table 3. It is evident that all blends have very good mechanical characteristics:

- Low values of Young's modulus (E)
- Tensile strength (σ_b)
- Residual elongation (ε_{res})
- Hardness

But their elongation at break (ε_b) is enough high. The blend based on 30 wt% ELPP with Mw = 136000 and crystallinity (K) 9% exhibits maximum σ_b and ε_b among TPVs containing 30 wt% ELPP. It was shown that mechanical characteristics of TPVs improve with an increase in the content of the ELPP. Subsequently, the system hardness grows. With a change in the type of ELPP, the elongations at break of TPVs can be varied without any influence on the elastic modulus except that in the case of the blend containing ELPP with lowest M_w and greatest K.

TABLE 3 Properties of TPVs with reduced hardness

TPV composition		ELPP		E, MPa	σ_b, MΠa	ε_b, %	ε_{res}, %	Hardness Shore A	MFI, g/10 min (T = 190°C)	
ELPP	**EPDM**	**M_w**	**K, %**						**2.16 kg**	**5.00 kg**
30	70	70000	36	6.2	2.7	530	6	43	no flow	no flow
30	70	82000	4	2.0	1.1	540	3	37	3.3	–*
40	60	82000	4	2.2	1.3	600	3	42	flow without load	flow without load
30	70	126000	16	2.2	1.5	490	3	40	flow without load	flow without load
30	70	136000	9	2.3	3.0	780	3	38	no flow	0.7
40	60	136000	9	2.6	4.8	960	3	43	1.7	12.9
50	50	136000	9	3.5	6.3	940	3	50	4.2	–*
30	70	168000	0	2.6	1.2	380	6	40	no flow	no flow
30	70	200000	17	2.4	1.5	590	3	38	no flow	no flow
30	70	235000	13	1.9	2.1	750	6	40	no flow	0.5–1.0

Note: *At this load, no measurements were taken.

The rheological characteristics of TPVs, in particular, the MFI, are key parameters determining their potential for processing [54]. As follows from the data in Table 3, none of the blends containing 30 wt% ELPP flow under a 2.16 kg load except for the composition based on ELPP with M_w = 82,000. As the load increases up to 5 kg, high viscosity flow is observed for some other types of TPVs. Note that an increase in the content of ELPP results in blend flowability, but simultaneously, the TPV hardness increases. The above facts are obviously indicative of the existence of the yield strength of TPVs, whose value depends considerably on the blend composition (the type of components, their ratio, the nature of the curatives, and so on) and structure [47].

Thus, the rheological behavior of TPVs with a reduced hardness is considerably influenced by the molecular weight characteristics of ELPP. An increase in the content of the elastomeric PP up to 40–50 wt% makes it possible to improve its rheological characteristics significantly. However, in this case, as mentioned above, the material hardness increases.

On the basis of the analysis of the data, it may be concluded that the use of ELPP as the main component of blends, along with the method of dynamic vulcanization, allows the production of TPVs with high elasticity:

- Low elastic moduli (below 4 MPa)
- High elongations at break (above 500%)

These materials demonstrate reduced hardness (35–50 Shore A units). However, a processing of TPVs with a high EPDM content (70 wt%) requires the addition of low crystallinity ELPPs with M_w = 80000–140000.

The developed TPVS are a new group of elastomer materials that have no analogs in Russia and will allow expansion of the assortment of polymer materials. The complex of properties of the obtained materials offers the prospect of their wide application in various areas:

- Construction
- The food industry
- Medicine
- Electrical engineering
- The automotive industry

17.7 CONCLUSION

The markets for TPVs are expected to grow at rates of over 6% per year, mainly in automotive, construction, healthcare, and consumer goods. The economic and design advantages of TPVs are that their consumption growth rate is nearly twice that of vulcanized rubber. The properties of current commercial TPVs are being constantly improved by modifying the base polymers and altering compounds, as well as by introducing new types of TPV. The TPV developments have focused on improving:

- Resilience and set (particularly at temperature)
- Softness
- Adhesion
- Oil/fluid
- Temperature
- Fire resistance
- Weathering and degradation
- Processing, and so on.

This review is intended to demonstrate of this new technology.

KEYWORDS

- **Dynamic vulcanization**
- **Mechanical properties**
- **Rheological behavior**
- **Structure**
- **Thermoplastic elastomers**
- **Thermoplastic vulcanizates**

REFERENCES

1. Xanthos, M. *Reactive Extrusion, Principles and Practice*, Munich, Hanser Publishers (1992).
2. Prut, E. V. and Zelenetskii A. N. *Russian Chem. Reviews*, **70**, 65 (2001).
3. Karger-Kocsis, J. *Polypropylene: Structure, Blends, and Composites* (Ed.) Chapman & Hall, London, **2** (1995).
4. E. P. Moore and Jr. (Eds.), *Polypropylene Handbook*, Hanser, Munich, Germany (1996).
5. Gessler, A. M. and Haslett, W. H. U.S. 3,037,954 (June 5, 1962).
6. Legge, N. R. *Rubber Chem. Technol.*, **62** (3), 529 (1989).
7. Fischer, W. K. U.S. 3,758,643 (September 11, 1972).
8. Coran, A. Y. and Patel, R. P. U.S. 4,104,210 (August 1, 1978).
9. Coran, A. Y. and Patel, R. P. U.S. 4,130,534 (December 19, 1978).
10. Coran, A. Y. and Patel, R. P. U.S. 4,130,535 (December 19, 1978).
11. Coran, A. Y. and Patel, R. P. *Rubber Chem. Technol.*, **53**, 141 (1980).
12. Coran, A. Y. and Patel, R. P. *Rubber Chem. Technol.*, **53**, 781 (1980).
13. Coran, A. Y. and Patel, R. P. *Rubber Chem. Technol.*, **54**, 91 (1981).
14. Coran, A. Y. and Patel, R. P. *Rubber Chem. Technol.*, **54**, 892 (1981).
15. Abdou-Sabet, S. and Fath, M. A. U.S. 4,311,628 (January 19, 1982).
16. Coran, A. Y. Thermoplastic Elastomers N. R. Legge, G. Holden, and H. G. Schroeder (Eds.), A Comprehensive Review, Hanser, Minich, Ch. 7, p. 133 (1987).
17. Coran, A. Y. Science and Technology of Rubber J. E. Mark, B. Erman, and F. R. Eirich (Eds.), Elsevier Academic Press, Amsterdam, Ch. 7, p. 321 (2005).
18. Moore, C. G., Porter, M., and Saville, B. The Chemistry and Physics of Rubber like Substances L. Bateman (Ed.). Maclaren, London, Ch. 15, p. 449 (1963).
19. Akiba, M. and Hashim, A. S. *Progr. Polym. Sci.*, **22**, 475 (1997).
20. Koning, C., van Duin, M., Pagnoulle, C., and Jerome, R. *Progr. Polym. Sci.*, **23**, 707 (1998).
21. Van Duin, M. and Souphanthong, A. *Rubber Chem. Technol.*, **68**, 717 (1995).
22. Van Duin, M. *Rubber Chem. Technol.*, **73**, 706 (2000).
23. Van Duin, M. *Kautsch. Gummi Kunstst.*, **55**, 150 (2002).
24. Abdou-Sabet, S., Puydak, R. C., and Rader, C. P. *Rubber Chem. Technol.*, **69**, 476 (1996).
25. Ellul, M. D., Tsoub, A. H., and Weiguo, Hu. *Polym.*, **45**, 3351 (2004).
26. Dickens, B. *J. Polym. Sci., Polym. Chem. (Ed.)*, **20**, 1065 (1982).
27. Speier, J. L. and Hook, D. E. US Patent 2,823,218 (February 11, 1958).

28. Smits, V. and Materne, T. The Eight International Conference on Thermoplastic Elastomers. 14-15. Berlin, Germany. Organized by Rapra Technology. Paper 4 (September 2005).
29. Singha, A. *Radiat. Phys. Chem.*, **60**, 453 (2001).
30. Drobny, J. G. *Radiation Technol. for Polym.*, CRC Press, Boca Raton (2003).
31. Bhowmick, A. K. and Vijaybaskar, V. *Rubber Chem. Technol.*, **79**, 402 (2006).
32. Schulze, U., Majumder, P. S., Heinrich, G., Stephan, M., and Gohs, U. *Macromol. Mater. Eng.*, **293**, 692 (2008).
33. Naskar, K., Gohs, U., Wagenknecht, U., and Heinrich, G., *eXPRESS Polym. Letters*, **3**, 677 (2009).
34. Thakur, V., Gohs, U., Wagenknecht, U., and Heinrich, G. *Polym. J.*, **44**, 439 (2012).
35. Abdou-Sabet, S. ACS Meeting. *PMSE Symp.,* **79**, 86 (1998).
36. J. Karger-Kocsis (Ed.) Polypropylene, Kluwer Academic, Dordrecht (1999).
37. Jang, B. Z., Uhlmann, D. R., and Vander Sande, J. B. *J. Appl. Polym. Sci.*, **29**, 3409 (1984).
38. Jang, B. Z., Uhlmann, D. R., and Vander Sande, J. B. *J. Appl. Polym. Sci.*, **29**, 4377 (1984).
39. Coppola, F., Greco, R., Martuscelli, E., Kammer, H. W., and Kummerlowe, C. *Polym.*, **28**, 47 (1987).
40. Hoppner, D. and Wendorff, J. H. *Colloid Polym. Sci.* **268**, 500 (1990).
41. Fortelny, I., Kovar, J., and Sikora, A., et al. *Angew. Macromol. Chem.*, **132**, 111 (1985).
42. Greco, R., Mancarella, C., Martuschelli, E., Ragosta, G., and Jinghua, Y. *Polym.*, **28**, 1929 (1987).
43. Dao, K. C. *J. Appl. Polym. Sci.*, **27**, 4799 (1982).
44. Medintseva, T. I., Erina, N. A., and Prut, E. V. *Polym. Sci.* **50A**, 647 (2008).
45. Binning, G., Rohrer, H., Gerber, Ch., and Weibel, E. *Phys. Rev. Lett.,* **49**, 57 (1982).
46. Yerina, N. and Magonov, S. *Rubber Chem. Technol.* **76**, 846 (2003).
47. Medintseva, T. I., Dreval, V. E., Erina, N. A., and Prut, E. V. *Polym. Sci.,* **45A**, 1213 (2003).
48. Prut, E. V., Erina, N. A., Karger-Kocsis, J., and Medintseva, T. I. *J. Appl. Polym. Sci.*, **109**, 1212 (2008).
49. RU 2 Patent No. 069 217, *Byull. Izobret.*, (32) (1996).
50. Ha, C. S. and Kim, S. C. *J. Appl. Polym. Sci.*, **37**. 317 (1989).
51. Naskar, K. and Noordermeer, J. W. M. *Rubber Chem. Technol.*, **77**, 955 (2004).
52. Medintseva, T. I., Kuptsov, S. A., Sergeev, A. I., and Prut, E. V. *Polym. Sci.,* **48**A, 941 (2006).
53. Prut, E. V., Petrushenko, D. V., Medintseva, T. I., and Shashkin, D. P. *Polym. Sci.*, **52A**, 139 (2010).
54. D. R. Paul and C. B. Bucknall (Eds.). Polymer Blends, Wiley, New York, (2000).
55. Han, P. K. and White, J. L. *Rubber Chem. Technol.,* **68**, 728 (1995).
56. Jayraman, K., Kolli, V. G., Kang, S. Y., Kumar, S., and Ellul, M. D. *J. Appl. Polym. Sci.* **93**, 113 (2004).
57. Sengers, W. G. F., Sengupta, P., Gotsis, A. D., Noordermeer, J. W. M., and Picken, S. J. *Polym.*, **45**, 8881 (2004).
58. Sengers, W. G. F., Wubbenhorst, M., Picken, S. J., and Gotsis, A. D. *Polym.*, **46**, 6391 (2005).
59. Shibryaeva, L. S., Korzh, N. N., Karpova, S. G., Popov, A. A., Tchepel, L. M., Medintseva, T. I., and Prut, E. V. *Russ. J. Chem. Phys.*, **23**, 78 (2004).
60. Medintseva, T. I., Kuptsov, S. A., Erina, N. A., and Prut, E. V. *Polym. Sci.*, **49A**, 42 (2007).
61. Prut, E. V., Medintseva, T. I., Kochanova, O. V., Erina, N. A., Zhorina, L. A., and Kuleznev, V. N. *J. Thermoplast. Comp. Mat.* (In print)

62. Bucknall, C. B. Toughened Plastics. *Appl. Sci.* London, 1977. Khimiya, Leningrad (1981).
63. Wu, S. *Polym.*, **26**, 1855 (1985).
64. Dubnikova, I. L., Oshmyan, V. G., and Gorenberg, A. Ya. *J. Mater. Sci.* **32,** 1613 (1997).
65. Dubnikova, I. L. and Oshmyan, V. G. *Polym. Sci.*, **40A**, 925 (1998).
66. Kompaniets, L. V., Erina, N. A., and Chepel, L. M. et al. *Polym. Sci.*, **39A**, 827 (1997).
67. Jain, A. K., Nagpal, A. K., Singhal, R., and Gupta, N. K. *J. Appl. Polym. Sci.* **78**, 2089 (2000).
68. Gupta, N. K., Jain, A. K., Singhal, R., and Nagpal, A. K. *J. Appl. Polym. Sci.,* **78,** 2104 (2000).
69. Medintseva, T. I., Kompaniets, L. V., Chadaev, D. V., and Prut, E. V. *Polym. Sci.,* **46A**, 279 (2004).
70. Medintseva, T. I., Dreval, V. E., and Prut, E. V. *Macromol. Symp.*, **233**, 78 (2006).
71. Kerner, E. *Proc. Phys. Soc., London.* **69B**, 808 (1956).
72. Uemura, S. and Takayanagi, M. *J. Appl. Polym. Sci.,* **10,** 113 (1966).
73. Davies, W. E. A. *J. Phys. D: Appl. Phys.*, **4,** 1325 (1971).
74. Coran, A. Y. and Patel, R. *J. Appl. Polym. Sci.*, **20,** 3005 (1976).
75. D. Nwabunma and T. Kyu (Eds.). Polymer Blends, Wiley, Hoboken, New Jersey (2008).
76. Prut, E. V. and Medintseva, T. I. 10th European Symposium on Polymer Blends, Dresden, Germany, p. 39 (March 7–10, 2010).
77. Prut, E. V. and Medintseva, T. I. *Gummi Fasern Kunsts.*, **64**, 434 (2011).
78. Prut, E., Medintseva, T., Solomatin, D., and Kuznetsova, O. *Macromol. Symp.,* (321–322), 59 (2012).
79. Medintseva, T., Erina, N., and Prut, E. *Macromol. Symp.*, **176**, 49 (2001).
80. Han, C. D. Rheology and Processing of Polymeric Materials, Oxford University Press, New York (2007).
81. Subhasish, Paul, and Kale, D. D. *J. Appl. Polym. Sci.,* **84**, 665 (2002).
82. Lopez Manchado, M. A., Biagiotti, J., and Kenny, J. M. *J. Appl. Polym. Sci.* **81,** 1 (2001).
83. Danesi, S. and Porter, R. S. *Polym.,* **19**, 448 (1978).
84. D'Orazio, L., Mancarella, C., Martuschelli, E., and Polato, F. *Polym.,* **32**, 1186 (1991).
85. Mighri, F., Huneault, M. A., Ajji, A., Ko, G. H., and Watanabe, F. *J. Appl. Polym. Sci.*, **82**, 2113 (2001).
86. Chung, O., Coran, A. Y., and White, J. L. *SPE ANTEC.*, **43**, 3455 (1997).
87. Goettler, L. A., Richwine, J. R., Wille, F. J. *Rubber Chem. Technol.*, **55,** 1448 (1982).
88. Zhorina, L. A., Kompaniets, L. V., Kanauzova, A. L., and Prut, E. V. *Polymer Science.*, **45A**, 606 (2003).
89. Bartenev, G. M. and Frenkel, S. Ya. Leningrad, Khimiya, *Physics of Polymers* (1990).
90. Dogadkin, B. A., Donzov, A. A., and Shershnev, V. A. Khimiya elastomerov, Moscow, Khimiya (1981).
91. Lattimer, R. P., Kinsey, R. A., and Layer, R. W. *Rubber Chem. Technol.*, **62,** 107 (1989).
92. Prut, E. V., Nedorezova, P. M., Klyamkina, A. N., Medintseva, T. I., Zhorina, L. A., Kuznetsova, O. P., Chapurina, A. V., and Aladyshev, A. M. *Polym. Sci.*, **55A,** 177 (2013).

CHAPTER 18

PROTECTION OF SYNTHETIC POLYMERS FROM BIODEGRADATION

ELENA L. PEKHTASHEVA and GENNADY E. ZAIKOV

CONTENTS

18.1 INTRODUCTION

This chapter includes the information about biodegradation and biodeterioration of plastics and materials on the base of plastics by different bacteria—*Penicillium, Aspergillus, Chaetomium, Fusarium, Alternaria, Trichoderma, Rhizopus,* and genus fungi. The contributors are discussing the problems of biostability for polyethylene, polypropylene, polystyrene, polyvinyl chloride (PVC) (rigid), polyamide, polyethylene terephthalate, and materials on the base of these polymers.

18.2 BIODEGRADATION AND PROTECTION OF PLASTICS

Materials based on polymers, which are in the viscoelastic or rubbery state at the molding stage and solid during operation are called plastics. The polymers are high-molecular compounds consisting of long chain molecules formed by multiple alterations of equal groups of atoms, linked with one another by chemical bonds. Along with polymers, plastics comprise fillers, plasticizers, dyes, stabilizers, and other additives. Synthetic polymer materials are widely used in virtually all fields of science and technology, in industry, construction, agriculture, and so on. They are more resistant to degradation by microorganisms than natural high-molecular compounds. The polymeric backbone of the macromolecule of synthetic high-molecular compounds is too large and strong to be directly consumed by bacteria or fungi. However, in some instances, even they are degraded by microorganisms [1-51].

Insects and rodents can also destroy polymers. The biological damages of plastics by insects and rodents are manifested in direct mechanical destruction of some parts, protective coatings and packing materials.

The destruction of plastic package and further insect and rodent colonization and propagation may happen in components of instruments and mechanisms, difficult to access for a man, which may be a safe ecological niche for animals. For many times, accumulation of animals and their metabolites in heavy duty parts of electric appliances caused short circuits and other failures. Most frequently, damages are caused by *Penicillium, Aspergillus, Chaetomium, Fusarium, Alternaria, Trichoderma, Rhizopus,* and so on genus fungi.

The mold fungi induce chemical (by metabolites) and mechanical (growing, mycelium hypha penetration into the material) damages of the materials. The extracellular enzymes and organic acids are basic chemical products of fungus metabolism, which induce damages of synthetic polymer materials via chemical degradation (hydrolysis, oxidation, and so on) of polymer macromolecules or low-molecular components (fillers, plasticizers, and so on).

Along with purely chemical degradation of polymeric materials, microorganism, and metabolites can cause changes of their physicochemical and electrophysical properties, as a result of swelling and cracking. Deterioration of decorative and other exterior properties of polymeric materials resulting biogrowth (mold spots occurrence, although in this instance performance of the article can be preserved) is possible.

The development of mold fungus culture of the polymer surface promotes water vapor condensation from the atmosphere, water accumulation, and only this circumstance can have an adverse effect on the change of polymer material properties. The chemical interaction of microorganism metabolites with a polymer or additives of synthetic material may result in changes of some physicochemical properties of the material. The properties of fungus unstable materials, such as strength, flexibility, dielectric parameters may decrease, insulating properties may deteriorate, color of stained surfaces may change, and so on.

The bacteria damage plastics less frequently, but their action may be insidious. In some cases, their presence can be hardly seen by the naked eye. The damage may be judged about by occurrence of foreign odor, staining, mucous, and so on. The various species and genus bacteria (*Pseudomonas, Bacillus*, and so on) participate in biodegradation of plastics. The bacteria adapt to synthetic polymers and degrade high-molecular compounds of different chemical structure to low-molecular fractions using various enzymes and metabolites [52-64].

Along with other materials, biodegradation of plastics usually happen simultaneously with their aging under the action of external physical and chemical factors of the environment (UV-radiation, water, temperature drops, and so on). Biodegradation and aging, both these processes complement and aggravate each other.

The degradation of plastics depends not only on the species and genus of impacting microorganisms. The following parameters affect the degradation rate of plastics—chemical structure of the polymer itself, its physical structure, molecular mass, molecular-mass fraction distribution, the presence and composition of plasticizers, fillers, stabilizers, and other additives.

There is a definite dependence between biodemage rate and chemical structure of the polymer. The biostability depends on the chemical origin, molecular mass, and supermolecular structure of the polymer [1-51].

Types of bonds $R - C_3$ and $R - CH_2 - R'$ are inaccessible or hardly accessible for microorganisms. In turn, unsaturated valences $R = CH_2$ and $R = CH - R'$, as well as bonds of polymers most sensitive to hydrolysis, such as acetal, amide, ether, and carbonylic or carboxylic ones, are easily accessible bonds to be broken by microorganisms.

The macromolecule size is the important factor defining polymer resistance to biodegradation. Polymers with high molecular mass are harder degradable by microorganisms, whereas, monomers or oligomers can be easily degraded by microorganisms.

The supermolecular structure of synthetic polymers is equally important factor affecting biodegradation. Compact dislocation of crystalline polymer structure fragments restricts their swelling in water and, simultaneously, prevents enzyme penetration into their structure. Hence, the effect of enzymes on both polymer backbone and biodegradable parts of macromolecule chain are restricted. The presence of defects in macro and microstructure and molecular heterogeneity promotes biodegradation.

18.2.1 BIODAMAGES OF THE MAIN COMPONENTS OF PLASTICS

Plastics are based on polymeric binders, which are polymeric resins. By type of polymeric resins, plastics are divided into thermoreactive and thermoplastic (with respect to the method of their hardening during production of the material), polyethylene, PVC, polyamide, and so on (with respect to chemical structure of the polymer) [52].

The carbochain polymers with the macromolecule backbone composed exclusively of carbon atoms (polyethylene, polypropylene, PVC, and so on) and heterochain polymers, which backbone, along with carbon, comprises oxygen, nitrogen, and other atoms (polyamides, polyurethanes, and so on), are distinguished.

The polymeric resins demonstrate different biostability, depending on chemical structure of macromolecule, polymeric chain length, side branches, and so on The general rule is increasing resistance of polymers to microbiological damage with the macromolecule chain length. All other conditions being equal, linear carbochain polymers are less bioresistant than branched or heterochain ones.

The influence of chemical structure on biostability of polymers was determined on the example of polyurethanes. For this purpose, over 109 specimens containing no admixtures, on which microscopic fungi might develop, were synthesized. It is found that polyurethanes with ether bond were damaged stronger than these with ester bond. The presence of ether bond simplifies degradation and consumption of polymer. It is also found that compounds are also decayed, which have a long carbon chain between ether bonds. The presence of three neighboring methyl groups also increases damaging of polyurethanes by microscopic fungi.

It is found that, for example, the microbiological stability of polymeric resins directly depends on molecular mass of the polymer and decreases in the presence of low-molecular fragments in the material. The same effect is observed in case of aging caused by light and heat.

Amorphous-crystalline phase transition of the polymer increases its biostability. Among polymeric resins with increased resistance to damage by mold fungi, the following should be mentioned—polyethylene, polypropylene, polystyrene, PVC (rigid), polyamide, and polyethylene terephthalate. Less fungus resistant polymers are polyvinyl acetate, polyvinyl alcohol, chlorosulfonated polyethylene, and so on.

Plasticizers are important components of plastics. They are mostly dicarboxylic and polycarboxylic aliphatic and aromatic acid esters. The plasticizer content may reach 30–50% of the plastic mass, therefore, to a great extent, biostability of the material depends on its biostability.

A dependence between biostability of organic plasticizers and the length and spatial configuration of the backbone is found—orthophthalic acid esters are most stable, whereas, para-, metha-, iso- terephthalic acid derivatives possess the lowest stability.

Plasticizers of ether type are hydrolyzed to short-chain bases and acids, which are utilized by microorganisms. This process may proceed at relatively low air humidity (50%) and temperature (+ 20°C). The consuming plasticizers and fillers as the nutrition source, microorganisms intensify aging of plastics.

The comparison of resistance to degradation by mold fungi of the most widespread plasticizers (phthalic and adipic acid ethers) indicates the highest stability of aromatic phthalic acid ethers rather than adipic acid ethers–aliphatic dicarboxylic acid. Aliphatic sebacinic acid ethers have low fungus resistance.

Fillers are important components of plastics. They generally represent inert solids injected to polymeric materials for the purpose of regulating mechanical and other properties. Injection of the filler also reduces the cost of materials and articles from plastics, increases their strength, electrical, and other properties.

The organic fillers (wood powder, cotton fibers, paper, and so on) are nutritive substrates for microorganisms and reduce fungal resistance of polymeric composites, whereas, inorganic fillers (asbestos, glass fiber, quartz powder, and kaolin) increase biostability.

However, if the binder impregnates organic filler well and is highly resistant to molds, fungal resistance of the polymeric material can be quite high. It is important to achieve maximum impregnation of the filler by resin in the technological process. Sometimes, this is achieved under vacuum. The mechanical processing of finished plastic products with non-bioresistant fillers without appropriate protection of processed areas by stable to fungi lacquer coating is unallowable. The dynamics of plastics damage depends on both chemical composition and physical structure.

Fungus mycelium can propagate in extremely thin cracks and pores, formed at the interphase and surface boundaries in the material. For instance, ethylene-vinyl-acetate copolymers were damaged with fungus hyphae development at the boundary of the polymer and starch grains. Meanwhile, the damage increased with vinyl acetate content in the material and was stimulated by adding starch, as the filler.

Polyethylene is a carbochain thermoplastic polymer, one of the most widely applied polymers of the polyolefin sequence. It is used for manufacture of films, protective coatings, insulating articles, containers, package materials, and so on. It has high dielectric properties and chemical stability.

The microbiological resistance of polyethylene is characterized by a property, general for all alkanes–the higher molecular mass, the higher biostability of the material. Polyethylene is usually damages from the surface, most degradable being the polymer with the molecular mass below 25000. The biostability of high density polyethylene is higher than biostability of low density polyethylene.

When exposed to soil in moderate climate, polyethylene products are assumed 8 year resistant to microbiological damages. In tropics, the effective life decreases.

Polyethylene surface overgrown by mold becomes rough and covered with mosaic black-brown spots.

Polystyrene is a carbochain thermoplastic polymer obtained by styrene polymerization in the presence of various initiators. It is water resistant and has high dielectric properties. Radio and electric equipment parts, insulating films, foamed plastic, and so on are produced from polystyrene. It is resistant to microorganisms—8 month exposure to a mixture of molds causes not damage.

Polyamides are heterochain thermoplastic polymers with - CO - NH – group in the backbone. They possess increased mechanical properties, good dielectric properties, however, they are low resistant to light and oxidant impacts. They are used in production of films, fibers, and various products.

Note that the in-depth study of biostable polyamide materials was preceded by quite detailed investigation of ε-caprolactam, which is the primary material for ε-polycaproamide synthesis, the latter being the most widespread kinds of fibrous polyamide materials.

In this case, the study of caprolactam degradation processes caused by microorganisms should be considered as a model for studies of microbial degradation of polycaproamide materials. It is found [65-75] that during biochemical sewage treatment caprolactam is completely degraded by microorganisms of active sewage silt.

Among 30 strains, only 4 species from *Eubacteriacede* family and *Bacillus, Bacterium* and *Pseudomonas* genus were found capable of propagating on a mineral medium containing caprolactam. Japanese scientists have extracted pure bacterium cultures of *Pseudomonas, Corynebacterium* and *Achromobacter* genus able to consume caprolactam as the unique source of carbon and nitrogen. It is author's opinion [65-75] that *Micrococcus* varians strain extracted from the soil is able to destroy caprolactam.

The complete hydrolysis of ε-caprolactam in concentration as high as 3 g/l by spore-forming bacteria from *Bacillus subtilis- mesentericus* group at their continuous cultivation proceeds within 24 hr.

Note also that caprolactam consumption by bacteria is clearly adaptive—when grown on nutritious media, bacteria quickly become unable to oxidize caprolactam.

Some microorganisms use caprolactam as the source of carbon and nitrogen, others require an additional source of carbon and use caprolactam as the carbon source exclusively.

A scheme of ε-caprolactam metabolism due to bacterial degradation is suggested [65-75]. This scheme includes ε-caprolactam hydrolysis to ε-aminocaproic acid by lactamase enzyme with further synthesis of L-ketoglutaric acid, which then is transaminated by transaminase enzyme to glutamic acid. Meanwhile, adipic acid half aldehyde is synthesized, which is then converted by dehydrogenase to adipic acid.

It is found that the general scheme of ε-caprolactam metabolism caused by bacteria is analogous to the one previously suggested by Japanese scientists. The first stage of caprolactam degradation is hydrolysis by amide bond to ε-aminocaproic acid. Similar to caprolactam, aminocaproic acid is far not normal cell metabolite. Nevertheless, its further transformation proceeds by way common for natural amino acids. Significant ammonia quantities in the culture fluid were detected. The ammonia excretion is associated with deamination of ε-aminocaproic acid. As a result of ε-caprolactam decay, volatile acetic and butyric acids and neutral products (aldehydes, ketones, and alcohols) are formed.

In the *Achromobacter guttatus* culture medium degrading caprolactam, glutamic acid, lysine, threonine, and methionine are identified. The laboratory examination for microbiological stability of film materials (polyamide-6, polyamide-6,6) on optimal nutritive medium indicates that these materials are bio-unstable. All samples demonstrate surface and through destruction of polymers. All samples demonstrate surface and through degradation of polymers. Polyamide-12 demonstrated high resistance to molds. Impact of some strains of fungi on polyamide films (PK-4 film) reduces strength to 80%.

The PVC is a carbochain thermoplastic polymer, one of the most widely applied to manufacture of rigid and plasticized polymers for pipe, sheet, film, shaped profile, fiber, protective coating, and so on production. This resin is combined well with many plasticizers. Flexible PVC is the base for artificial leather production, widely used in aircraft and automobile industry, agricultural engineering, personal protection equipment production, production of shoes, light industry products, and so on. The biostability of plasticized PVC significantly depends on biostability of applied plasticizers, stabilizers, and so on. The plasticized PVC film loses strength, resulting microorganism impact, which is also accompanied by mass reduction and rigidity increase. Different film staining (red, orange, and pink) is observed and light transmittance is reduced.

A large body of research point at connection of observable changes within sufficient biostability of plasticizers and the rate of their migration from the volume of flexible PVC. The rigid PVC is more biostable in relation to bacteria and molds. After 8-year tests in the soil, pipes from rigid PVC have not considerably decreased their physicomechanical properties, although the material itself is able to support fungus growth.

Polycarbonates are heterochain thermoplastic polymers synthesized by polycondensation of carbonic acid ethers and dioxy compounds. They possess increased mechanical properties, water and weather resistance, are good dielectrics and physiologically innocent. They are used in production of electronics, watches, refrigerators, films, dishware, and so on.

The study of microbiological resistance of one of polycarbonates has shown that molds are able to propagate on the material surface at 100% air humidity and temperature of + 30°C. Polycarbonate film is also not exactly fungus-resistant. This should be considered at its application in electrical appliances, for medical purposes and packaging. These are resistant to the impact of bacteria.

Polyurethanes are heterochain thermoplastic polymers with urethane group in the macromolecule. It is characterized by multiple applications in technology. Flexible and rigid foamed plastics, elastomers, fibers, films, adhesives, lacquers, solid, and flexible covers are produced from polyurethanes.

As compared with polyolefins, polyurethane-based synthetic polymeric materials possess lower resistance to fungal action.

The comparative study of microbiological stability of polyurethane polymers synthesized from polyethers and polyesters shows that polymers from polyesters are more damageable by mold fungi than polymers from polyethers.

Polyurethane protective coats of aviation aluminum fuel tanks and metal structures are highly damaged by fungi down to exfoliation and full degradation.

Along with the structure and composition of plastics, their biostability is significantly affected by environmental conditions—high air humidity, increased temperature, difference in day and night temperatures. The water vapor condensation and moisture accumulation on the surface material promote propagation of microorganisms. Some plastics, already affected just by high water content, change their properties. This effect is added by chemical corrosion induced by microorganism metabolites, which results in deterioration of properties and quality of the products.

The microorganisms of various groups deteriorate mechanical, hygienic, and esthetic properties of plastics. The pigments from microorganisms stain plastics—grey, green, violet or pink spots occur, bleaching or pitting of the surface is also possible.

Sometimes, damages happen on the surface and are only manifested in mycelium growth, which can then be removed and, consequently, will cause no considerable effect on entire material or article performance. In other cases, biodamages may be deeper, when along with the appearance, physicochemical, physicomechanical, and other properties of materials change, for example, the change of viscosity, strength, hardness, insulating, and other properties is observed.

In connection with plastic articles use under conditions suggesting high impact of microorganisms, the problem of polymeric material protection is extremely urgent. At present, over 3000 compounds possessing biocidal properties are described. However, no antiseptics complying with the requirement to them are found yet.

Despite the fact that the majority of known biocides are probed on plastics, they have no practical application. This is due to specificity of plastics manufacture. When manufactured and processed, plastics are subject to impact of high temperatures allowable just for few biocides.

Along with thermal stability, biocides shall be chemically stable: this means that biocides must not interact with other components of plastics and, at the same time, be compatible with the plastic (the polymer and all its components).

One of the requirements imposed on biocides for plastics is a wide spectrum of antimicrobial action at low concentrations, because high biocide concentrations may decrease physicomechanical and electrical properties of articles from plastics.

Moreover, biocide must be harmless and shall not be washed off during operation. Since a number of specific requirements are imposed on the articles from plastics, for example, electric resistance and dielectric properties, biocides must be non-polar compounds.

All these requirements limit the number of biocides used for plastics protection. Among biocides for plastics, during the years the following substances are used—salicyl anilide, copper 8-oxyquinolate, 2-oxydiphenyl, 4-nitrophenol, sodium pentachlorophenolate, and so on The following biocides are also known—trilan (4,5-trichlorobenzoxazolinone), cymid (dichloromaleic cyclohexylimide), and some organoarsenic and organotin compounds.

Trilan, cymid, and epoxar (an arsenic-containing compound) have shown good results as biocides for fungus-resistant films from PVC and artificial leather for technical purposes used in articles, which are to be delivered to tropics. For example, addition of 1–2% of cymid to artivicial leather films made of PVC provides long-term preservation of strength, good appearance, and other properties of the material under the most severe conditions. The strength of material not protected by trilan, under the same conditions, is reduced by 15–30% within 3 months.

The advantage of epoxar is that along with biocidal properties, it is able to improve light and thermal stability of polymeric materials, that is the universal stabilizer.

Sometimes, fungicidal treated polymeric materials allow a trick for complex technical issues. The films containing biocides were successfully used for lining bottoms of channels, pools, and other hydraulic works. Such protective coatings were not overgrown by microorganisms and algae, had increased lifetime, prevented water leakage to soil, and even prevented, to a definite extent, water against "contamination" by microorganisms and algae.

The use of biocides in plastics can be intended for both preventing them against biodegradation and hygienic purposes. For example, in some medical institution biocidal plastics have proved themselves in production of plastic handles, water-closet seats, some parts of medical equipment, film articles, antiseptic linings for bassinets, and so on.

The hygienic purpose articles made from polyethylene, shock-resistant polystyrene and other plastics fungicidal treated by hexachlorophene (1–2%) retain antiseptic properties after a year of operation in hospitals and prevent contagion, whereas, under the same conditions, common, biocide unprotected materials can be the infection foci.

18.3 BIODEGRADABLE POLYMERIC MATERIALS

The waste polymeric materials, extremely slow degradable under natural conditions, are considerable source of environmental pollution. The single-use plastic package, films, and packing materials are of special hazard. That is why, along with solving the problem of increasing plastics durability, special types of polymers with regulated lifetime are designed and manufactured. The adverse feature of these polymers is their ability to retain consumer properties during the entire operation period and only then be subject to biological and physicochemical transformations.

Owing to chemical structure and high molecular mass, the most of volume plastics types (polyethylene, polypropylene, polystyrene, and PVC) degrade very slowly. The soil microorganisms might degrade these polymers, if their long macromolecules would split into shorter fragments [1-51].

There are several approaches to creation of biodegradable polymers:

- The creation of photodegradable polymers, which are able to degrade under natural conditions to low-molecular fractions due to special additives present in them. Further on, these fractions are degraded by soil microflora,
- The designing of polymeric composites containing, along with high-molecular backbone, organic fillers, which are the nutrition for microorganisms (starch, cellulose, pectin, amylase, and so on),
- The creation of polymers with the structure similar to natural polymers,
- The polymers can be synthesized by biotechnology methods.

One of directions in creation of photodegradable polymers is inclusion of chromophores into polymeric backbone, which provide UV-light absorption by polymers that induces their degradation.

At present, studies associated with development of composites containing, along with high-molecular backbone, organic fillers, which are the nutritive medium for microorganisms are of interest. Meanwhile, beside material degradation associated with filler destruction by bacteria, an effect of additional degradation stipulated by the features of filled polymer structure is observed. It is known that the filler may be accumulated in lesser ordered polymer zones. Moreover, macromolecule packing density in the interface layer of the "polymer-filler" system is approximately twice lower, than in the rest volume of disordered phase of the polymer. Therefore, when the filler is destroyed by bacteria, microorganism access to less stable in relation to biodegradation of a part of polymer becomes simpler [52].

CURIOUS FACTS

American Warner–Lambert Company has designed the first polymeric material Novolon, which only consists of starch and water and is completely biodegradable. This polymer can be processed by traditional methods, and mechanically, is intermediate between polystyrene and polyethylene.

Archer Daniels Midland Company (USA) has designed a polyethylene-based concentrate of Polyclean trademark used for production of biodegradable films. The concentrate contains 40% of starch and an oxidizing additive: starch quantity in the final product is 5–6%. The oxidizing component acts as the catalyst both in the light and in darkness. The starch degradation makes access of microorganisms and oxygen to polymer surface simpler, that is a specific synergic effect is observed.

ICI Americas Inc. Company produces a heat-sensitive plastic, which is naturally degraded. It has properties similar to polypropylene. This is linear polyether (poly-3-hydroxybutyram-3-hydroxyvalerate), which is produced by sugar fermentation with the help of Alcfligenes eutrophys bacteria. This material is degraded by microorganisms, inhabiting in the soil, sewage and at water body beds.

Of great interest are studies aimed at creation of materials, which are not only biodegradable, but are also produced from recoverable biological resource, given that traditional raw material sources for the polymer synthesis are limited. These polymers are mostly based on starch, products of bacterial fermentation of sugar, heat-sensitive plastics based on animal starch with added petrochemical products.

Polymers may also be turned biodegradable by breeding specific microorganism strains able to degrade polymers. This direction is only successful in relation to polyvinyl alcohol yet. Japanese scientists have extracted *Pseudomonas* bacteria from the soil, which produce an enzyme cleaving polyvinyl alcohol. After degradation, polymer fragments are completely consumed by these bacteria. Therefore, *Pseudomonas* bacteria are added to activate sewage sludge for more complete cleaning from this polymer.

It is found [68] that regularly alternating co-polyamides containing L-amino caproic acid as one of the components are easily biodegradable, and glycine and L-amino caproic acid co-polyamide, degraded by bacteria and fungi in several weeks have been synthesized.

Polyamide synthesized from benzyl malonic acid and hexamethylene diamine is highly biodegradable. Such polymers [68] may be used as hypoallergic surgical threads, capsules for seedlings at recovery of forests, nonwoven materials and films used at mature tree moving as a single-use package. Other fields of application are also suggested—fishery, protection against floods, and so on.

KEYWORDS

- **Chemical degradation**
- **Dielectric parameters**
- **Plasticizers**
- **Rodent colonization**
- **Supermolecular structure**

REFERENCES

1. Rotmistrov, M. N., Gvozdyak, P. I., and Stavskaya, S. S. *Microbial degradation of synthetic organic compounds* (in Russian), Naukova Dumka Publishing House, Kiev, p. 224 (1975).
2. Blagnik, R. and Newly, B. *Microbiological corrosion* (in Russsian), Khimiya (Chemistry) Publishing House, Leningrad, p. 222 (1965).
3. Yermilov, I. A. Theoretical and practical bases of microbial degradation of chemical fibers (in Russian), Nauka (Science) Publishing House, Moscow, p. 248 (1991).
4. Wolfe, L. A. and Meos, A. I. *Fibers for special applications* (in Russian), Khimiya (Chemistry) Publishing House, Moscow, p. 224 (1971).
5. Bugorkov, V. S., Ageev, T. A., and Galperin, V. M. *The basic directions for creating photopolymer materials and biodestruktiruemyh* (in Russian), *Plastics Journal*, N 9, pp. 48–51 (1991).
6. Legonkova, O. A. and Sukharev, L. A. *Polymer biological resistance to biodegradation* (in Russian), RadioSoft Publishing House, Moscow, p. 272 (2004).
7. Perepelkin, K. E. The present and future man-made fibers (in Russian), LegPromBuzines Publishing House, Moscow, № 8 (22), № 9 (23), № 10 (24), № 11 (25), № 12 (26) (2000).
8. Emanuel, N. M. and Buchachenko, A. L. "*Chemical physics of degradation and stabilization of polymers*", VSP International Science Publ., Utrecht, p. 354 (1982).
9. Bochkov, A. F. and Zaikov, G. E. "*Chemistry of the glycosidic bonds. Formation and cleavage*", Pergamon Press, Oxford, p. 210 (1979).
10. Razumovskii, S. D. and Zaikov, G. E. "*Ozone and its reactions with organic compounds*", Elsevier, Amsterdam, p. 404 (1984).
11. Emanuel, N. M., Zaikov, G. E., and Maizus, Z. K. "*Oxidation of organic compounds. Medium effects in radical reactions*", Pergamon Press, Oxford, p. 628 (1984).
12. Afanasiev, V. A. and Zaikov, G. E. "*In the realm of catalysis*", Mir Publishers, Moscow, p. 220 (1979).
13. Moiseev, Yu. V. and Zaikov, G. E. "*Chemical resistance of polymers in reactive media*", Plenum Press, New York, p. 586 (1987).
14. Zaikov, G. E., Iordanskii, A. L., and Markin, V. S. "*Diffusion of electrolytes in polymers*", VNU Science Press, Utrecht, p. 328 (1988).
15. Minsker, K. S., Kolesov, S. V., and Zaikov, G. E. "*Degradation and stabilization of polymers on the base of vinylchloride*", Pergamon Press, Oxford, p. 526 (1988).
16. Aseeva, R. M. and Zaikov, G. E. "*Combustion of polymer materials*", Karl Hanser Verlag, Munchen, p. 389 (1986).
17. Popov, A. A., Rapoport, N. A., and Zaikov, G. E. "*Oxidation of stressed polymers*", Gordon & Breach, New York, p. 336.

18. Bochkov, A. F., Zaikov, G. E., and Afanasiev, V. A. "*Carbohydrates*", VSP Science Press VB, Zeist-Utrecht, p. 154 (1991).
19. Afanasiev, V. A. and Zaikov, G. E. "*Physical methods in chemistry*", Nova Science Publ., New York, p. 180 (1992).
20. Todorov, I. N., Zaikov, G. E., and Degterev, I A. "*Bioactive compounds: biotransformation and biological action*", Nova Science Publ., New York, p. 292 (1993).
21. Roubajlo, V. L., Maslov, S. A., and Zaikov, G. E. "*Liquid phase oxidation of unsaturated compounds*", Nova Science Publ., New York, p. 294 (1993).
22. Iordanskii, A. L., Rudakova, T. E., and Zaikov, G. E. "*Interaction of polymers with bioactive and corrosive media*", VSP International Publ., Utrecht, p. 298 (1994).
23. "*Degradation and stabilization of polymers. Theory and practice*", G. E. Zaikov (Ed.), Nova Science Publ., New York, p. 238 (1995).
24. Polishchuk, A. Ya. and Zaikov, G. E. "*Multicomponent transport in polymer systems*", Gordon & Breach, New York, p. 231 (1996).
25. Minsker, K. S. and Berlin, A. A. "*Fast reaction processes*", Gordon & Breach, New York, p. 364 (1996).
26. Davydov, E. Ya., Vorotnikov, A. P., Pariyskii, G. B., and Zaikov, G. E. "*Kinetic pecularities of solid phase reactions*", John Willey & Sons, Chichester (UK), p. 150 (1998).
27. Aneli, J. N., Khananashvili, L. M., and Zaikov, G. E. "*Structuring and conductivity of polymer composites*", Nova Science Publ., New York, p. 326 (1998).
28. Gumargalieva, K. Z. and Zaikov, G. E. "*Biodegradation and biodeterioration of polymers. Kinetical aspects*", Nova Science Publ., New York, p. 210 (1998).
29. Rakovsky, S. K. and Zaikov, G. E. "*Kinetics and mechanism of ozone reactions with organic and polymeric compounds in liquid phase*", Nova Science Publ., New York, p. 345 (1998).
30. Lomakin, S. M. and Zaikov, G. E. "*Ecological aspects of polymer flame retardancy*", VSP International Publ., Utrecht, p. 158 (1999).
31. Minsker, K. S. and Zaikov, G. E. "*Chemistry of chlorine-containing polymers: synthesis, degradation, stabilization*", Nova Science Publ., New York, p. 198 (2000).
32. Jimenez, A. and Zaikov, G. E. "*Polymer analysis and degradation*", Nova Science Publ., New York, p. 287 (2000).
33. Kozlov, G. V. and Zaikov, G. E. "*Fractal analysis of polymers*", Nova Science Publ., New York, p. 244 (2001).
34. Zaikov, G. E., Buchachenko, A. L., and Ivanov, V. B. "*Aging of polymers, polymer blends and polymer composites*", Nova Science Publ., New York, **1**, p. 258 (2002).
35. Zaikov, G. E., Buchachenko, A. L., and Ivanov, V. B. "*Aging of polymers, polymer blends and polymer composites*", Nova Science Publ., New York, **2**, p. 253 (2002).
36. Zaikov, G. E., Buchachenko, A. L., and Ivanov, V. B. "*Polymer aging at the cutting adge*", Nova Science Publ., New York, p. 176 (2002).
37. Semenov, S. A., Gumargalieva, K. Z., and Zaikov, G. E. "*Biodegradation and durability of materials under the effect of microorganisms*", VSP International Science Publ., Utrecht, p. 199 (2003).
38. Zaikov, G. E. "*Burning, destruction and stabilization of polymers*" (in Russian), Scientific Basis and Technology Publishing House, St. Petersburg, p. 422 (2008).
39. Mikitaev, A. K., Kozlov, G. V., and Zaikov, G. E. "*Polymer nanocomposites. A variety of structural forms and applications* (in Russian)", Nauka (Science) Publishing House, Moscoe, p. 279 (2009).
40. Mikitaev, A. K., Kozlov, G. V., Zaikov, G. E. "*Fractal analysis of the gas transport in polymers*" (in Russian), Nauka (Science) Publishing House, Moscow, p. 199 (2009).

41. Rakovsky, S. K. and Zaikov, G. E. "*Interaction of ozone with chemical compounds. New frontiers*", Rapra Technology, London, p. 486 (2009).
42. Naumova, R. P. *Conversion study of caprolactam by bacteria* (in Russian), N. A. Plate (Ed.), Kazan State Technological University Publishing House, Kazan, pp.67–69 (1964).
43. Naumova, R. P., Zakharova, N. G., and Zakharova, S. J. Degradation of synthetic lactam and ε-amino acids by microorganisms, cleaning industrial waste water. Microbiological methods of dealing with environmental pollution (in Russian), N. A. Plate (Ed.), Nauka (Science) Publishing House (Pushchino Branch), Pushchino-town,–pp.70–72 (1979).
44. Kato, K. and Fukumura, T. Bacterial breakdown of ε-caprolactam, *Chem. and Industr. journal*, N 23.pp. 1146 (1962).
45. Kinoshita, S., Kobayashi, E., and Okada, H. Degradation of ε-caprolactam by Achromobacter guttatus KF 71, *Journal of Fermentation Technology*, **51**, pp. 719–725 (1973).
46. Uemura, T. Autolytic enzyme assocated with cell walls of Bacillus subtilis, *Journal of Biol. Chem.*, **241**(15), pp.3462–3467 (1966).
47. Roy, A. Destruction of caprolactam and hexamethylenediamine bacterial groups Bac.subtilis-mesentericus, Thesis, Institute of Colloid and Water Chemistry, USSR Academy of Sciences, Kiev, p. 162 (1975).(in Russian)
48. Roy, A. The destruction of caprolactam microorganisms in continuous cultivation. Scientific fundamentals of water treatment, N. A. Plate (Ed.), Naukova Dumka (Publishing House of Ukrain Academy of Sciences), Kiev, (2), p.152–156 (1976) (in Russian).
49. Shevtsova, I. I. The destruction of microorganisms caprolactam (in Russian), Bulletin of Kiev State University, Division of Biology, Kiev, (11), p.149–152 (1969).
50. Fucumura, T. Bacterial Breakdown of ε-caprolactam and its cyclic oligomers, Plant and cell. *Physiology journal*, **7,** N1. pp. 93–104 (1966).
51. Naumova, R. P. and Belov, I. S. The transformation of aminocaproic acid in bacterial destruction of caprolactam (in Russian), *Biochemistry journal*, № 33, p. 946 (1968).
52. Biodamage: Textbook. Manual for biological faculties of universities (in Russian), V. D. Ilyicheva (Ed.), High School Publishing House, Moscow, p. 352 (1987).
53. Pekhtasheva, E. L. Topical issues of biological damage (in Russian), Nauka (Science) Publishing House, Moscow, p. 265 (1983).
54. Neverov, A. N. Actual problems of biological damage and protection of materials, components and structures, Science, Scientific Council for biological damage of the USSR Publishing House, Moscow, p. 256 (1989).(in Rus.)
55. Neverov, A. N. *Biodamage and methods for assessing the biological stability of materials*(in Russian), Scientific Council for biological damage of the USSR Publishing House, Moscow, pp. 140 (1988).
56. Neverov, A. N. *Biodamage, methods of protection* (in Russian), The Scientific Council for biological damage of the USSR Publishing House, Poltava, p. 182 (1985).
57. Kanev, I. G. *Biological damage of industrial materials* (in Russian), Nauka (Science) Publishing House, Leningrad, p. 232 (1984).
58. *Biological problems of environmental materials* (in Russian), A. N. Neverov (Ed.), Scientific Council on the biological damage of Sciences Publishing House, Penza, p. 108 (1995).
59. Methods for determining the biological stability of materials, N. A. Plate (Ed.), Science Council of biological damage of the USSR Publishing House, Moscow, p. 202 (1979).
60. *Microorganisms and lower plants - the destroyers of materials and products* (in Russian), N. A. Plate (Ed.), Nauka (Science) Publishing House, Moscow, p. 225 (1979).
61. *First Conference on Biological Damage*, N. A. Plate (Ed.), Nauka (Science) Publishing House, Moscow, p. 226 (1978).

62. Problems of biological damage to materials. Environmental aspects, A. N. Neverov (Ed.), Scientific Council of biological damage of the USSR Publishing House, Moscow, p. 124 (1988). (in Rus.)
63. *Environmental problems of biodegradation of industrial, construction materials and industrial wastes* (in Russian), A. N. Neverov (Ed.), Scientific Council on Problems of biological damage Publishing House, Penza, p. 192 (2000).
64. Fourth Conference on Biological Damage (in Russian), N. A. Plate (Ed.), Scientific Council of biological damage of the USSR Publishing House, Nizhnii Novgorod-city, p. 99 (1991).
65. Andreyuk, E. I., Bilan, V., and Koval, E. Z. *Microbial corrosion and its agents* (in Russian), Naukova Dumka, (Publishing House of Ukrain Academy of Sciences), Kiev, p. 258 (1980).
66. Blagnik, R. and Newly, B. *Microbiological corrosion* (in Russian), Khimiya (Chemistry) Publishing House, Moscow, p. 222 (1965).
67. Shkut, Y. T. and Kostylev, A. F. *Histology and microbiology skins* (in Russian), Light Industry Publishing House, Moscow, p. 150 (1980).
68. Martha, V., Kutukova, K. S., and Moszkowski, Sh. D. *Microbiology skins* (in Russian), Gizlegprom Publishing House, Leningrad, p. 318 (1936).

CHAPTER 19

FIRE AND POLYMERS

S. M. LOMAKIN, P. A. SAKHAROV, and G. E. ZAIKOV

CONTENTS

19.1 INTRODUCTION

In this chapter we focused on the development of environmentally friendly intumescent/char forming coatings for wood and polymeric materials. For this purpose we have developed the method of oxidation of raw materials—polysaccharides, seeds, and lignin by oxygen in the presence of catalyst. The main products of oxidation of such substrates are the salts of polyoxy and polyphenoxy acids. The efficiency of fireproofing action of novel surface protecting intumescent coating for wood and polymeric materials based on modified renewable raw materials was studied. The flammability tests of wood and polymers samples treated by oxidized raw materials confirm their high-performance fire protection.

Today, as well as it is a lot of centuries back, prevention and extinguishing of fires is one of the global problems stands in front of humanity. In the world annually thousands of the new reagents appears directed on the decision of this problem. But as we can see, the problems such as, for example, extinguishing of forest fires or creation of low inflammable polymeric materials is far from the solution.

The combustion of natural and polymer materials, like the combustion process of any other fuel material, is a combination of complex physical and chemical processes, which include the transformation of initial products. This whole conversion process may be divided on stages, with specific physical and chemical processes occurring in each of these stages. In contrast to the combustion of gases, the combustion process of condensed substances has a multi-phase character. Each stage of the initial transformation of a substance correlates with a corresponding value (combustion wave) with specific physical and chemical properties (state of aggregation, temperature range, concentration of the reacting substances, kinetic parameters of the reaction and so on).

The char-forming materials often swell and intumesce during their degradation (combustion), and the flame-retardant approach is to promote the formation of such intumescent char. The study of new polymer flame retardants has been directed at finding ways to increase the tendency of plastics to char when they are burned [1, 2].

There is a strong correlation between char yield and fire resistance. This follows because char is formed at the expense of combustible gases and because the presence of a char inhibits further flame spread by acting as a thermal barrier around the unburned material. The tendency of a polymer to char can be increased with chemical additives and by altering its molecular structure. It has been studied polymeric additives (polyvinyl alcohol systems) which promote the formation of char [3-5].

These polymeric additives usually produce a highly conjugated system - aromatic structures which char during thermal degradation and/or transform into cross-linking agents at high temperatures.

The char, which is formed in the process of thermal degradation, can play several roles in fire retardancy. The formation of char in and of itself has a significant effect on the degradation because char formation must occur at the expense of other reactions that may form volatiles, thus, char formation may limit the amount of fuel available. An example of this occurs in cellulose, which may degrade either by a series of dehydration reactions that yield water, carbon dioxide, and char, or by a process in which levoglucosan is produced, which eventually leads to the formation of volatiles.

It is believed that the temperature at the surface of a burning polymer is close to the temperature at which extensive thermal degradation occurs (usually 300–600°C). The bottom layer of char, near the polymer surface, is at the same temperature, whereas, the upper surface, exposed to the flame, can be as hot as 1500°C [6].

Therefore, fire retardancy chemistry is concerned with chars, which may be produced at temperatures between 300 and 1500°C [7].

Char-forming materials often swell and intumesce during their degradation (combustion), and the flame-retardant approach is to promote the formation of such intumescent char. Intumescent systems, on heating, give a swollen multicellular char capable of protecting the underlying material from the action of the flame. The fire-protective coatings with intumescent properties have been in use for 50 years, whereas incorporation of intumescent additives in polymeric materials is a relatively recent approach [8-10].

On burning, these additives develop the foamed char on the surface of degraded material. The suggested mechanism of fire retardancy assumes that the char acts as a physical barrier against heat transmission and diffusion of oxygen toward the polymer and of combustible degradation products of the polymer toward the flame. Thus, the rate of pyrolysis of the natural and polymeric materials is expected to decrease below flame-feeding requirements, which leads to flame extinguishment. The intumescent behavior resulting from a combination of charring and foaming of the surface of the burning polymers is being widely developed for fire retardancy because it is characterized by a low environmental impact.

Apparently, intumescent coatings are the most effective methods of protecting wood and natural materials from fire. When heated, they form a thick, porous carbonaceous layer. This provides an ideal insulation of the protected surface against an excessive increase in temperature and oxygen availability, thus preventing thermal decomposition which plays a decisive role in retarding the combustion process. Under the heating, intumescent compounds expand up to 200 times their volume, and, in some cases, to even 200 times [11]. Intumescent systems such as paints, lacquers, mastics, and linings must contain ingredients which, when heated to high temperatures, will form large amounts of non-flammable residues. These residues, under the influence of emitted gases produce foam with good insulating properties [11, 12, 13].

The heating of cellulosic materials and wood at 105–110°C results in removal of moisture. Reactions occurring at this stage proceed slowly and are mostly endothermic. After exceeding the above temperature, a slow thermal decomposition of the components of natural polymers begins and at 150–200°C gas products of the decomposition start to be released.

At 160°C decomposition of lignin begins. The lignin under thermal degradation yields phenols from cleavage of ether and carbon–carbon linkages, resulting in more char than in the case of cellulose. Most of the fixed carbon in charcoal originates from lignin. At 180°C hydrolysis of low molecular weight polysaccharides (hemicelluloses) begins [13]. Thermal stability of hemicelluloses is lower than that of cellulose and they release more incombustible gases and fewer tarry substances [12] The most abundant gaseous products contain about 70% of incombustible CO_2 and about 30% of combustible CO. Depending on availability of oxygen, subsequent reactions can be exothermic or endothermic. In the temperature range of 220–260°C exothermic reactions begin. They are characterized by evolution of gaseous products of decomposition, release of tarry substances, and appearance of ignition areas of hydrocarbons with low boiling points. Cellulose decomposes in the temperature range between 260 and 350°C, and it is primarily responsible for the formation of flammable volatile compounds [12].

At temperatures about 275–280°C, accelerated release of considerable quantities of heat begins and increased amounts of liquid and gaseous products (280–300°C), including methanol and acetic acid are formed. The amount of evolving carbon monoxide and dioxide decreases, mechanical slackening of wood structure proceeds and ignition occurs. Mass loss of wood reaches about 39%. [12] Tar begins to appear at 290°C. The release of gases still increases and rapid formation of charcoal takes place. This reaction is highly exothermic and proceeds at 280–320°C. Secondary reactions of pyrolysis become predominant result an increased amount of gaseous products. Combustion proceeds in the gas phase at a small distance from the surface rather than on the wood surface itself. The ignition of wood occurs at 300–400°C depends on its origin and lignin content [13].

At the final stage of combustion of wood (above 500°C) the formation of combustible compounds is small and the formation of charcoal increases. Charcoal makes an insulating layer which hinders heat transfer, thus preventing the temperature from increasing to the pyrolysis of wood occurs.

The mechanisms of flame retardancy depend on the character of the action of flame retardants chemical compounds present in fire retardants. In comprehensive review on wood flame retardant R. Kozlowski and M. Wladyka-Przybylak proposed two general groups of fire retardants for wood—additive and reactive [12]. The additive compounds are those whose interaction with a substrate is only physical in its nature, whereas, reactive compounds interact chemically with cellulose, hemicellulose or lignin [12, 14]. The applied additive compounds include—mono and diammonium phosphate, halogenated phosphate esters, phosphonates, and inorganic compounds such as antimony oxide and halogens, ammonium salts (ammonium bromide, ammonium fluoroborate, ammonium polyphosphate, and ammonium chloride), amino resins (compounds used for their manufacture are dicyandiamide, phosphoric acid, formaldehyde, melamine and urea), hydrated alumina, stannic oxide hydrate, zinc chloride, and boron compounds (boric acid, borax, zinc borate, triammonium borate, ethyl, and methyl borates) [12].

One of the most effective methods of protecting wood from fire is the use of fire retardant intumescent coatings. When heated, they form a thick, porous carbonaceous layer. This provides an ideal insulation of the protected surface against an excessive increase in temperature and atmospheric oxygen, thus preventing thermal decomposition which plays a decisive role in retarding the combustion process.

Studies of new flame retardant systems for the protection of natural polymers and wood can be directed to chemical modification of natural polymers and wood and the use of more efficient intumescent systems and fire protectors. Here we represent the results of researches of use of non-polluting reagents from renewed vegetative raw materials as highly effective flame retardants intumescent/char-forming systems for cellulosic materials and wood which can find wide applications due to their cheapness and simple technology of production.

Earlier it has been found that starch, cellulose, and lignin can be easy oxidized by oxygen in the presence of copper salts and alkali. The basic products of oxidation of polysaccharides and lignin are salts of polyoxyacids [15]. Such salts can be used as components of washing powders, components of boring solutions, gums for use in building and so on. But unexpectedly it was found that the salts of polyoxyacids are also diminishing the burning of different materials [16]. In the present work the data on properties of the oxidized polysaccharides and a lignin, used as flame retardants are presented.

19.2 OXIDIZED POLYSACCHARIDES AND LIGNIN

The methods of oxidative modifying of polysaccharides with receiving of polyacids as main products are widely used in practice due to availability of initial raw material and high consumer properties of oxidized polysaccharides. The salts of polyacids are widely used as water soluble glues in production of paper, cardboard, in processes of materials dressing, as components of drilling agents, and so on. However, as well as for a lot of other processes of organic compounds oxidation as oxidizing agent of polysaccharides the hypochlorites and periodates [17, 18], hydrogen peroxide and not gaseous oxygen are used till recently that is connected with low activity of oxygen in processes of polysaccharides oxidation.

In the presence of copper complexes and bases not only simple in their structure alcohols and ketones but also polysaccharides (starches, dextranes, and cellulose) may be oxidized by oxygen with high rates. The high rates of polysaccharides oxidation by oxygen exceeding oxidation rates by hypochlorites and other oxidizing agents are reached at temperatures 40–90°C [19].

The oxidation of polysaccharides with the greatest efficiency proceeds in the presence of copper and alkali salts. In Figure 1 kinetic curve of oxygen absorption (curve 1), NaOH consumption (curve 2) and changes of viscosity of initial gel (curve 3) during at of 10% of potato starch gel oxidation are presented. As it is seen from Figure 1 at the first 10–20 min of starch oxidation viscosity of initial gel decreases in hundreds times and final (after 3–4 hr) viscosity of a solution comes is near to viscosity of water. The molecular weight of an initial polysaccharide at first minutes decreases slightly [20].

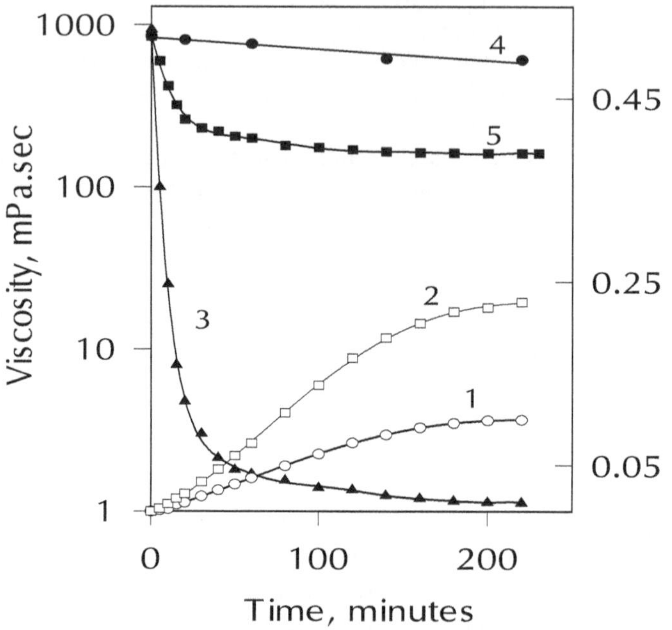

FIGURE 1 The kinetic curves of O_2 (1) and NaOH (2) consumption and change of viscosity of initial gel in the course of starch oxidation in the presents of cupper ions (3) and change of viscosity of gel in the absence of catalyst (4). Curve 5–change of the viscosity of initial gel of starch (5 g starch and 0.5 g NaOH in 50 ml H_2O) under inert atmosphere after addition of 1 g dry powder of deep oxidized starch. $[CuCl_2]_o = 5.10^{-3}$ M, 5 g potato starch, 0.5 g NaOH, 50 ml H_2O, 338 K.

Without copper salt the oxidation of starch does not proceed with measureable rate and viscosity of gel practically does not change (curve 4). But an addition of oxidized starch to the native gel of starch without oxygen drastically (in ten time) diminishes the viscosity of initial gel (curve 5).

Below the most probable formula of the oxidized polysaccharide (starch) is presented. The chemical formula of the oxidized polysaccharides includes unmodified α-D-glucopyranosyl cycles and the oxidized links of polysaccharides containing carboxyl groups [20].

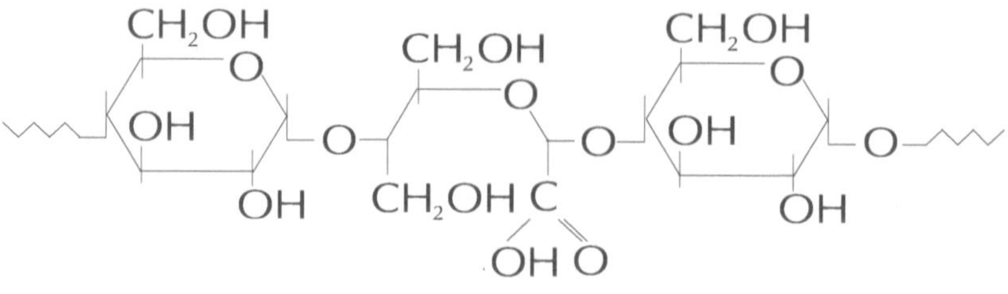

Copper salts are the most effective catalysts of oxidation of polysaccharides in the presence of alkali. For lack of the catalyst starch and cellulose practically are not oxidized at temperatures below 373 K. However, lignin which contains many of high reactive phenolic groups (under high rates) is oxidized by oxygen and without the catalyst with measurable rates.

As well as under oxidation of polyols with low molecular weights [21], the anion forms of polysaccharides reacts with Cu^{2+} ions with $[Cu^{2+}...A^{-}]$ adducts formation (A^{-}–deprotonated polysaccharide). The function of catalyst (Cu^{2+} ions) is to activate deprotonated forms of substrate to oxygen. It is possible to suppose that the role of bivalent copper ions is the oxidation of anion form of substrate and followed interaction of intermediate radicals or anion radicals with O_2. However, high-molecular polysaccharides with a low content of end aldehyde groups in anaerobic medium are redox inactive and the rate of reduction of Cu^2 to Cu^+- ions (due to electron transfer from substrate anion form to Cu^{2+}) proceeds with rates in hundred times lower than the rate of oxygen adsorption during the process of polysaccharides oxidation in alkaline mediums. Apparently, just direct interaction of oxygen with anions A^- in coordination sphere of copper ion leads to formation of hydroxycarboxylates as the main primary products of oxidation. Absorption of oxygen is completely stagnated after neutralization of introduced alkali by polyoxyacids those formed during the oxidation process. Thus, by varying of amount of introduced alkali we may change the degree of polysaccharides oxidation into polyacids. This fact allows us to change the final products viscosity, bonding ability, solubility in water, and so on.

As initial raw materials for receiving of salts of polyoxyacids not only starch can be used, but any of other starch-containing raw materials—corns of maize, oats, rice, and so on, including ill-conditioned raw materials (grain-crops affected by various fungus diseases, waste of rice slashing, waste of mill houses, and so on). This fact enables to decrease the prime cost of the final product significantly. Moreover, it was observed the similar to the polysaccharides behavior of lignin under oxidation condition in alkali media with/without the catalyst. The lignin is amongst the most abundant biopolymers on earth. It is estimated that the planet currently contains 3×10^{11} metric tons of lignin with an annual biosynthetic rate of approximately 2×10^{10} tons [22]. It constitutes approximately 30% of the dry weight of softwoods and about 20% of the weight of hardwoods [23]. It is absent from primitive plants such as algae, and fungi which lack a vascular system and mechanical reinforcement. The presence of lignin within the cellulosic fibre wall, mixed with hemicelluloses, creates a naturally occurring composite material which imparts strength and rigidity to trees and plants.

Lignin is a random copolymer consisting of phenylpropane units having characteristic side chains. It slightly crosslinks and takes an amorphous structure in the solid state. The molecular motion is observed as glass transition by thermal, viscoelastic, and spectroscopic measurements. The hydroxyl group of lignin plays a crucial role in interaction with water. Lignin is usually considered as a polyphenolic material having an amorphous structure, which arises from an enzyme-initiated dehydrogenative polymerization of *p*-coumaryl, coniferyl, and sinapyl alcohols. The basic lignin structure is classified into only two components, one is the aromatic part and the other is the C3 chain. The only usable reaction site in lignin is the OH group, which is the case for both phenolic and alcoholic hydroxyl groups.

The lignin is one of the most important bio-resources for the raw material of the synthesis of environmentally compatible polymers. They are derived from renewable resources such as trees, grasses, and agricultural crops. About 30% of wood constituents are lignin. These are nontoxic and extremely versatile in performance. The production of lignin as a by-product of pulping process in the world is over 30 million tons per year [23].

According to the widely accepted concept, lignin may be defined as an amorphous, three-dimensional polyphenolic polymer arising from an enzyme-mediated dehydrogenative polymerization of three phenylpropanoid monomers, coniferyl, sinapyl, and p-coumaryl alcohols. Basically, three major structures of lignin, 4-hydroxyphenyl (1), guaiacyl (2), and syringyl (3) structures are conjugated to produce a lignin polymer in the process of radical-based lignin biosynthesis.

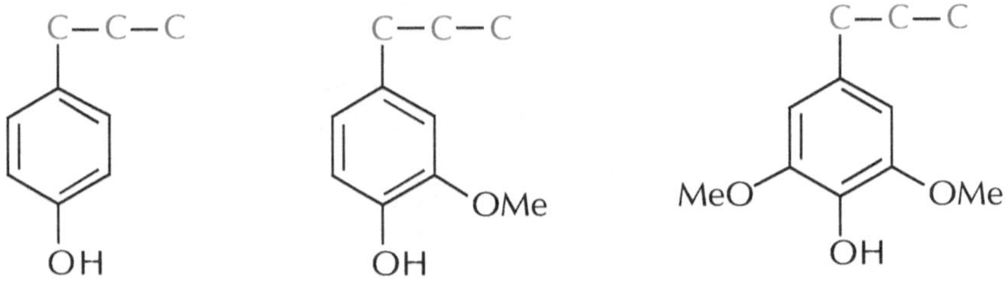

(1) (2) (3)

The results of estimation of efficiency of fire protecting action of coverings based on oxidized starch-containing raw materials with different amounts of applied reagent were carried out by standard method (ASTM E136-09 Standard Test Method for Behavior of Materials in a Vertical Tube Furnace) (Figure 2). It is obvious that all pine samples treated with "one-layer covering" of oxidized starch reagent (OSR) - 100 g/m^2 provide good fire-protection activity (II group) to reach a rank of hardly-inflammable wood. In the case of "multi-layer" covering, 3 ÷ 4 covering, (OSR consumption 300 ÷ 400 g/m^2) mass losses at fire testing are significantly decreased allow us to obtain the I group of fire-protection–hardly inflammable wood.

19.3 MODE OF ACTION OF OXIDIZED POLYSACCHARIDES (INTUMESCENCE BEHAVIOR)

The study of thermal decomposition of oxidized starch at dynamic heating in the range of temperatures 25–950°C has shown, that formation of the made foam coke occurs at the very first stage at temperature 150–280°C as a result of synchronous processes of reduction of viscosity of polymer at its transition from glass forming form to viscous-fluid and chemical reactions of decarboxylation and dehydration. Making foam agents are gaseous products of decomposition–water and a carbon dioxide. The reactions of intermolecular dehydration promote the formation of the sewed spatially-mesh structure and stabilization and hardening of the made foam coke. Figure 2 shows the formation of intumescent coke from oxidized starch under the heating at 400°C in air.

FIGURE 2 Formations of coke after heat treatment of a metal plate with covering oxidized starch.

The combustibility study of pinewood samples treated with aqueous solution of oxidized starch has shown high efficiency of flame retardant action of this intumescent additive during combustion process.

The fire tests results of plain pinewood samples and the pine samples treated with aqueous solution of oxidized starch according to ASTM E136-09 (Standard Test Method for Behavior of Materials in a Vertical Tube Furnace) (Figure 3, and Figure 4).

FIGURE 3 Photograph of initial pinewood samples before fire test.

FIGURE 4 Photograph of pinewood samples treated with aqueous solution oxidized starch flame retardant after fire tests (ASTM E136-09).

The effect of fireproof action of high-molecular oxidized starch flame retardant is connected with formation on a surface of wood sufficient foam coke layer. This foam coke shows excellent heat-shielding barrier properties, complicates heat access to the wood surface and hinders the evolution of combustible gaseous products of decomposition of wood.

Figure 5 shows the actual mass loss pinewood samples treated with aqueous solution of different oxidized polysaccharides after fire tests (ASTM E136-09).

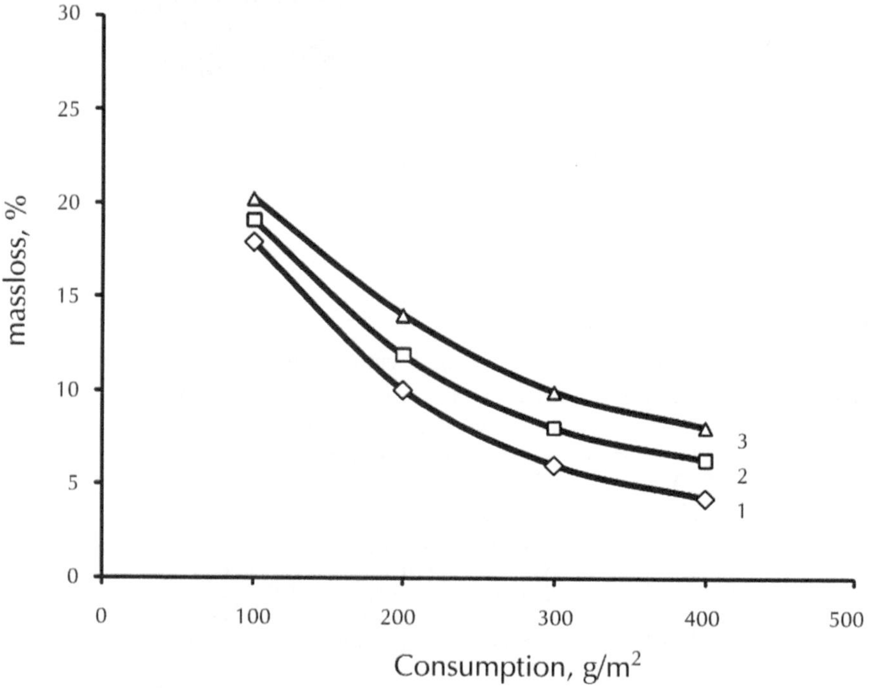

FIGURE 5 Dependence of mass losses of wood samples on amount of applied fire-protecting coverings based on modified polysaccharides—(1)–oxidized starch (high degree of oxidation), (2)–oxidized rice (high degree of oxidation), and (3)–oxidized starch (average degree of oxidation).

The combustibility of pristine pinewood sample and the treated by oxidized lignin and starch ones was also evaluated by the Mass Loss Calorimeter (ISO 13927) at an incident heat flux of 35 kW/m^2 condition [24]. The mass loss rates (MLR) of pinewood samples acquired at 35 kW/m^2 are presented in Figure 6.

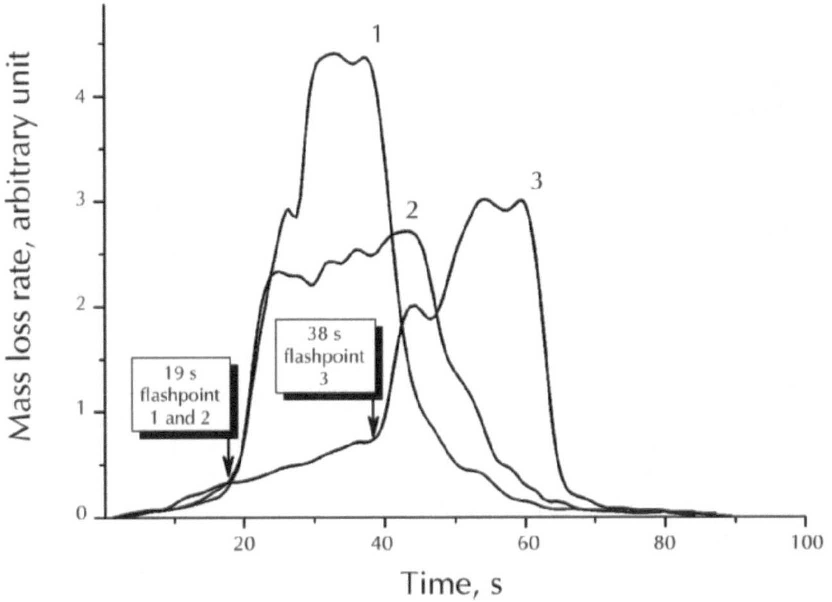

FIGURE 6 Mass loss rate versus time for plain pinewood sample–(1), pinewood sample treated with of oxidized lignin (15% by wt.)–(2) and pinewood sample treated with oxidized starch (15% by wt.)–(3).

It is clearly seen that under conditions of initial surface combustion, the mass loss rate of the pinewood samples treated with of oxidized lignin and starch (15% by wt.) and the pinewood sample treated with oxidized starch and its rate are noticeably lower than an adequate value for the neat pinewood samples. The peak of MLR of the pinewood sample treated with of oxidized lignin is 41% lower than that of pure pinewood sample, while that of the pinewood sample treated with oxidized starch is 33% lower. Due to predominant solid-phase mode of action of oxidized lignin and starch flame retardants the MLR curves are supposed to be similar to the rate of heat release (RHR) curves, so the reduction of the MLR is evidently the primary factor responsible for the lower RHR of the pinewood samples. An improvement in flame resistance of the pinewood samples treated with of oxidized lignin over the pinewood sample happens as a result of the char formation providing a transient protective barrier of oxidized lignin (as polyphenolic carbonizing structure), whereas, the pinewood sample treated with oxidized starch indicates the strong intumescent action. An additional observation obtained from the MLR plots is the substantial increase in induction period time of self-ignition (flashpoint) of the pinewood sample treated with oxidized starch as compared with plain pinewood sample and the pinewood sample treated with oxidized lignin due to intumescent behavior of oxidized starch (figure 6).

KEYWORDS

- **Char-forming materials**
- **Degradation**
- **Extensive thermal degradation**
- **Flammability test**
- **Polymeric additives**

REFERENCES

1. Lomakin, S. M., Brown, J. E., Breese, R. S., and Nyden, M. R., *Polymer Degrad. And Stab.*, **41**, 229–243 (1993).
2. Factor, A. Char Formation in Aromatic Engineering Polymers, in Fire and Polymers, G. L. Nelson (Ed.), ACS Symposium Series 425, ASC, Washington DC, pp. 274–287 (1990).
3. Lomakin, S. M., Zaikov, G. E., and Artsis, M. I. *Intern. J. Polym. Mater.*, **32**(1–4), 173–202 (1996).
4. Lomakin, S. M., Zaikov, G. E., Artsis, M. I., Ruban, L. V., and Aseeva, R. M., *Oxidation Comm.*, **18**(2), 105–112 (1995).
5. Lomakin, S. M., Zaikov, G. E., and Artsis, M. I. *Intern. J. Polym. Mater.*, **26**(3–4), 187–194 (1994).
6. Aseeva, R. M. and Zaikov, G. E. *Combustion of Polymeric Materials* (in Russian). Moscow, Nauka, pp. 84–135 (1981).
7. Levchik, S. and Wilkie, Ch. A. Char Formation, Chapter 6 in *Fire Retardancy of Polymeric Materials*. A. F. Grand and Ch. A. Wilkie (Eds.), Marcel Dekker, Inc. New York, pp. 171–216 (2000)
8. Camino, G. *Flame retardants: Intumescent systems*. In G Pritchard (Ed.), Plastics Additives, London, Chapman Hall, pp. 297–306 (1998).
9. Camino, G. and Delobel, R. Intumescence, Chapter 7 in *Fire Retardancy of Polymeric Materials*. A. F. Grand and Ch. A. Wilkie (Eds.), Marcel Dekker, Inc. New York, 217–244 (2000).
10. Camino, G. and Lomakin, S. Intumescent materials In: A. R. Horrocks and D. Price (Eds.), *Fire Retardant Materials*. CRC Press, New York, Washington DC, 318–335 (2001).
11. Vandersall, H. L. Intumescent coating systems, their development and chemistry *J Fire Flamm*, 2 April, pp. 97–140 (1971).
12. Kozlowski, R. and Wladyka-Przybylak, M. Natural polymers, wood and lignocellulosic materials In A. R. Horrocks and D. Price (Eds.), *Fire Retardant Materials*. CRC Press, New York, Washington DC, 293–317 (2001).
13. Hurst, N. W. and Jones, T. A. A review of products evolved from heated coal, wood and PVC, *Fire Mater*, **9**(1), 1–9 (1985).
14. Lewin, M., Atlas, S. M., and Pearce, E. M. *Flame-Retardant Polymeric Materials*, New York and London, Plenum Press, **1**, (1975).
15. Sakharov, A. M. In Chemical and Biochemical Physics: New Frontiers, Zaikov, G. E. (Ed.), Nova Science Publishers, 113 (2006).
16. Sivenkov, A. B., Serkov, B. B., Aseeva, R. M., Sakharov, A. M., Sakharov, P. A., and Skibida, I. P. Fire & Explosion Safety (in Russian). **1**, 39 (2002).
17. Floor, M., Kieboom, A. P. G., and Bekhum, H. v. *Starch*. **41**, 303 (1989).

18. Santacesria, E., Trully, F., Brussani, G. F., Gelosa, D., and Di Serio, M. (1994).
19. C. Skibida I.P., Sakharov An.M., Sakharov Al.M. (1993). EP 91122164.6, Carbohydrate Polymers, 23, 35.
20. Sakharov, A. M., and Skibida, I. P. *Chem.Phys.*, **20**, 101 (2001).
21. Sakharov, A. M., Silakhtaryan, N. T., and Skibida, I. P. *Kinetics and Catalysis*, **37**, 393 (1996).
22. *Cereal Straw as a Resource for Sustainable Biomaterials and Biofuels: Chemistry, Extractives, Lignins, Hemicelluloses, and Cellulose*. Run-Cang-Sun (Ed.), Elsevier, p. 292 (2010).
23. Hatakeyama, H. and Hatakeyama, T. Lignin Structure, Properties, and Applications. In *Biopolymers - Lignin, Proteins, Bioactive Nanocomposites*. A. Abe, K. Dušek and Sh. Kobayashi (Eds.), Springer, Heidelberg, Dordrecht, London, New York, pp. 2–64(2010).
24. [24] Standard Test Method for Screening Test for Mass Loss and Ignitability of Materials (ASTM E2102). American Society for Testing and Materials, Philadelphia, PA.

INDEX

E

F

G

H

I

J

K

L

M

R

S

T

U

V

W

X

Y

Z